Damage Diagnosis and Safety Assessment of Bridge Structures under Multiple Hazards

Damage Diagnosis and Safety Assessment of Bridge Structures under Multiple Hazards

Editors

Kai Wei
Mingjin Zhang
Jian Zhong
Yutao Pang

MDPI

Basel • Beijing • Wuhan • Barcelona • Belgrade • Novi Sad • Cluj • Manchester

Editors

Kai Wei
Department of Bridge Engineering
Southwest Jiaotong University
Chengdu
China

Mingjin Zhang
Department of Bridge Engineering
Southwest Jiaotong University
Chengdu
China

Jian Zhong
Department of Civil Engineering
Hefei University of Technology
Hefei
China

Yutao Pang
Department of Civil Engineering
China University of Geosciences
Wuhan
China

Editorial Office
MDPI
St. Alban-Anlage 66
4052 Basel, Switzerland

This is a reprint of articles from the Special Issue published online in the open access journal *Sustainability* (ISSN 2071-1050) (available at: www.mdpi.com/journal/sustainability/special_issues/Damage_Diagnosis_and_Safety_Assessment).

For citation purposes, cite each article independently as indicated on the article page online and as indicated below:

Lastname, A.A.; Lastname, B.B. Article Title. *Journal Name* **Year**, *Volume Number*, Page Range.

ISBN 978-3-0365-9763-8 (Hbk)
ISBN 978-3-0365-9762-1 (PDF)
doi.org/10.3390/books978-3-0365-9762-1

Contents

About the Editors

Kai Wei

Kai Wei is a Professor and Doctoral Supervisor within the Department of Bridge Engineering at Southwest Jiaotong University. He is the recipient of the National Excellent Youth Fund, the Young Talent of the Sichuan "Tianfu Emei" Program, and a Member of the Sichuan May Fourth Youth Medal Collective. He has long been engaged in theoretical and applied basic research on bridge hydrodynamics, presided over three projects of the National Natural Science Foundation of China, three sub-projects of regional joint fund and national key research and development projects, and led several scientific and technological research projects of some major bridges, including Changtai Yangtze River Bridge, Hangzhou Bay High-speed Railway Bridge, Xihoumen Rail-cum-road Bridge, etc. He serves as a Section Board Member of the journal *Sustainability*, an Editorial Board Member of the *Journal of Civil and Environmental Engineering*, and a Young Editorial Board Member of the *China Journal of Highway and Transport*. He has published one monograph and more than one hundred journal articles. He was also the winner of the Special Prize in the 2022 Science and Technology Award of the China Highway and Transportation Society, and he received second prize in the 2017 Shanghai Science and Technology Awards.

Mingjin Zhang

Mingjin Zhang is a Professor, Doctoral Supervisor, and recipient of the National Ten Thousand Talents Programme for young outstanding individuals. Mingjin Zhang is mainly engaged in the wind-resistant research of bridges. He has hosted more than 4 national scientific research projects and over 10 wind-resistant research projects for long-span bridges in mountainous areas. He has published 50 SCI journal papers and received three first prizes for scientific and technological progress at the provincial level. He has served as a Young Editorial Board Member of the journals of Southwest Jiaotong University and Acta Aerodynamica Sinica, and served as a reviewer for more than 20 academic journals both domestically and internationally.

Jian Zhong

Jian Zhong, Ph.D. is an Associate Professor and Doctoral Supervisor at Hefei University of Technology, and he also serves as the Deputy Director of the Road and Bridge Research Institute, as well as a member of the Higher Education Teaching Review and Evaluation Committee of the Ministry of Education. He obtained his bachelor's degree from Zhejiang University, doctoral degree from Tongji University Bridge Seismic Research Group, and another doctoral degree from the Georgia Institute of Technology. He is engaged in bridge seismic research, including risk analysis of near/cross fault bridges, seismic safety of urban bridge network lifelines, artificial intelligence, and seismic resilience. He has hosted projects such as the National Natural Science Foundation's general program and the Youth Fund. He has published 57 SCI/EI journal papers, published nearly 40 SCI journal papers as a first author/corresponding author in internationally renowned journals such as ES and ASCE JSE (including 36 papers in Zone 1/2, 7 highly cited ESI papers, and 1 hot paper), authorized/published 32 national invention patents, developed 5 software copyrights, and participated in the compilation of 1 industry standard. He has served as a Youth Editorial Board Member for three international journals and a Guest Editor for two SCI journals.

Yutao Pang

Yutao Pang, Ph.D. is an Associate Professor at China University of Geosciences (Wuhan). He obtained his bachelor's degree from China University of Geosciences (Wuhan), and another doctoral degree from Tongji University Bridge Seismic Research Group. He was engaged in a one-year postdoctoral research fellowship at Polytechnique Montréal. He is also engaged in bridge seismic research, including seismic fragility analysis, resilience analysis of bridges, and artificial intelligence. He has hosted projects such as the National Natural Science Foundation's Youth Fund. He has published 40 SCI/EI journal papers, published nearly 30 SCI journal papers as a first author/corresponding author in internationally renowned journals such as ES and ASCE JSE (3 highly cited ESI papers), authorized 10 national invention patents, and participated in the compilation of 1 industry standard.

Preface

The states of bridges decline over time due to degradation processes, such as creep, corrosion, and cyclic loading. During its life cycle, a bridge faces the great threat of multiple hazards, such as winds, earthquakes, collisions, scour, floods, and waves. Structural damage in existing bridges affects structural safety and weakens sustainability. Therefore, advanced damage diagnosis techniques and safety assessment methods are needed to achieve structural safety and sustainability.

This Special Issue, titled "Damage Diagnosis and Structural Safety Assessment of Bridges Under Multiple Hazards", brings together recent developments in damage diagnosis methods, testing techniques, dynamic behavior, damage mechanisms, and risk assessment approaches. This reprint includes all thirteen papers published within this Special Issue, which are summarized as follows:

Zhang et al. investigate the relationship between theoretical ultimate span length and rise-span ratio for arch bridges. Wu et al. present an analytical solution to analyze aerodynamic forces on a streamlined box girder with coupled vibration. Wu et al. propose a numerical method for conformal mapping and performance prediction of box girder bridges under wind. Chen et al. investigate the seismic behavior of concrete-filled steel tubular-laced piers through simplified numerical models to improve efficiency. Zeng et al. propose a new kind of concrete-filled double-skin steel tube T-joint based on hysteresis experiments. Wang et al. study pounding effects on prefabricated frame bridges based on pounding forces. An et al. study the impact of structural collision caused by separation conditions on the failure mode of the block. An et al. deduce the possible vertical separation contact condition of a bridge based on the near-field vertical seismic spectrum. Xiao et al. propose a new method to survey scenarios of offshore projects with elevation transmission, monitoring operations, and trigonometric leveling based on dynamic compensation. Xiao et al. design a new structure for a combined survey platform under strong and deep-water conditions based on vibration reduction and isolation. Tang et al. propose a new grid pile foundation to improve the reliability of foundations. Wang et al. investigate the distribution of sunshine temperature fields in bridges based on numerical simulation and field testing. Yang et al. enhance the strength and ductility of concrete through hybrid fiber-reinforced self-consolidating concrete.

These articles present up-to-date research on the damage diagnosis and safety assessment of bridges under multiple hazards. The Guest Editors believe that the reprint of this Special Issue will provide important theoretical and technical support for bridge damage diagnosis and safety assessment and promote cooperation between design, construction, research, and teaching.

Kai Wei, Mingjin Zhang, Jian Zhong, and Yutao Pang
Editors

sustainability

Article

Theoretical Analysis of Ultimate Main Span Length for Arch Bridge

Xianxiong Zhang [1], Zhuozhang Deng [2], Genshen Fang [2,*] and Yaojun Ge [2]

1 Poly Changda Engineering Co., Ltd., Guangzhou 510620, China
2 State Key Laboratory of Disaster Reduction in Civil Engineering, Tongji University, Shanghai 200092, China
* Correspondence: 2222tjfgs@tongji.edu.cn

Abstract: The advancement of construction techniques and high-performance sustainable materials enables the increase of span length for arch bridge. It is of great importance to study the theoretical ultimate span length of arch bridge. Based on the parabolic and catenary arch axes, the analytical solutions of ultimate span length of arch bridge are solved using theoretical derivation accounting for the strength, in-plane stability and out-plane stability conditions, respectively. Then, the use of high-performance concrete, reactive powder concrete and high-strength steel is considered to study the relationship between theoretical ultimate span length and rise-span ratio as well as material strength for concrete and steel arch bridges. The results show that the theoretical ultimate span length derived by catenary arch axis is smaller by about 2–6% than that obtained by parabolic arch axis, but the difference is insignificant. When the rise-span ratio is 1/5, the theoretical ultimate span length for concrete arch bridge using R200 reactive powder concrete can reach 2000 m (2161 m for catenary arch axis and 2099 m for parabolic arch axis) while the main span of steel arch bridge using Q690 high-strength steel can be longer than 2500 m (2948 m for catenary arch axis and 2865 m for parabolic arch axis).

Keywords: arch bridge; ultimate span length; theoretical analysis; high-performance material; strength; stability

check for updates

Citation: Zhang, X.; Deng, Z.; Fang, G.; Ge, Y. Theoretical Analysis of Ultimate Main Span Length for Arch Bridge. *Sustainability* **2022**, *14*, 17043. https://doi.org/10.3390/su142417043

Academic Editors: Kai Wei, Mingjin Zhang, Jian Zhong and Yutao Pang

Received: 29 October 2022
Accepted: 15 December 2022
Published: 19 December 2022

1. Introduction

Long-span arch bridge is one of competitive types of bridge to cross rivers and canyons due to its favorable durability and mechanical performance [1]. Recent years, many long-span arch bridges were constructed in China and rewritten the world record of the main span. For steel arch bridges, Chaotianmen Bridge with a main span of 552 m set a world record in 2009 [2]. For concrete arch bridge, the construction of the Tianelongtan Bridge in Guangxi Province, China will break the record of the span length of arch bridge to the value of 600 m [3]. Table 1 summarizes 15 world records of the main span length during the development of arch bridge, including the year of the completion, years of the record, material of the bridge etc. Compared with the rapid development of suspension bridge and cable-stayed bridge, the main span length of arch bridge is almost at a standstill for a long time. It can be noted that since the completion of the first 500 m-main-span-level arch bridge (Bayonne Bridge with the main span of 504 m in the USA) in 1931 [4], the span length has only been increased to 575 m in past about 90 years after the completion of Pingnan 3rd Bridge in 2020 [5]. However, the span capacity of arch bridge is not well examined, especially with the rapid development of advanced construction techniques and new materials. The span length of arch bridge could be continuously broken in the future. The study of ultimate main span length of arch bridge has become a hotspot in the bridge engineering community [6–10].

Table 1. World record of the main span length for arch bridge.

No.	Year of Completion	Years of Record	Bridge Name	Main Span (m)	Rate of Increase	Material	Country
1	605	695	Zhaozhou Bridge	37.5	——	Stone	China
2	1300	41	Maddalena Bridge [11]	38	1.3%	Stone	Italy
3	1341	15	Diable Bridge	45	18.4%	Stone	Italy
4	1356	21	Castelvecchio Bridge	49	8.9%	Stone	Italy
5	1377	500	Trezzo Bridge *	72	46.9%	Stone	Italy
6	1877	7	Maria Bridge	160	122.2%	Cast iron	Portugal
7	1884	2	Garabit Bridge	165	3.1%	Cast iron	France
8	1886	12	Dom Luís Bridge	172.5	4.5%	Cast iron	Portugal
9	1898	18	Upper Steel Bridge *	256	48.4%	Steel truss	USA
10	1916	15	Hell Gate Bridge	298	16.4%	Steel truss	USA
11	1931	46	Bayonne Bridge [2]	504	69.1%	Steel truss	USA
12	1977	26	New River Gorge Bridge	518	2.8%	Steel truss	USA
13	2003	6	Lupu Bridge [12]	550	6.2%	Steel box	China
14	2009	11	Chaotianmen Bridge [4]	552	0.4%	Steel truss	China
15	2020	——	Pingnan 3rd Bridge [3]	575	4.2%	CFST *	China

* Note: Trezzo Bridge collapsed in 1416; Upper Steel Bridge collapsed in 1938; CFST: Concrete Filled Steel Tube.

Some pioneering studies on the ultimate span of arch bridge have been performed in past several decades, as summarized in Table 2. As can be seen, the steel, concrete or CFST are customarily employed for the arch. The catenary and parabolic arch rib axes are usually treated as the reasonable arch axes under dead load with the rise-to-span ratios of 1/3~1/5. The ultimate span length is almost positively relevant to the material strength and steel arch bridge is considered to have a larger span capacity than that of concrete. The ultimate span length for concrete arch bridge is about 600 m when the C100 or lower grade of concrete is utilized. Comparatively, except Tang [8] and Järvenpää and Jutila [13], the steel arch bridge is able to reach 1000 m when the Q460 or lower grade if steel is employed. As for the CFST arch bridge, Wang [14] proposed an optimization method for solving the ultimate span length based on the response surface method, founding that the maximum span length of the CFST arch bridge can reach up to 821 m if the Q420 steel and C80 concrete were utilized.

Table 2. Summary of study on ultimate span of arch bridge.

Researcher	Year	Material	Arch Axis	Rise-Span Ratio	Material Grade	Ultimate Span (m)	Method	Condition
Xia [15,16]	2005	Concrete	Parabolic Catenary	1/4 1/4	C60 C60	481 451	Theoretical analysis	Strength and in-plane stability, solid-web rectangular arch section. An ultimate span reduction coefficient of 0.75 is used for only considering the self-weight of the main arch.
		Steel	Parabolic Catenary	1/4 1/4	Q345 Q345	916 860		
Li [17]	2007	Concrete Steel	Catenary Catenary	1/5 1/5	C100 Q460	590 660	Numerical simulation of arch	Strength and stability conditions
Wang [18]	2012	Steel	Parabolic Catenary	1/4 1/5.5	Q420 Q420	866 818	Numerical simulation of arch	Strength and stability conditions
Zhao [19]	2017	Concrete	Parabolic	1/4	C60	540	Theoretical analysis	Strength and in-plane stability conditions. An ultimate span reduction coefficient of 0.62 is used for only considering the self-weight of the main arch.
Tang [8]	2017	Steel	Catenary	1/5.5	—	5000	Theoretical analysis	Strength, the allowable stress of steel is 420 MPa
Wang [14]	2019	CFST	Catenary	1/4	Q420/C80	821	Numerical simulation of full bridge	Strength, stiffness and stability conditions
Järvenpää and Jutila [13]	2019	Steel	Parabolic Catenary	1/2.31 1/2.96	— —	6250 8284	Theoretical analysis	Strength condition, the allowable stress of steel is 500 MPa

With the advancement of concrete-filled steel tube and stiff skeleton concrete arch bridge technology as well as the development of new materials, such as high-performance steel and ultra-high-performance concrete, the main span length of arch bridge is expected to make a breakthrough. Among them, the concrete arch bridge has natural rationality in the application of materials. With the application of ultra-high-performance concrete, the spanning capacity of concrete arch bridge increases constantly. And the stiff skeleton construction method provides a reasonable solution for the construction of long-span concrete arch bridge. It can be said that the concrete arch bridge has a very potential to continuously break the span record. As for the steel arch bridge, although it has no economic advantage, it is proved to have a strong spanning capacity according to the theoretical analysis. Chen and Liu [20] believes that it is feasible to build a 3300 m main span steel arch bridge. Some trial design schemes of super-span arch bridges have also been proposed to validate the feasibility. Čandrlić and Radić [21] studied the applicability of reactive powder concrete (RPC) to the construction of 1000 m main span concrete arch bridge. Zheng et al. [22] performed a feasibility study on the construction of 700 m main span concrete filled steel tube arch bridge. Shao et al. [23,24] proposed a new system of super long-span steel UHPC composite truss arch bridge and proved its feasibility to construct the 800 m and 1000 m main span arch bridges.

Although there are some theoretical and numerical studies on investigating the ultimate span length of long-span arch bridges, some other conditions, such as the out-of-plane stability and the application of recent high-performance sustainable materials are not well considered. To further investigate the theoretical ultimate span length of arch bridge, this study performs the theoretical analysis of ultimate span length of arch bridges using parabolic and catenary arch axes, respectively considering the strength, in-plane stability and out of plane stability conditions. The relationships between the theoretical ultimate span length and the rise-span ratio as well as the material strength are comparatively studied. The ultimate span lengths for concrete and steel arch bridges are finally obtained with respect to different material grades and rise-span ratios.

2. Parabolic Arch Axis

2.1. Strength Condition

As shown in Figure 1, a half-span arch is subjected to the uniformly distributed load with the density of q (N/m). l (m) and f (m) are the span length and rise of the arch, respectively. M (N·m), H_g (N) and Q (N) are moment, axial force, and shear force at the apex of arch, respectively. By setting the origin of the coordinate at the apex of arch, the parabolic curve of the arch axis is formulated as:

$$y = 4fx^2/l^2 \tag{1}$$

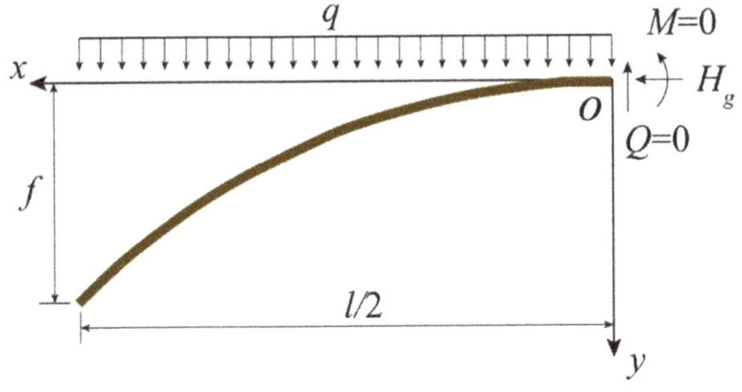

Figure 1. Parabolic arch axis.

When the dead load is assumed to be the uniformly distributed load as shown in Figure 1, the dead load compression line coincides with the parabolic arch axis. That is, there is only axial compression force in the arch under the action of dead load, and there is no bending moment and shear force. At the apex of arch, the horizontal compression force under dead load can be solved as:

$$H_g = \frac{\sum M_j}{f} = \frac{\int_0^{l/2} q(l/2 - x)dx}{f} = \frac{ql^2}{8f} = \frac{ql}{8n} \tag{2}$$

where $\sum M_j$ is the bending moment at the abutment of the arch due to the half-span dead load, $n = f/l$ is the ratio of rise to span. The axial force at the abutment of the arch can be easily obtained:

$$N_j = \frac{H_g}{\cos\theta_j} = \frac{H_g}{1/\sqrt{1+\tan\theta_j^2}} = \frac{ql}{8n}\sqrt{1+16n^2} \tag{3}$$

where θ_j (°) is the angle between the tangent direction of the arch axis at the abutment and horizontal direction. Assume that the proportion of the self-weight of the arch to the total load is λ. The arch is assumed to have a uniform cross section with the area of A (m^2) and apparent density of γ (N/m^3). The density of the dead load is therefore estimated by $q = \gamma A$. The design value of material compressive strength of the arch is f_d (MPa). Based on the General Code for Design of Highway Bridges and Culverts (JTG D60-2015) [25], the structural importance factor of 1.1 and partial safety factor of permanent action of 1.2 are considered. The strength condition of theoretical ultimate span length for arch bridge based on parabolic arch axis is then derived as:

$$1.1 \times 1.2 \times \frac{N_j}{A} = 1.32 \times \frac{\gamma l}{8n}\sqrt{1+16n^2} \le \lambda f_d \tag{4}$$

or

$$l \le l_{\max 1} = \frac{8 \cdot \lambda \cdot n \cdot f_d}{1.32 \cdot \gamma \cdot \sqrt{1+16n^2}} = \frac{6.06\lambda n}{\sqrt{1+16n^2}} \cdot \frac{f_d}{\gamma} \tag{5}$$

It can be seen that the value of $n/\sqrt{1+16n^2}$ increases with n, suggesting that the ultimate span length increases with the rise-span ratio in strength condition.

2.2. Stability Conditions

2.2.1. In-Plane Stability

Although the parabolic arch only bears axial compressive force without bending moment under the action of uniform load, the axial compressive force varies along the arch axis. Meanwhile, the variation of the curvature of the arch results in that the coefficients of the equilibrium differential equation are not constant. The theoretical solution of the equation is usually unavailable. In engineering applications, the effective calculation length S_0 of the arch is often used to approximately describe the critical compressive force. Similar to the centrally compressed member, the critical compressive force of the in-plane stability (normally used as the critical compressive force at the cross section of the 1/4 span) can be calculated by [26]:

$$N_{\text{cr1}} = \frac{\pi^2 EI_z}{l_z^2} = \frac{\pi^2 EI_z}{(0.36 l_a)^2} \tag{6}$$

where l_z (m) is the calculation length of in-plane stability of the arch, $l_z = 0.36l_a$ for unhinged arch, in which l_a is the arc length of the arch axis, E (MPa) is the modulus of elasticity of the material used by the arch, I_z (m^4) is the moment of inertia of the cross section with respect to the vertical direction. The arc length of the arch axis can be approximately calculated using the first two terms of the Taylor's series expansion:

$$l_a = 2\int_0^{l/2}\sqrt{1+\left(\frac{8n}{l}x\right)^2}dx \approx 2\int_0^{l/2}\left[1+\frac{1}{2}\left(\frac{8n}{l}x\right)^2\right]dx = l\left(1+\frac{8}{3}n^2\right) \tag{7}$$

Under the action of uniform load, the axial compressive force at the cross-section of 1/4 span is expressed as:

$$N_{l/4} = \frac{H_g}{\cos\theta_{l/4}} = \frac{H_g}{1/\sqrt{1+\tan\theta_{l/4}^2}} = \frac{ql}{8n}\sqrt{1+4n^2} \tag{8}$$

where $\theta_{l/4}$ (°) is the angle between the tangent direction of the arch axis at 1/4 span and horizontal direction. By setting a stability safety coefficient of φ = 4.0, the in-plane stability condition of theoretical ultimate span length for arch bridge based on parabolic arch axis is then derived as:

$$l^3 \le \frac{2\lambda\pi^2}{0.36^2} \times \frac{n}{\sqrt{1+4n^2}(1+8n^2/3)^2} \times \frac{E}{\gamma} \times \frac{I_z}{A} \tag{9}$$

where λ is the ratio of slenderness. To calculate the I_z/A in Equation (9), a box cross section with the width of b (m), height of h (m) and thickness of plate of t (m) is introduced. The high-order terms with respect to t can be ignored when calculating the moment of inertia since the value of t is relatively much smaller than b and h. The ratio of the moment of inertia and area of the section can be calculated by:

$$\frac{I_z}{A} = \frac{3bh^2 + h^3}{12(b+h)} \tag{10}$$

$$\frac{I_y}{A} = \frac{3hb^2 + b^3}{12(b+h)} \tag{11}$$

where I_y (m^4) is the moment of inertia with respect to the lateral direction. The ratio between the height of the arch cross section and span length h/l is usually at the range of 1/50~1/100. The width of the section to the span length ratio b/l is at the range of 1/20~1/30. If h/l = 1/50 and b/l = 1/30, the in-plane stability condition can be rewritten as:

$$l \le l_{\max 2} = \frac{0.0114\lambda n}{\sqrt{1+4n^2}(1+8n^2/3)^2} \cdot \frac{E}{\gamma} \tag{12}$$

As can be seen, the value of $n/\left[\sqrt{1+4n^2}(1+8n^2/3)^2\right]$ increases first before decreasing with n at the range of $0 < n < 1$ and has the maximum of 0.17 at n = 0.29.

2.2.2. Out-of-Plane Stability

Similar to the element subjected to the axially loaded compression, the critical compressive force related to the out-of-plane stability of the arch can be expressed as:

$$N_{cr2} = \frac{\pi^2 EI_y}{l_y^2} = \frac{\pi^2 EI_y}{\left(\zeta \cdot \frac{l}{2}\left(\frac{1}{4n}+n\right)\right)^2} \tag{13}$$

where l_y (m) is the calculation length of out-of-plane stability of the arch, $l_y = \zeta \cdot \frac{l}{2}\left(\frac{1}{4n}+n\right)$ for unhinged arch, in which ζ is the calculation length coefficient of out-of-plane stability, as listed in Table 3.

Table 3. Calculation length coefficient of out-of-plane stability.

n	1/3	1/4	1/5	1/6	1/7	1/8	1/9	1/10
ζ	1.167	0.962	0.797	0.576	0.495	0.452	0.425	0.406

Based on the in-plane stability condition of $N_{l/4} \leq \lambda N_{\text{cr2}}/\varphi$ and the safety factor of φ = 4.0, the out-of-plane stability condition of theoretical ultimate span length for arch bridge based on parabolic arch axis is then derived as:

$$l \leq l_{\max 3} = \frac{0.2050\lambda n^3}{\zeta^2(1+4n^2)^{\frac{5}{2}}} \cdot \frac{E}{\gamma} \tag{14}$$

where the value of ζ increase with n, but $n^3/(1+4n^2)^{\frac{5}{2}}$ increase first before decreasing with n at the range of $0 < n < 1$ and has the maximum at n = 0.61.

3. Catenary Arch Axis

3.1. Strength Condition

When the dead load density (gravity per unit length) over the arch is continuously distributed and gradually increases from the apex to the abutment of the arch and has an approximate linear relationship with the arch axis, its reasonable arch axis is a catenary. The shape of the catenary is not only related to the rise-span ratio, but also depends on the arch axis coefficient. The force of the catenary arch can be optimized by adjusting the arch axis coefficient so that has a relatively strong adaptability to non-uniform loads. The catenary is the most commonly used arch axis for log-span arch bridges. As shown in Figure 2, a half-span arch with the span length of l (m) and rise of f (m) is subjected to the non-uniform load. The density of the load at the abutment and apex of the arch are q_j (N/m) and q_d (N/m), respectively. M (N·m), H_g (N) and Q (N) are moment, axial force and shear force at the apex of arch, respectively. By setting the origin of the coordinate at the apex of arch, the load density at the location of x is expressed as:

$$q_x = q_d\left(1 + \frac{m-1}{f}y\right) \tag{15}$$

where $m = q_j/q_d$ is the arch axis coefficient. The catenary arch axis is formulated as:

$$y = \frac{f}{m-1}(\cosh k\xi - 1) \tag{16}$$

where $\xi = 2x/l$, $k = \ln\left(m + \sqrt{m^2-1}\right)$. When the dead load is continuously distributed and gradually increased from the apex to the abutment of the arch and has an approximately linear relationship with the arch axis, the dead load compression line coincides with the catenary arch axis. That is, there is only axial compression force in the arch under the action of dead load, and there is no bending moment and shear force. At the apex of arch, the horizontal compression force under dead load can be solved as:

$$H_g = \frac{\sum M_j}{f} = \frac{\int_0^{l/2} q_x(l/2-x)dx}{f} = \frac{q_d l}{4nk^2}(m-1) \tag{17}$$

The compressive force at the abutment of the arch is:

$$N_j = \frac{H_g}{\cos\theta_j} = \frac{H_g}{1/\sqrt{1+\tan\theta_j^2}} = \frac{q_d l}{4nk^2}(m-1)\sqrt{1+4n^2k^2\frac{m+1}{m-1}} \tag{18}$$

Similar to the parabolic arch axis in Section 2.1, assume that the proportion of the self-weight of the arch to the total load is λ. The arch is assumed to have a uniform cross section with the area of A (m^2) and apparent density of γ. The strength condition of theoretical ultimate span length for arch bridge based on catenary arch axis is derived as:

$$l \leq l_{\max 1} \frac{4\lambda nk^2 f_d}{1.32\gamma(m-1)\sqrt{1+4n^2k^2\frac{m+1}{m-1}}} = \frac{3.03\lambda nk^2}{(m-1)\sqrt{1+4n^2k^2\frac{m+1}{m-1}}} \cdot \frac{f_d}{\gamma} \quad (19)$$

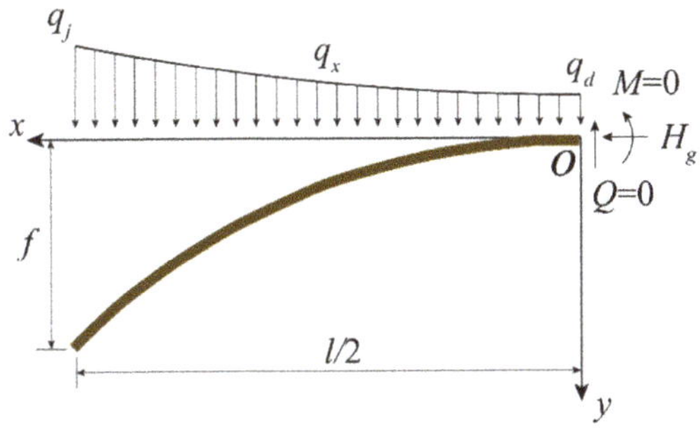

Figure 2. Catenary arch axis.

3.2. Stability Conditions

3.2.1. In-Plane Stability

Similar to the centrally compressed member, the critical compressive force of the in-plane stability of catenary arch (normally used as the critical compressive force at the cross section of the 1/4 span) can be calculated by using Equation (6). The arc length of the catenary arch axis can be approximately calculated by:

$$l_a = 2\int_0^{l/2} \sqrt{1+\left(\frac{2nk}{(m-1)}\sinh k\xi\right)^2} dx = 2\int_0^{l/2} \sqrt{1+\frac{4n^2k^2}{(m-1)^2}\left(\sinh k\frac{2x}{l}\right)^2} dx \quad (20)$$

where the integral equation is like $\int \sqrt{1+a\cdot\sinh^2(bx)dx}$, which is the second type of the elliptic integral if $a \neq 1$. When a = 1, or $4n^2k^2 = (m-1)^2$:

$$\cos\theta_j = \frac{1}{\sqrt{1+\tan\theta_j^2}} = \frac{1}{\sqrt{1+\sinh k}} = \frac{1}{m} \quad (21)$$

That means $m = 1/\cos\theta_j$ when $4n^2k^2 = (m-1)^2$. The load density at the abutment of the arch is expressed as:

$$q_j = m \cdot q_d = \frac{\gamma A}{\cos\theta_j} \quad (22)$$

By introducing an element of the abutment of the arch with the arch length of ds, Equation (22) can be rewritten as:

$$q_j dx = \gamma A \frac{dx}{\cos\theta_j} = \gamma A ds \quad (23)$$

where $dx = ds\cos\theta_j$ is the horizontal projection length of the element. Equation (23) suggests that the dead load density at the abutment of the arch is exactly equal to the self-weight of the arch with a uniform cross-section at abutment. Then, the arch axis is just the reasonable arch axis under the action of self-weight of the arch. That is, the reasonable arch axis coefficient can be determined by the equation of $2nk = m - 1$. By substituting $k = \ln\left(m + \sqrt{m^2 - 1}\right)$ into this equation produces:

$$\frac{m-1}{\ln\left(m + \sqrt{m^2 - 1}\right)} = 2n \tag{24}$$

Equation (24) can be used to determine a reasonable arch axis coefficient m based on the rise-span ratio n. The arc length of the catenary arch axis is therefore calculated by:

$$l_a = 2\int_0^{l/2} \sqrt{1 + \left(\text{sinh}k\frac{2x}{l}\right)^2}\,dx = \frac{l}{k}\text{sinh}k \tag{25}$$

The axial compressive force at the cross-section of 1/4 span is expressed as:

$$N_{l/4} = \frac{H_g}{\cos\theta_{l/4}} = \frac{H_g}{1/\sqrt{1 + \tan\theta_{l/4}^2}} = \frac{q_d l}{2k}\cosh\frac{k}{2} \tag{26}$$

Based on the in-plane stability condition of $N_{l/4} \le \lambda N_{cr2}/\varphi$ and the safety factor of $\varphi = 4.0$ as well as the cross-section property in Equation (10), the in-plane stability condition of theoretical ultimate span length for arch bridge based on catenary arch axis is then derived as:

$$l \le l_{\max 2} = \frac{0.0029\lambda k^3}{\text{sinh}^2 k \cdot \cosh\frac{k}{2}} \cdot \frac{E}{\gamma} \tag{27}$$

3.2.2. Out-of-Plane Stability

Similar to Equation (13), the critical compressive force related to the out-of-plane stability of the catenary arch can be expressed as:

$$\frac{q_d l}{2k}\cosh\frac{k}{2} \le \frac{\lambda}{4.0} \times \frac{\pi^2 E I_y}{\zeta^2 \cdot \frac{l^2}{4}\left(\frac{1}{4n} + n\right)^2} \tag{28}$$

or

$$l \le l_{\max 3} = \frac{0.0512\lambda k n^2}{\zeta^2 \cosh\frac{k}{2}(1 + 4n^2)^2} \cdot \frac{E}{\gamma} \tag{29}$$

4. Theoretical Ultimate Span Length

Based on above theoretical derivation, the strength, in-plane stability and out-of-plane stability conditions of the parabolic and catenary arch axes are summarized listed in Table 4, respectively. The theoretical ultimate span length of arch bridges $l_{\max}$ (m) is expressed as

$$l_{\max} = \min\{l_{\max 1}, l_{\max 2}, l_{\max 3}\} \tag{30}$$

Table 4. Ultimate span length in different conditions.

Arch Axis	Strength $l_{\max 1}$	In-Plane Stability $l_{\max 2}$	Out-of-Plane Stability $l_{\max 3}$
Parabola	$\frac{6.06\lambda n}{\sqrt{1+16n^2}} \cdot \frac{f_d}{\gamma}$	$\frac{0.0114\lambda n}{\sqrt{1+4n^2}(1+8n^2/3)^2} \cdot \frac{E}{\gamma}$	$\frac{0.2050\lambda n^3}{\zeta^2(1+4n^2)^{\frac{5}{2}}} \cdot \frac{E}{\gamma}$
Catenary	$\frac{3.03\lambda n k^2}{(m-1)\sqrt{1+4n^2k^2\frac{m+1}{m-1}}} \cdot \frac{f_d}{\gamma}$	$\frac{0.0029\lambda k^3}{\text{sinh}^2 k \cdot \cosh\frac{k}{2}} \cdot \frac{E}{\gamma}$	$\frac{0.0512\lambda k n^2}{\zeta^2 \cosh\frac{k}{2}(1+4n^2)^2} \cdot \frac{E}{\gamma}$

To further discuss the ultimate span length, the concrete and steel arch bridges with different materials are introduced and compared.

4.1. Concrete Arch Bridge

As suggested by the Specifications for Design of Highway Reinforced Concrete and Prestressed Concrete Bridges and Culverts (JTG 3362-2018) [27] and Technical Specification for Reactive Powder Concrete Structures (DBJ43/T 325-2017) [28], the design values of compressive strength f_d and modulus of elasticity E for concrete are listed in Table 5. The apparent density of concrete γ = 26.0 kN/m^3.

Table 5. Design values of compressive strength f_d and modulus of elasticity E for concrete.

Grade	C60	C80	R100	R120	R140	R160	R180	R200
f_d (MPa)	26.5	34.6	48.0	58.0	68.0	77.0	87.0	97.0
E ($\times 10^4$ MPa)	3.60	3.80	4.00	4.29	4.52	4.71	4.86	5.00

By substituting the design values of different grades of concrete in Table 5 into the theoretical solutions in Table 4, the ultimate span lengths of parabolic and catenary arch bridges with respect to the strength, in-plane stability and out-of-plane stability conditions can be readily obtained, as shown in Figures 3 and 4. The final theoretical ultimate span length is shown in Figure 5.

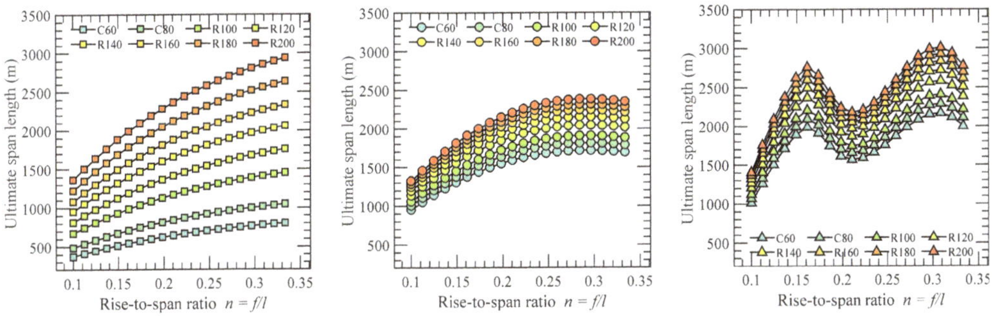

(a) (b) (c)

Figure 3. Ultimate span length of parabolic concrete arch bridge: (**a**) l_{max1}; (**b**) l_{max2}; (**c**) l_{max3}.

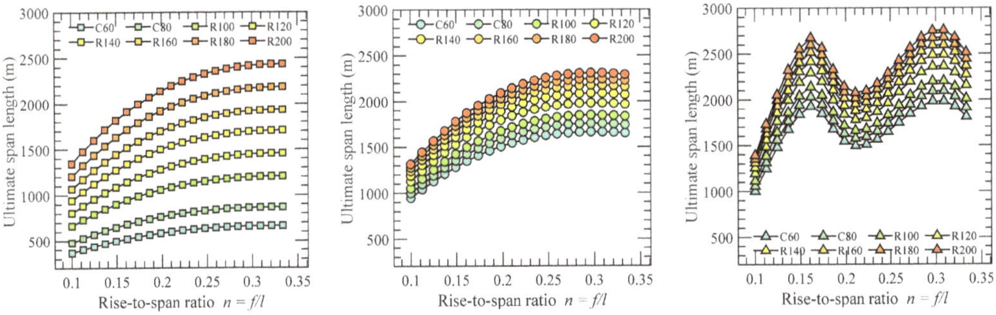

(a) (b) (c)

Figure 4. Ultimate span length of catenary concrete arch bridge: (**a**) l_{max1}; (**b**) l_{max2}; (**c**) l_{max3}.

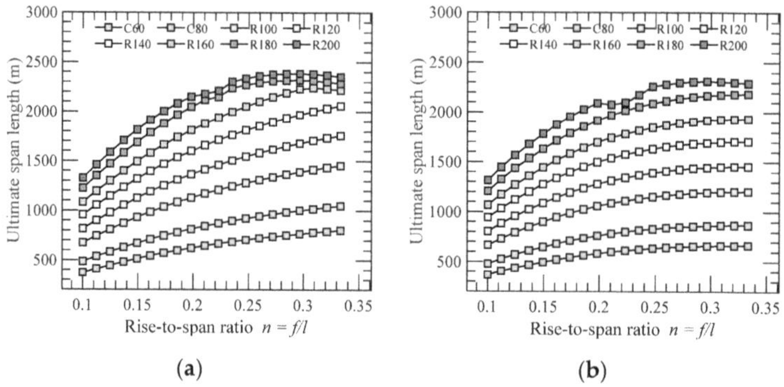

Figure 5. Ultimate span length of concrete arch bridge: (**a**) Parabolic arch axis; (**b**) Catenary arch axis.

As can be seen, the ultimate span length of arch bridge is mainly controlled by the strength condition as the increase of the grade of concrete. The stability condition starts to affect the result when the strength of the concrete reaches up to the grade of R160 for parabolic arch and R180 for catenary arch. The stability condition completely dominates the ultimate span length when the grade of R200 concrete is employed. Specifically, when the rise-span ratio is 1/5, the ultimate span lengths for parabolic and catenary arch axes are about 627 m and 586 m, respectively if the C60 concrete is used. This value can be increased to be 2160 m and 2099 m, respectively if the ultra-high-performance concrete with the grade of R200 is utilized.

4.2. Steel Arch Bridge

The high-performance steel has been well examined and applied in some real bridge structures. For example, the Q500q grade steel was utilized by Husutong Yangtze River Bridge completed in July 2020, which is the second longest span cable-stayed bridge. The Q690q grade steel was applied to the Jianghan 7th Arch Bridge with the main span of 408 m. As suggested by the Specifications for Design of Highway Steel Bridge (JTG D64-2015) [29] and the Structural Steel for Bridge (GB/T 714-2015) [30], the design strength f_d of steel is listed in Table 5. The modulus of elasticity $E = 2.06 \times 10^5$ MPa. The apparent density of concrete $\gamma = 78.5$ kN/m^3.

By substituting the design strength of different steel grades in Table 6 into the theoretical solutions in Table 4, the ultimate span lengths of parabolic and catenary arch bridges with respect to the strength, in-plane stability and out-of-plane stability conditions can be readily obtained, as shown in Figures 6 and 7. The final theoretical ultimate span length is shown in Figure 8.

Table 6. Design strength of steel f_d.

Grade	Q345	Q370	Q420	Q460	Q500	Q550	Q620	Q690
f_d (MPa)	265	285	325	365	380	420	460	520

Similar to the concrete arch bridge, the ultimate span length of steel arch bridge is also mainly controlled by the strength condition as the increase of the grade of steel. The stability condition starts to affect the result when the strength of the steel reaches up to the grade of Q420 for parabolic arch and Q460 for catenary arch. The stability condition completely dominates the ultimate span length when the grade of Q500 steel is employed. Specifically, when the rise-span ratio is 1/5, the ultimate span lengths for parabolic and catenary arch axes are about 2077 m and 1942 m, respectively if the Q345 steel is used. This value can be increased to be 2948 m and 2865 m, respectively if the high-performance steel with the grade of Q690 is utilized.

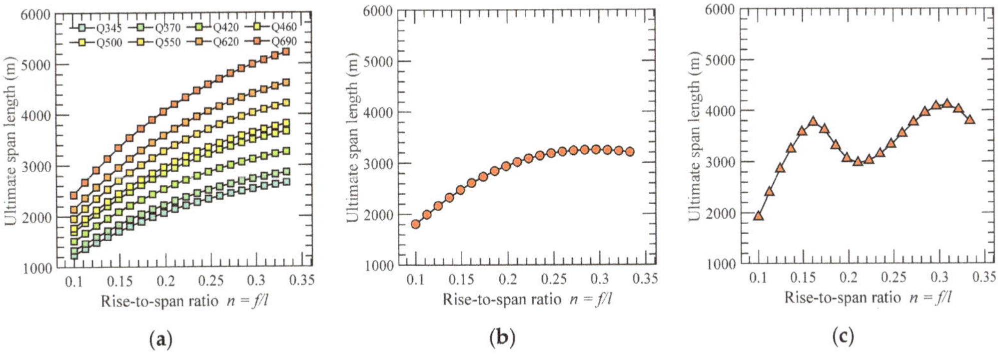

Figure 6. Ultimate span length of parabolic steel arch bridge: (**a**) l_{max1}; (**b**) l_{max2}; (**c**) l_{max3}.

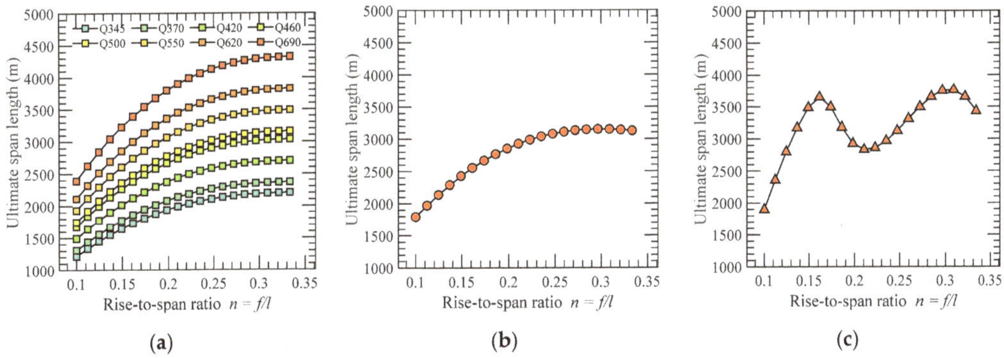

Figure 7. Ultimate span length of steel arch bridge: (**a**) l_{max1}; (**b**) l_{max2}; (**c**) l_{max3}.

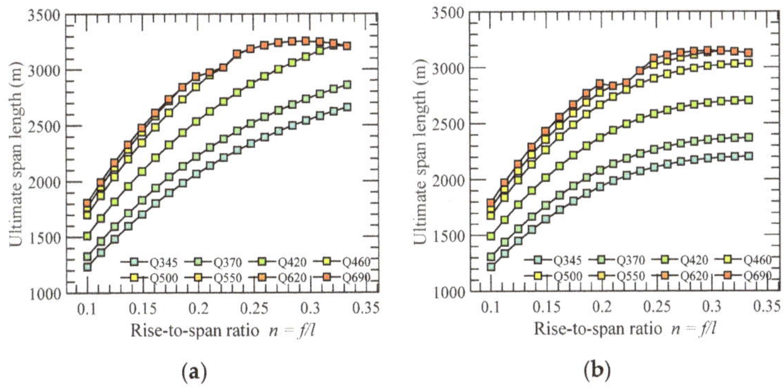

Figure 8. Ultimate span length of steel arch bridge: (**a**) Parabolic arch axis; (**b**) Catenary arch axis.

For comparison purpose, Tables 7 and 8 list the theoretical solutions of ultimate span length for concrete and steel arch bridges using parabolic and catenary arch axes, respectively when the rise-span ratio is 1/5. As can be seen, the results based on catenary arch axis are slightly smaller by about 2–6% than that of parabolic arch axis. This is because that the uniformly distributed load is assumed by the parabolic arch axis, which neglects the increment of dead load at abutment compared to the load at the apex of the arch.

Table 7. Ultimate span length for concrete arch bridge at $n = 1/5$.

Grade	C60	C80	R100	R120	R140	R160	R180	R200
Parabolic arch axis l_{max} (m)	627	819	1136	1372	1609	1822	2058	2161
Catenary arch axis l_{max} (m)	586	766	1062	1283	1505	1704	1925	2099

Table 8. Ultimate span length for steel arch bridge at $n = 1/5$.

Grade	Q345	Q370	Q420	Q460	Q500	Q550	Q620	Q690
Parabolic arch axis l_{max} (m)	2077	2233	2547	2860	2948	2948	2948	2948
Catenary arch axis l_{max} (m)	1942	2089	2382	2675	2785	2865	2865	2865

5. Conclusions and Discussion

Based on the theoretical derivation of strength, in-plane stability and out-of-plane stability conditions, the theoretical solutions of ultimate span length for arch bridge using the parabolic and catenary arch axes are achieved, respectively. They are applied to estimate the ultimate span lengths of concrete and steel arch bridges with various rise-span ratios using different grades of material. The results show that the theoretical ultimate span length based on catenary arch axis is slightly smaller than that of parabolic arch axis. When the rise-span ratio of the bridge is 1/5, the theoretical ultimate span of concrete arch bridge can be longer than 2000 m if the ultra-high-performance concrete with the grade of R200 is utilized. As for steel arch bridge, the ultimate span length can be longer than 2500 m if the high-performance steel with the grade of Q690 is utilized.

The findings of ultimate span length in this study provide the basis to better understanding the length capacity of arch bridge when advanced new materials are utilized. It is worth mentioning that only the arch structure under the action of dead load is analyzed in this study. The design of a real long-span arch bridge is much more complicated. The effects of some other extreme loadings, such as wind [31–33], earthquake [34–37] and vehicle etc. could dominate the design of the arch. Moreover, such other issues, including instability of spandrel column for concrete deck arch bridges [38], strength of hangers for through arch bridge, local buckling instability etc. should be considered during the design of a real arch bridge. The difficulty of construction and rapid increase of cost are also great challenges for the completion of these very long-span arch bridges.

Author Contributions: All four co-authors contributed to the completion of this article. Conceptualization, Z.D. and Y.G.; methodology, Z.D. and G.F.; software, X.Z. and Z.D.; validation, G.F. and Y.G.; formal analysis, X.Z. and Z.D.; investigation, Z.D.; resources, Y.G.; data curation, Z.D.; writing—original draft preparation, G.F.; writing—review and editing, X.Z. and G.F.; visualization, Z.D.; supervision, Y.G.; project administration, G.F. and Y.G.; funding acquisition, G.F. and Y.G. All authors have read and agreed to the published version of the manuscript.

Funding: This research was funded by the National Natural Science Foundation of China (52108469, 51978527), the Shanghai Pujiang Program (20PJ1413600) and the Fundamental Research Funds for the Central Universities (22120220577).

Institutional Review Board Statement: Not applicable.

Informed Consent Statement: Not applicable.

Data Availability Statement: Any interested parties can contact the corresponding author directly via email for information on the data.

Conflicts of Interest: The authors declare no conflict of interest.

References

1. Hu, N.; Dai, G.L.; Yan, B.; Liu, K. Recent development of design and construction of medium and long span high-speed railway bridges in China. *Eng. Struct.* **2014**, *74*, 233–241. [CrossRef]
2. Cheng, J. Optimum design of steel truss arch bridges using a hybrid genetic algorithm. *J. Constr. Steel Res.* **2010**, *66*, 1011–1017. [CrossRef]
3. Liu, J.P.; Chen, B.C.; Li, C.; Zhang, M.J.; Mou, T.M.; Tabatabai, H. Recent application of and research on concrete arch bridges in China. *Struct. Eng. Int.* **2022**. [CrossRef]
4. Thrall, A.P.; Billington, D.P. Bayonne Bridge: The Work of Othmar Ammann Master Builder. *J. Bridge Eng.* **2008**, *13*, 635–643. [CrossRef]
5. Zheng, J.L.; Du, H.L.; Mu, T.M.; Liu, J.; Qin, D.; Mei, G.; Tu, B. Innovations in design, construction, and management of Pingnan Third Bridge—The largest-span arch bridge in the world. *Struct. Eng. Int.* **2022**, *32*, 134–141. [CrossRef]
6. Nazmy, A.S. Stability and lode-carrying capacity of three-dimensional long-span steel arch bridges. *Comput. Struct.* **1997**, *65*, 857–868. [CrossRef]
7. Salonga, J.; Gauvreau, P. Comparative study of the proportions, form, and efficiency of concrete arch bridges. *J. Bridge Eng.* **2014**, *19*, 04013010. [CrossRef]
8. Tang, M.C. Super-long span bridges. *Struct. Infrastruct. Eng.* **2017**, *13*, 722–730. [CrossRef]
9. Ge, Y.J. Challenge and development of long-span arch bridges in statics, dynamics and aerodynamics. In Proceedings of the ARCH 2019: 9th International Conference on Arch Bridges, Porto, Portugal, 2–4 October 2019.
10. Zheng, J.L.; Wang, J.J. Concrete-Filled Steel Tube Arch Bridges in China. *Engineering* **2018**, *4*, 143–155. [CrossRef]
11. Azzara, R.M.; Falco, A.D.; Girardi, M.; Pellegrini, D. Ambient vibration recording on the Maddalena Bridge in Borgo a Mozzano (Italy): Data analysis. *Ann. Geophys.* **2017**, *60*, S0441. [CrossRef]
12. Lin, Y.P.; Zhang, Z.H.; Ma, B.; Zhou, L. Lupu arch bridge, Shanghai. *Struct. Eng. Int.* **2004**, *14*, 24–26. [CrossRef]
13. Järvenpää, E.; Jutila, A. Ultimate spans and optimal rise relations of steel arches. In Proceedings of the IABSE Congress, New York, NY, USA, 4–6 September 2019.
14. Wang, X.C. Study on the reasonable arch axis of long-span concrete-filled steel tubular arch bridge. Master's Thesis, Chongqing Jiaotong University, Chongqing, China, 2019.
15. Xia, M.; Liu, H. Disquisition of limitation span of arch bridge. *Transp. Sci. Technol.* **2005**, *12*, 39–42.
16. Xia, M. Analysis of ultimate carrying capacity for long-span fixed arch bridge with thrust. Ph.D. Thesis, Tongji University, Shanghai, China, 2005.
17. Li, X.H.; Chen, B.C. Development of long span arch bridges. *World Bridge* **2007**, *1*, 9–12.
18. Wang, J. The Study of Ultimate Spans for Steel Arch Bridges. Master's Thesis, Chongqing Jiaotong University, Chongqing, China, 2012.
19. Zhao, H.D. Analysis on the ultimate span of railway concrete deck type arch bridges. *High Speed Railw. Technol.* **2017**, *8*, 29–32.
20. Chen, B.C.; Liu, J.P. Review of construction and technology development of arch bridges in the world. *J. Traffic Transp. Eng.* **2020**, *20*, 27–41.
21. Čandrlić, V.; Radić, J.; Gukov, I. Research of Concrete Arch Bridges up to 1000 m in Span. In Proceedings of the 4th International Conference on Arch Bridge, Advances in Assessment, Structural Design and Construction, Barcelona, Spain, 17–19 November 2004.
22. Zheng, J.; Wang, J.J.; Mou, T.M.; Feng, Z.; Han, Y.; Qin, D. Feasibility study on design and construction of concrete filled steel tubular arch bridge with a span of 700 m. *Strateg. Study Chin. Acad. Eng.* **2014**, *16*, 33–37.
23. Shao, X.D.; He, G. Conceptual design and feasibility study of an 800 m scale steel-UHPC composite truss arch bridge. *China J. Highw. Transp.* **2020**, *33*, 73–82.
24. Shao, X.D.; He, G.; Shen, X.J.; Zhu, P.; Chan, Y. Conceptual design of 1000 m scale steel-UHPFRC composite truss arch bridge. *Eng. Struct.* **2021**, *226*, 111430. [CrossRef]
25. CCCC Highway Consultants Co., Ltd. *General Specification for Design of Highway Bridges and Culverts: JTG D60-2015*; China Communication Press: Beijing, China, 2015.
26. Xiang, H.F.; Liu, G.D. *Stability and Vibration of Arch Structures*; China Communication Press: Beijing, China, 1991.
27. CCCC Highway Consultants Co., Ltd. *Specifications for Design of Highway Reinforced Concrete and Prestressed Concrete Bridges and Culverts: JTG 3362-2018*; China Communication Press: Beijing, China, 2018.
28. Hunan University. *Technical Specifications for Reactive Powder Concrete Structures: DBJ43/T 325-2017*; China Architecture & Building Press: Beijing, China, 2017.
29. CCCC Highway Consultants Co., Ltd. *Specifications for Design of Highway Steel Bridges: JTG D64-2015*; China Communication Press: Beijing, China, 2015.
30. China Iron and Steel Association. *Structural Steel for Bridge: GB/T 714-2015*; Standards Press of China: Beijing, China, 2015.
31. Fang, G.S.; Pang, W.; Zhao, L.; Xu, K.; Cao, S.; Ge, Y. Tropical-Cyclone-Wind-Induced Flutter Failure Analysis of Long-Span Bridges. *Eng. Fail. Anal.* **2022**, *132*, 105933. [CrossRef]
32. Hu, X.N.; Fang, G.S.; Yang, J.Y.; Zhao, L.; Ge, Y. Simplified models for uncertainty quantification of extreme events using Monte Carlo technique. *Reliab. Eng. Syst. Saf.* **2023**, *230*, 108935. [CrossRef]

33. Fang, G.S.; Wei, M.M.; Zhao, L.; Xu, K.; Cao, S.; Ge, Y. Site- and building height-dependent design extreme wind speed vertical profile of tropical cyclone. *J. Build. Eng.* **2022**, *62*, 105322. [CrossRef]
34. Wang, X.W.; Mazumder, R.K.; Salarieh, B.; Salman, A.M.; Shafieezadeh, A.; Li, Y. Machine learning for risk and resilience assessment in structural engineering: Progress and future trends. *J. Struct. Eng.* **2022**, *148*, 03122003. [CrossRef]
35. Pang, Y.; Wang, X. Cloud-IDA-MSA conversion of fragility curves for efficient and high-fidelity resilience assessment. *J. Struct. Eng.* **2021**, *147*, 04021049. [CrossRef]
36. Zhong, J.; Ni, M.; Hu, H.; Yuan, W.; Yuan, H.; Pang, Y. Uncoupled multivariate power models for estimating performance-based seismic damage states of column curvature ductility. *Structures* **2022**, *36*, 752–764. [CrossRef]
37. Khedmatgozar Dolati, S.S.; Mehrabi, A.; Khedmatgozar Dolati, S.S. Application of Viscous Damper and Laminated Rubber Bearing Pads for Bridges in Seismic Regions. *Metals* **2021**, *11*, 1666. [CrossRef]
38. Khedmatgozar Dolati, S.S.; Mehrabi, A. Review of available systems and materials for splicing prestressed-precast concrete piles. *Structures* **2021**, *30*, 850–865. [CrossRef]

sustainability MDPI

Article

An Analytical Solution for Unsteady Aerodynamic Forces on Streamlined Box Girders with Coupled Vibration

Lianhuo Wu [1], Mingjin Zhang [1,*], Fanying Jiang [1], Zelin Zhou [2] and Yongle Li [1]

1 Department of Bridge Engineering, Southwest Jiaotong University, Chengdu 610031, China
2 China 19th Metallurgical Corporation, Chengdu 610031, China
* Correspondence: zhang-minjin@swjtu.edu.cn

Abstract: This paper is a contribution to analyzing the aerodynamic forces on a streamlined box girder (SBG) with coupled vibration in a potential flow. The key enabling step was to assume that the normal velocity of the airflow at an arbitrary point on the surface of the SBG was equal to the normal velocity of the surface motion. The aerodynamic drag force, lift force, and pitching moment were expressed as functions of the motion state of the SBG and the SBG's shape-related parameters. To investigate the applicability of this force model, the analytical solution at various angles of attack was compared with a numerical simulation in a viscous flow. The results imply that the amplitude of the analytical lift force and pitching moment agree well with the numerical results under the angles of attack of 0° and ±3°. Furthermore, the analytical drag force effectively predicts the second-order phenomenon resulting from the multiplication of the vertical and torsional vibration velocities. As a consequence, the present analytical solution provides an effective method for analyzing the aerodynamic forces acting on SBGs with coupled vibration.

Keywords: aerodynamic forces; streamlined box girder; coupled vibration; potential flow; second-order component

check for updates

Citation: Wu, L.; Zhang, M.; Jiang, F.; Zhou, Z.; Li, Y. An Analytical Solution for Unsteady Aerodynamic Forces on Streamlined Box Girders with Coupled Vibration. *Sustainability* **2023**, *15*, 7312. https://doi.org/10.3390/su15097312

Academic Editor: Firoz Alam

Received: 6 March 2023
Revised: 18 April 2023
Accepted: 26 April 2023
Published: 27 April 2023

1. Introduction

The shift towards environmentally friendly and sustainable solutions has become a ubiquitous aspect of modern life. Over the course of several decades, it has played a significant role in shaping the design, construction, and maintenance of bridges. Wind can cause significant vibrations on bridges, including vortex-induced vibration, flutter, and buffet [1,2], which have the potential to damage the bridge's structure and compromise its long-term sustainability. Therefore, it is crucial to conduct an aerodynamic force study on bridges to ensure their wind resistance safety.

The analysis of unsteady aerodynamic forces and vibrations has been a subject of significant interest in the field of engineering [3–5], particularly in the design and analysis of streamlined box girder (SBG) bridges. The increasing demand for long-span bridges makes the bridge structure increasingly sensitive to the wind load [6,7], which can result in remarkable structural vibrations [8–10] and even the collapse of the bridge [11]. The potential flow theory can provide a useful framework for understanding the basic principles of fluid flow and the resulting forces on structures. It has been extensively used to design airfoils. However, the application of the potential flow theory to the aerodynamic forces acting on SBGs has not been thoroughly studied. This study proposes an analytical solution for unsteady aerodynamic forces acting on SBGs with coupled vibration in a potential flow and investigates its applicability.

A model coupled via aerodynamic derivatives has long been widely used to describe aerodynamic forces [12,13]. Through flutter derivatives and the motion state of the main girder, the aerodynamic drag force, lift force, and pitching moment can be determined. The flutter derivatives can be obtained by measuring the aerodynamic force at a limited number of wind speeds via the wind tunnel test [14–16] and a computational fluid dynamics (CFD) study [17–20]. The classical Scanlan model is linearized. The frequency of an unsteady force is equal to the coupled vibration frequency. It is not able to explain the high-order unsteady force phenomenon, which was discovered through wind tunnel tests and CFD [21,22]. The common step to consider the high-order component is to add the high-order term in the aerodynamic force model. However, there are few explanations for the generation of high-order components.

The potential flow theory has been widely used in predicting the aerodynamic forces on an airfoil. Von Karman and Sears [23] and Theodorsen [24] presented the general theory of aerodynamic forces on an oscillating thin airfoil. Edwards et al. [25] derived the unsteady aerodynamic loads due to the arbitrary motions of a thin wing. Sun and Wang [26] developed a numerical model for calculating the unsteady flow of an elliptical circulation-control airfoil. Stangfeld et al. [27] compared the aerodynamic loads acting on a relatively thick NACA 0018 airfoil (a symmetric airfoil with a thickness-to-chord ratio of 18% and a maximum thickness located at 30% of the chord length) with pitching and heaving vibrations. Excellent correspondence was found between the experimentally obtained aerodynamic forces and potential theory. Catlett et al. [28] investigated the unsteady pressures and forces on airfoils with an oscillating motion. The unsteady lift forces and pitching moment of the thin airfoils agreed well with the CFD results. In addition, the potential theory was also applied to evaluate the aerodynamic loads on a railway train [29,30] and cylinders [31]. All the aforementioned works indicated the good performance of the potential flow theory in the calculation of aerodynamic forces on various types of structures. However, it is difficult to find research studying the application of the potential flow theory in bridge engineering.

Studying the analytical solution for aerodynamic forces is of great significance in enhancing the sustainability of bridge structures. However, the real flow is too complex to find an explicit expression to calculate the aerodynamic force. Therefore, this paper primarily concentrates on the analytical aerodynamic forces of an SBG with coupled vibration based on the potential flow theory. This solution is expected to offer insights into the underlying aerodynamic principles, paving the way for more precise models of aerodynamic performance in future studies. Meanwhile, the distinction between the aerodynamic forces in a potential flow and a viscous flow was analyzed.

To investigate the unsteady aerodynamic forces on the SBG, the present paper is organized as follows: Section 2 renders the complex potential on the surface of the SBG and in the whole flow domain. The analytical solution of the aerodynamic forces is derived in Section 3. In Section 4, to study the applicability of the potential flow solution in bridge engineering, the analytical solution is compared with the numerical results in a viscous flow. Finally, the conclusions are summarized.

2. Complex Potential

2.1. Stream Function on the Surface of an SBG

The complex potential $w(z)$ is an analytic function in an ideal flow whose real part is the velocity potential ϕ and whose imaginary part is the stream function ψ, as presented in Equation (1), where i is the imaginary unit. Thus, to obtain the aerodynamic force of a moving SBG, it is essential to determine the complex potential of the SBG.

$$w(z) = \phi + i\psi \tag{1}$$

Assume that an SBG is in an arbitrary coupled vibration of its horizontal velocity u_x, vertical velocity u_y, and angular velocity ω, as displayed in Figure 1. The coordinate of a point P on the SBG relative to the center of mass of the SBG is (x_p, y_p); then, the velocity of point P can be expressed as follows:

$$\begin{cases} u_{px} = u_x + \omega y_p \\ u_{py} = u_y - \omega x_p \end{cases} \tag{2}$$

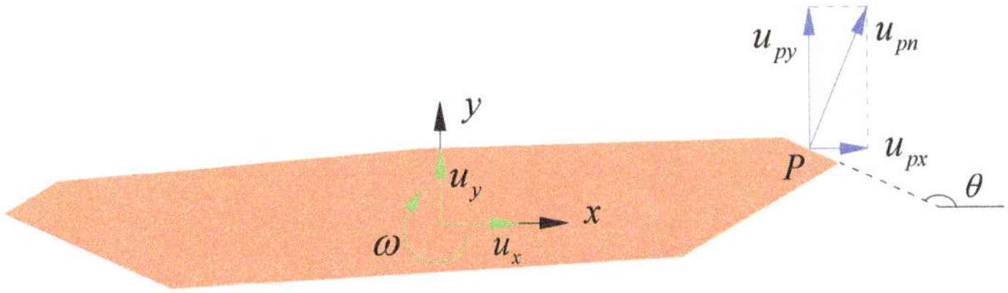

Figure 1. SBG in arbitrary coupled motion.

Let the angle between the surface of point P and the positive direction of the x-axis be θ; then, the normal velocity u_{pn} at point P becomes:

$$u_{pn} = \left(u_x + \omega y_p\right)\sin\theta - \left(u_y - \omega x_p\right)\cos\theta \tag{3}$$

According to the geometrical condition, we have:

$$\sin\theta = \tfrac{dy}{dc} \quad , \quad \cos\theta = \tfrac{dx}{dc} \tag{4}$$

where c is the surface of SBG; dx, dy, and dc denote the infinitesimal increase in x, y, and c.

Since the air cannot penetrate through the surface of the SBG, the normal velocity of the airflow at point P is equal to the normal velocity of the SBG motion at point P. Therefore, the normal velocity u_{pn} can be expressed as follows:

$$u_{pn} = \frac{\partial\psi}{\partial c} \tag{5}$$

Substituting Equations (4) and (5) into Equation (3) leads to:

$$\frac{\partial\psi}{\partial c} = (u_x + \omega y)\frac{dy}{dc} - \left(u_y - \omega x\right)\frac{dx}{dc} \tag{6}$$

where $\partial\psi$ and ∂c represent the partial differentials of ψ and c, respectively. Integrating Equation (6), the stream function ψ can be given as follows:

$$\psi = u_x y - u_y x + \frac{1}{2}\omega\left(x^2 + y^2\right) + Const \tag{7}$$

where $Const$ means a constant.

2.2. Complex Potential in the Fluid Domain

To obtain the complex potential, we define a function:

$$f(z,\overline{z}) = \overline{u}z + \frac{1}{2}i\omega z\overline{z} \tag{8}$$

in which $z = x + iy$, $u = u_x + iu_y$; $\overline{z}$ and $\overline{u}$ are complex conjugate of z and u, respectively. Then, the imaginary part of Equation (8) is equal to Equation (7) without the constant term, i.e.,

$$2i\psi = f(z,\overline{z}) - \overline{f}(z,\overline{z}) = \overline{u}z - u\overline{z} + i\omega z\overline{z} \tag{9}$$

where $\overline{f}$ is the complex conjugate of f. Assume that $z = f(\zeta)$ is the conformal mapping function from a unit circle in the ζ-plane to the SBG in the z-plane. The point on the edge of the unit circle ζ_u can be written as:

$$\zeta_u = e^{i\theta} \quad , \quad \overline{\zeta}_u = e^{-i\theta} = \tfrac{1}{\zeta_u} \tag{10}$$

where $\overline{\zeta}_u$ is the complex conjugate of ζ_u. Introducing $z = f(\zeta)$ and Equation (10) into Equation (9) leads to:

$$2i\psi = \overline{u}f(\zeta_u) - uf\left(\frac{1}{\zeta_u}\right) + i\omega f(\zeta_u)f\left(\frac{1}{\zeta_u}\right) = B(\zeta_u) \tag{11}$$

in which $B(\zeta_u)$ is the surface function of SBG, expressed by ζ_u. The complex potential of the unit circle satisfies:

$$w(\zeta_u) - w\left(\frac{1}{\zeta_u}\right) = 2i\psi \tag{12}$$

Therefore, the relationship between the complex potential on the surface of the unit circle and the surface function can be exhibited as:

$$w(\zeta_u) - w\left(\frac{1}{\zeta_u}\right) = B(\zeta_u) \tag{13}$$

Here, the complex potential on the surface will be extended to the full fluid domain. The surface function can be written as:

$$B(\zeta_u) = B_1(\zeta_u) + B_2(\zeta_u) \tag{14}$$

where $B_1(\zeta_u)$ is the negative exponential term, and $B_2(\zeta_u)$ is the non-negative exponential term. Let Equation (14) be divided by $2\pi i(\zeta_u - \zeta)$. Integrating both sides of the equation on the surface of the unit circle arrives at:

$$\oint_C \frac{B(\zeta_u)}{2\pi i(\zeta_u - \zeta)}d\zeta_u = \frac{1}{2\pi i}\oint_C \frac{B_1(\zeta_u)}{\zeta_u - \zeta}d\zeta_u + \frac{1}{2\pi i}\oint_C \frac{B_2(\zeta_u)}{\zeta_u - \zeta}d\zeta_u = \frac{1}{2\pi i}\oint_C \frac{w(\zeta_u)}{\zeta_u - \zeta}d\zeta_u - \frac{1}{2\pi i}\oint_C \frac{w(1/\zeta_u)}{\zeta_u - \zeta}d\zeta_u \tag{15}$$

On the basis of the Cauchy integral formula [32]:

$$\begin{array}{ll} \frac{1}{2\pi i}\oint_C \frac{w(\zeta_u)}{\zeta_u-\zeta}d\zeta_u = w(\zeta) & , \quad \frac{1}{2\pi i}\oint_C \frac{w(1/\zeta_u)}{\zeta_u-\zeta}d\zeta_u = 0 \\ \frac{1}{2\pi i}\oint_C \frac{B_1(\zeta_u)}{\zeta_u-\zeta}d\zeta_u = B_1(\zeta) & , \quad \frac{1}{2\pi i}\oint_C \frac{B_2(\zeta_u)}{\zeta_u-\zeta}d\zeta_u = 0 \end{array} \tag{16}$$

Then the complex potential of the SBG becomes:

$$w(\zeta) = B_1(\zeta) \tag{17}$$

The conformal mapping from a unit circle in the ζ-plane to the SBG in the z-plane can be expressed in the Laurent series [33] as follows:

$$z = \sum_{k=1}^{n} c_k\zeta^{2-k} \tag{18}$$

in which $c_k = a_k + ib_k$, a_k, and b_k and are real coefficients to be determined [34]. Introducing Equation (18) into Equation (11), the surface function of SBG is then determined as:

$$B(\zeta) = \overline{u}\sum_{k=1}^{n} c_k\zeta^{2-k} - u\sum_{k=1}^{n} \overline{c}_k\overline{\zeta}^{2-k} + i\omega\left(\sum_{k=1}^{n} c_k\zeta^{2-k}\right)\left(\sum_{k=1}^{n} \overline{c}_k\overline{\zeta}^{2-k}\right) \tag{19}$$

According to Equation (17), the complex potential of the SBG with coupled vibration is the negative exponential term of the surface function. Therefore, the complex potential can be obtained from Equation (19):

$$w(\zeta) = \overline{u}\sum_{k=3}^{n} c_k\zeta^{2-k} - u\overline{c}_1\frac{1}{\zeta} + i\omega\sum_{k=1}^{n-1}\sum_{j=k+1}^{n} \overline{c}_k c_j\zeta^{k-j} \tag{20}$$

3. Aerodynamic Forces

3.1. Pressure Function

Due to the low density of the airflow, the resultant pressure induced by the variation in the relative elevation of the SBG is deemed negligible. This implies that any minute alterations in the SBG's height are unlikely to exert a noteworthy influence on the ambient airflow, and the associated pressure can therefore be justifiably disregarded. In light of the unsteady Bernoulli equation [35], the pressure equation of the ideal incompressible fluid can be expressed as [28]:

$$p+\rho\frac{\partial\phi}{\partial t}+\frac{1}{2}\rho V^2 = Const(t) \tag{21}$$

where p is the aerodynamic pressure; ρ is the air density; ϕ is the potential function; $V = (u_x, u_y)$ is the vector of wind velocity; *Const*(t) means a constant.

To obtain the pressure function, two coordinate systems are established. As shown in Figure 2, $x^*O^*y^*$ is a moving coordinate system that moves with the SBG whose x^*-axis is always parallel to the long axis of the SBG. xOy is a static coordinate system. α is the angle between the two coordinate systems. The velocity component is represented by u_x^* and u_y^* in the $x^*O^*y^*$ coordinate system and u_x and u_y in the xOy coordinate system, respectively. Their relationships are then expressed as follows:

$$\begin{cases} u_x^* = u_x\cos\alpha + u_y\sin\alpha \quad , \quad u_y^* = -u_x\sin\alpha + u_y\cos\alpha \\ dx^* = dx\cos\alpha + dy\sin\alpha \quad , \quad dy^* = -dx\sin\alpha + dy\cos\alpha \end{cases} \tag{22}$$

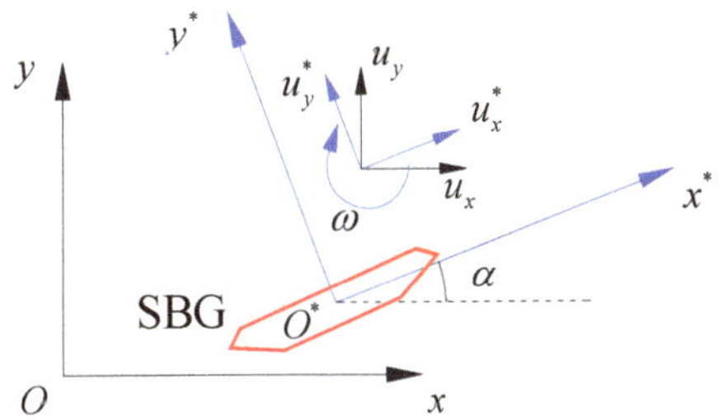

Figure 2. Moving and static coordinate system.

The potential function in the $x^*O^*y^*$ coordinate system is defined as $\phi^* = \phi^*(x^*, y^*, t^*)$ $= \int\left(u_x^*dx^* + u_y^*dy^*\right)$. Introducing Equation (22) into the above formula, it can be found that $\phi^* = \phi$, i.e., the potential functions in the two coordinate systems are equal. The partial derivative of ϕ with respect to time t is:

$$\frac{\partial\phi}{\partial t} = \lim_{\Delta t\to 0}\frac{\phi^*\left(x_{t+\Delta t}^*, y_{t+\Delta t}^*, t+\Delta t\right) - \phi^*\left(x_t^*, y_t^*, t\right)}{\Delta t} \tag{23}$$

In the $x^*O^*y^*$ coordinate system, the location of (x^*, y^*) at time $t + \Delta t$ can be expressed as:

$$x_{t+\Delta t}^* = x_t^* - (u_x + \omega y_t^*)\Delta t \quad , \quad y_{t+\Delta t}^* = y_t^* - (u_y - \omega x_t^*)\Delta t \tag{24}$$

Substituting Equation (24) into Equation (23), we arrive:

$$\frac{\partial\phi}{\partial t} = \frac{\partial\phi^*}{\partial t} - V_0\cdot\nabla^*\phi^* \tag{25}$$

Since

$$V_0\cdot\overline{V}_0 = ((u_x+\omega y)+i(u_y-\omega x))\cdot\overline{((u_x+\omega y)+i(u_y-\omega x))} = (u-i\omega z)(\overline{u}+i\omega\overline{z}) \tag{26}$$

Finally, introducing Equations (25) and (26) into Equation (21), the pressure function p can finally be obtained as follows:

$$p = -\rho\frac{\partial\phi}{\partial t} + \frac{1}{2}\rho(u - i\omega z)(\overline{u} + i\omega\overline{z}) - \frac{1}{2}\rho e^{i2\beta}\left(\frac{dw(z)}{dz} - (\overline{u} + i\omega\overline{z})\right)^2 \tag{27}$$

where β is the angle between the vector formed by a point in the flow field and the coordinate origin and the x-axis.

3.2. Drag Force and Lift Force

On the basis of the complex variable theory, it is known that $d\overline{z} = dze^{-i2\beta}$. Integrating the pressure function Equation (27) on the surface of the SBG, the aerodynamic force of the SBG can be acquired:

$$F_x - iF_y = -i\oint_C pd\overline{z} = -i\oint_C pe^{-i2\beta}dz \tag{28}$$

where F_x is the aerodynamic drag force of SBG, and F_y is the aerodynamic lift force of SBG. Substituting Equation (27) into Equation (28), the complex expression of the aerodynamic force becomes:

$$\begin{aligned} F_x - iF_y = &-\tfrac{1}{2}i\rho\oint_C (u - i\omega z)(\overline{u} + i\omega\overline{z})d\overline{z} + \tfrac{1}{2}i\rho\oint_C \left(\tfrac{dw(z)}{dz}\right)^2 dz \\ &+\tfrac{1}{2}i\rho\oint_C (\overline{u} + i\omega\overline{z})^2 dz - i\rho\oint_C \tfrac{dw(z)}{dz}(\overline{u} + i\omega\overline{z})dz + i\rho\oint_C \tfrac{\partial\phi}{\partial t}d\overline{z} \end{aligned} \tag{29}$$

To simplify Equation (29), we can define the function $f(z, \overline{z})$ as:

$$\begin{cases} \frac{\partial f}{\partial x} = \frac{\partial f}{\partial z}\frac{\partial z}{\partial x} + \frac{\partial f}{\partial \overline{z}}\frac{\partial \overline{z}}{\partial x} = \frac{\partial f}{\partial z} + \frac{\partial f}{\partial \overline{z}} \\ \frac{\partial f}{\partial y} = \frac{\partial f}{\partial z}\frac{\partial z}{\partial y} + \frac{\partial f}{\partial \overline{z}}\frac{\partial \overline{z}}{\partial y} = i\left(\frac{\partial f}{\partial z} - \frac{\partial f}{\partial \overline{z}}\right) \end{cases} \tag{30}$$

According to Stokes' theorem [36], the integral along the surface of SBG can be expressed as an area fraction:

$$\oint_C fdr = \int_S e_z \times \nabla f dS = e_z \times \int_S \left(\frac{\partial f}{\partial x}e_x + \frac{\partial f}{\partial y}e_y\right)dS \tag{31}$$

where S is the inner region of SBG, and $\boldsymbol{e}_x$, $\boldsymbol{e}_y$, and $\boldsymbol{e}_z$ are the unit vectors along the coordinate axis. Considering that $\boldsymbol{e}_y = \boldsymbol{e}_z \times \boldsymbol{e}_x = i\boldsymbol{e}_x$, $\boldsymbol{dr} = \boldsymbol{e}_x dx + \boldsymbol{e}_y dy = \boldsymbol{e}_x dz$, Equation (31) becomes:

$$\oint_C fdz = i\int_S \left(\frac{\partial f}{\partial x} + i\frac{\partial f}{\partial y}\right)dS \tag{32}$$

Introducing Equation (32) into Equation (30), we have:

$$\begin{cases} \oint_C f(z, \overline{z})dz = 2i\int_S \frac{\partial f}{\partial \overline{z}}dS \\ \oint_C f(z, \overline{z})d\overline{z} = -2i\int_S \frac{\partial f}{\partial z}dS \end{cases} \tag{33}$$

Assume that the area of the SBG is A, and z_c is the geometric center of the SBG, which can be expressed as $z_c = \frac{1}{A}\int_S zdS$. Then, in the light of Equation (33), the first and third terms in Equation (29) become:

$$\begin{aligned} &-\tfrac{1}{2}i\rho\oint_C (u - i\omega z)(\overline{u} + i\omega\overline{z})d\overline{z} = -\tfrac{1}{2}i\rho \cdot 2i\int_S i\omega(\overline{u} + i\omega\overline{z})dS = i\rho\omega A(\overline{u} + i\omega\overline{z}_c) \\ &\tfrac{1}{2}i\rho\oint_C (\overline{u} + i\omega\overline{z})^2 dz = \tfrac{1}{2}i\rho \cdot 2i\int_S 2i\omega(\overline{u} + i\omega\overline{z})dS = -2i\rho\omega A(\overline{u} + i\omega\overline{z}_c) \end{aligned} \tag{34}$$

According to the definition of the complex potential function, it is known that $\oint_C \phi d\bar{z} = \frac{1}{2}(\oint_C w d\bar{z} + \oint_C \overline{w} d\bar{z})$. Thus, the expression of the resultant force of the SBG expressed by the complex potential function can finally be attained as follows:

$$\begin{aligned} F_x - iF_y = & -i\rho\omega A(\overline{u} + i\omega\bar{z}_c) + \tfrac{1}{2}i\rho\oint_C \left(\tfrac{dw(z)}{dz}\right)^2 dz - i\rho\overline{u}\oint_C dw(z) \\ & +\rho\omega\oint_C \bar{z}dw(z) + \tfrac{1}{2}i\rho\tfrac{\partial}{\partial t}\left(\oint_C w(z)d\bar{z} + \oint_C \overline{w}(z)d\bar{z}\right) \end{aligned} \tag{35}$$

By substituting Equations (18) and (20) into Equation (35) and then integrating it, the aerodynamic drag force and lift force are represented as follows:

$$\begin{aligned} F_x &= -\rho A\omega u_y + \mathrm{Re}(F) \\ F_y &= \rho A\omega u_x - \mathrm{Im}(F) \end{aligned} \tag{36}$$

where Re(F) and Im(F) are the real part and imaginary part of F, which is expressed as:

$$\begin{aligned} F = & \rho\omega 2\pi i\left(\overline{u}\sum_{k=3}^{n}(2-k)\bar{c}_k c_k + uc_1c_3 + i\omega\sum_{k=3}^{n}\sum_{j=1}^{n-k+2}\bar{c}_k\bar{c}_j c_{j+k-2}(-2-k)\right) - \pi\rho\left(\frac{\partial\overline{u}}{\partial t}\left(c_1^2 + \sum_{k=3}^{n}(k-2)c_k\bar{c}_k\right)\right. \\ & \left. -\frac{\partial u}{\partial t}2c_1c_3 + i\frac{\partial\omega}{\partial t}\left(-\sum_{k=3}^{n-1}\bar{c}_1\bar{c}_k c_{k-1} + \sum_{k=3}^{n}\sum_{j=k-1}^{n}(k-2)\bar{c}_k\bar{c}_{j-k+2}c_j + \sum_{k=1}^{n-1}c_1c_k\bar{c}_{k+1} - \sum_{k=3}^{n-1}\sum_{j=k}^{n-1}(k-2)\bar{c}_k c_j\bar{c}_{j-k+2}\right)\right) \end{aligned} \tag{37}$$

3.3. Pitching Moment

By performing the integration of the moment of surface pressure on the geometric center of the SBG, the aerodynamic pitching moment of the SBG can be expressed as:

$$M = -\mathrm{Re}\left(\oint_C pzd\bar{z}\right) \tag{38}$$

By incorporating the mathematical expressions represented by Equations (18), (20), and (27) into Equation (38), a more comprehensive and refined equation can be obtained. Subsequently, integrate Equation (38). The pitching moment can then be expressed as follows:

$$\begin{aligned} M = \mathrm{Re}\Bigg(& -\rho\omega uA\bar{z}_C + 2\pi i\rho\overline{u}\left(\overline{u}c_1c_3 + i\omega\sum_{k=1}^{n-1}c_1\bar{c}_k c_{k+1}\right) + 2\pi\rho\omega\Bigg(u\sum_{k=1}^{n-1}c_k\bar{c}_{k+1}\bar{c}_1 + \overline{u}\sum_{l=3}^{n}(2-l)c_1\bar{c}_{l-1}c_l \\ & \left. +\overline{u}\sum_{k=2}^{n-1}\sum_{l=3}^{n-k+2}(2-l)c_k\bar{c}_{k+l-2}c_l\right) + \pi\rho i\frac{\partial}{\partial t}\Bigg(\overline{u}\sum_{j=3}^{n-1}(j-2)c_1\bar{c}_j c_{j+1} + \overline{u}\sum_{k=2}^{n-1}\sum_{j=k+1}^{n}(j-2)c_k\bar{c}_j c_{j-k+2} \\ & -u\sum_{k=2}^{n-1}(k-1)c_k\bar{c}_{k+1}\bar{c}_1 + i\omega\sum_{l=1}^{n-2}\sum_{j=3}^{n-l+1}(j-2)c_1\bar{c}_j\bar{c}_l c_{j+l-1} + i\omega\sum_{k=2}^{n-1}\sum_{l=1}^{n-1}\sum_{j=k+1}^{dyn}(j-2)c_k\bar{c}_j\bar{c}_l c_{j-k+l} \\ & +u\sum_{k=4}^{n}\sum_{j=3}^{k-1}(j-2)c_k\bar{c}_j\bar{c}_{k-j+2} - \overline{u}\sum_{k=4}^{n}(k-3)c_1c_k\bar{c}_{k-1} - u\sum_{k=2}^{n-1}\bar{c}_1c_k\bar{c}_{k+1} + \overline{u}c_1\bar{c}_1c_2 \\ & \left.\left. +i\omega\sum_{k=1}^{n-1}\sum_{j=k+1}^{n}\bar{c}_1c_{j-k+1}c_k\bar{c}_j - i\omega\sum_{k=4}^{n-1}\sum_{l=1}^{n-1}\sum_{j=dyx}^{k-1}(j-2)c_k\bar{c}_j c_l\bar{c}_{k-j+l}\right)\right) \end{aligned} \tag{39}$$

where $dyn = \min\{n, n-l+k\}$, $dyx = \max\{3, l+k-n\}$.

4. Comparison with the Viscous Flow

Unlike the assumed inviscid fluid, the real airflow is viscous. Therefore, to verify the applicability of the potential flow solution of Equations (37) and (39) in bridge engineering, the unsteady aerodynamic forces of three SBG shapes in the viscous mean wind flow are investigated via CFD simulation. The three SBGs are shown in Figure 3, in which B1 is the deck of the Great Belt East Bridge [37]. The differences between the three SBGs are their heights, which represent different degrees of bluntness. The conformal mapping coefficients of the three boxes in Equation (18) are listed in Table 1, based on the direct iteration conformal mapping method by Wu et al. [27].

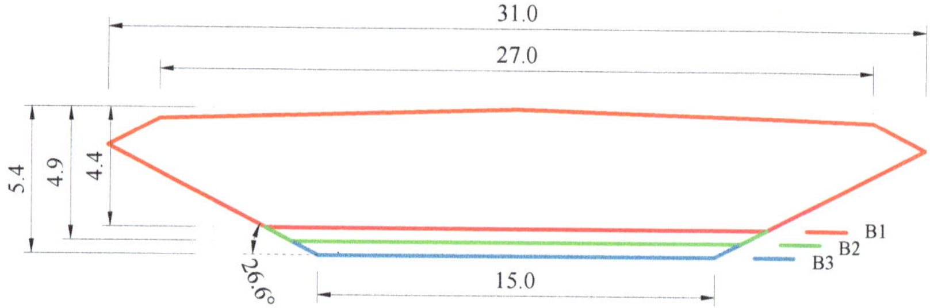

Figure 3. Sketch map of SBG shapes (unit: m).

Table 1. Conformal mapping coefficients.

SBG	c_1	c_2	c_3	c_4	c_5	c_6	c_7	c_8	c_9	c_{10}
B1	8.79	−0.57i	6.45	0.45i	−0.07	0.16i	0.16	−0.09i	0.09	−0.04i
B2	8.86	−0.70i	6.30	0.59i	−0.01	0.16i	0.18	−0.12i	0.06	−0.02i
B3	8.93	−0.82i	6.16	0.72i	0.07	0.13i	0.19	−0.14i	0.03	0.01i

4.1. Numerical Model Verification

The aerodynamic forces of SBGs in a viscous flow are computed via numerical simulation. To ensure the validity of the numerical model, the simulation of B3 (i.e., the Great Belt East Bridge) was conducted first. A hybrid grid scheme of the structured grids and the unstructured grids with a total grid number of 291,202 was adopted, as presented in Figure 4. To simulate the motion of the SBG, the computational domain was divided into three regions, namely, the rigid mesh zone, dynamic mesh zone, and fixed mesh zone. The rigid mesh zone moved along with the SBG to ensure the quality of the mesh near the SBG. The normal spacing next to the SBG wall was 5×10^{-6} m. The height and width of the computational domain were 12 B and 18 B, respectively, in which B is the width of the SBG. The inlet boundary of the computational domain was set to the velocity inlet. The right boundary of the computational domain was set to the pressure outlet. The top and bottom boundaries were set as symmetry boundaries.

The ANSYS Fluent program was adopted to compute the aerodynamic forces of the SBG. Turbulence was modeled using the SST k-ω viscous scheme, and the PISO scheme was selected to solve the pressure–velocity coupling. The least-squares cell-based scheme was used to compute the gradient, while the second-order interpolation scheme was utilized for pressure approximation. Furthermore, the second-order upwind scheme was implemented for other spatial discretization terms. The dimensionless time step was set to 0.001 s to ensure accuracy.

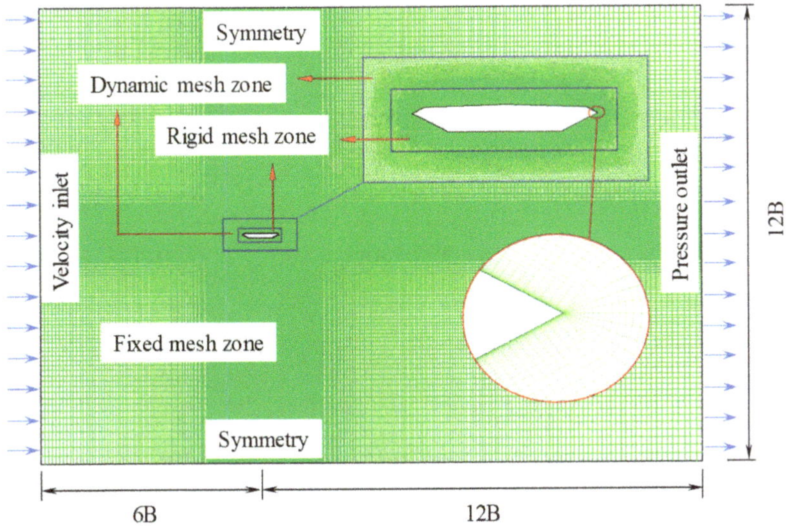

Figure 4. Numerical simulation computational domain and mesh grid.

For the sake of reliable numerical results, the numerical model must be validated. The forced single-degree-of-freedom vertical and pitching harmonic vibrations were carried out. The amplitude of the vibration was set at 0.02 B for vertical motion and 3° for pitching motion. The vibration frequency was 2.0 Hz for both the vertical and pitching vibrations. A comparison of the flutter derivatives is exhibited in Figure 5, for which the wind tunnel test was conducted by Reinhold et al. [38]. The flutter velocity is presented in Table 2. The results indicate that the flutter derivatives obtained from the numerical simulation are consistent with those from the wind tunnel test. The flutter critical wind velocity and the flutter frequency have errors of only 0.3% and 3.7%, respectively, compared to the experimental values. This suggests that the present numerical model demonstrates good accuracy in computing the aerodynamic forces of the SBG. Since B2 and B3 used the same mesh scheme and numerical settings as B1, it is reasonable to assume that their CFD-generated aerodynamic forces are also reliable.

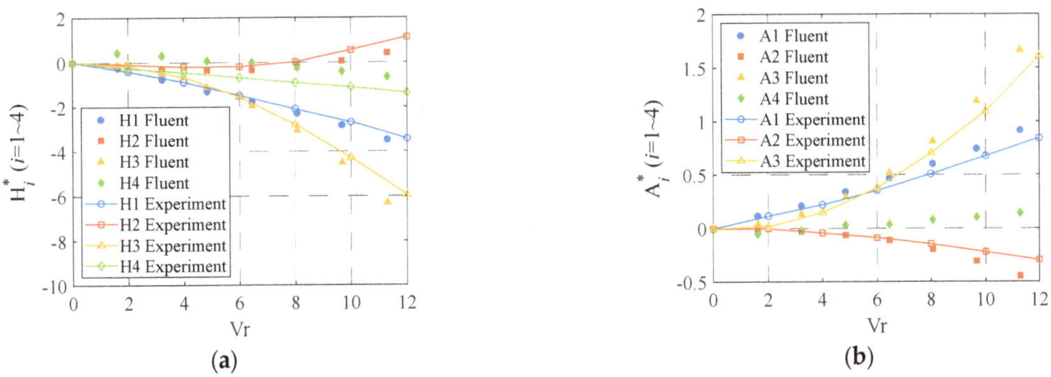

Figure 5. Comparison of flutter derivatives: (**a**) H_i^* (i = 1~4); (**b**) A_i^* (i = 1~4).

Table 2. Comparison of flutter critical velocity and frequency.

Method	Wind Tunnel Test [38]	Discrete Vortex Method [39]	Present Simulation
Flutter critical velocity (m/s)	37.6	35.3	37.5
Flutter frequency (Hz)	0.165	0.165	0.159

4.2. Aerodynamic Forces Comparative Analysis

The dimensionless aerodynamic coefficients C_d, C_l, and C_m are utilized to describe the aerodynamic forces on SBGs. The expressions of the aerodynamic coefficients are displayed in Equation (40), where V = 10.0 m/s is the wind velocity, and H means the height of the SBGs.

$$C_d = F_x/\left(0.5\rho V^2 H\right) \quad ; \quad C_l = F_y/\left(0.5\rho V^2 B\right) \quad ; \quad C_m = M/\left(0.5\rho V^2 B^2\right) \tag{40}$$

Regarding the action of the wind load, assume that the SBGs are in the coupled vibration of $u_x = -V + A_x 2\pi f \cos(2\pi f t)$, $u_y = -A_y 2\pi f \cos(2\pi f t)$, and $\omega = A_m 2\pi f \cos(2\pi f t)$, where A_x = 0.01 B, A_y = 0.02 B, A_m = 3°, B = 31.0 m, f = 0.2 Hz, and t denotes time series. Figure 6 illustrates a comparison of the aerodynamic drag force, lift force, and pitching moment of three SBGs in both potential and viscous flows. The results reveal that the evolution law of aerodynamic force with time via the potential solution is close to the viscous flow. Since there is no flow separation in a potential flow, the drag force in a potential flow is smaller than the drag force in a viscous flow. However, the potential solution reproduces the second-order component of the drag force. The second-order effect was also observed before [16,22] and cannot be achieved by the Scanlan model [13], as expressed in Equation (41), where P_i^* (i = 1,3,5,6) represent the flutter derivatives related to the drag force. The second-order component is generated by $-\rho A\omega u_y$ and $\mathrm{Re}\left(\rho\omega 2\pi i\left(\overline{u}\sum_{k=3}^{n}(2-k)\overline{c}_k c_k + u c_1 c_3\right)\right)$ in Equation (36), indicating that the second-order component is primarily caused by the multiplication of the vertical vibration velocity u_y and the pitching vibration velocity ω.

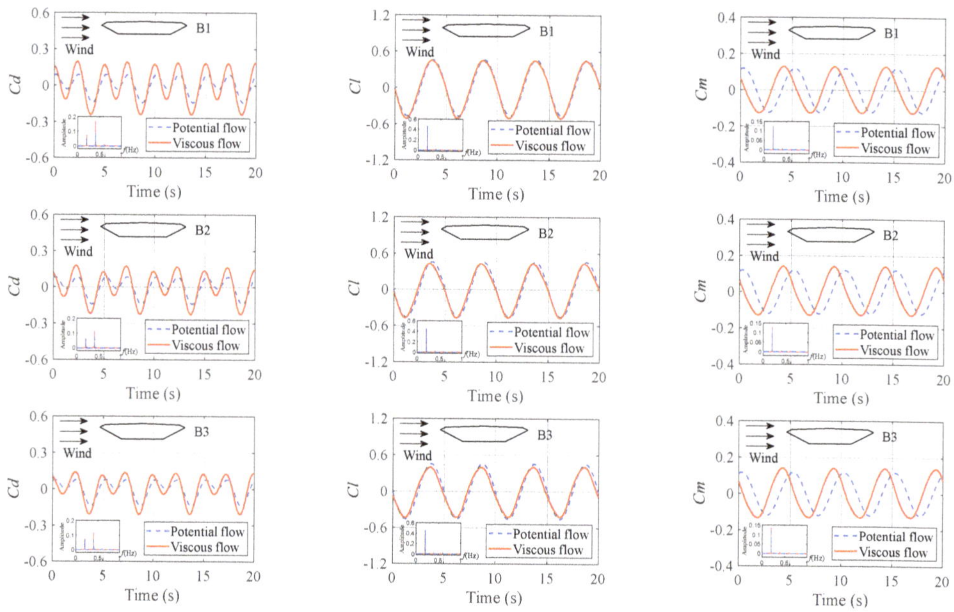

Figure 6. Comparison of aerodynamic forces.

For the lift force and pitching moment, the frequency is almost the same as the coupled vibration frequency *f*. There are also second-order components for the lift force and pitching moment in Equations (36) and (39), but they are too slight to be neglected. A similar phenomenon was reported before by Chen et al. [21]. The amplitudes of the lift force and the pitching moment of the potential flow solution are close to those of the viscous flow. However, in most cases, the aerodynamic force in a potential flow is smaller than the aerodynamic force in a viscous flow. Furthermore, the aerodynamic forces in a potential flow will lag behind those in a viscous flow. The phase lag of the pitching moment is the most salient observation, with a value of approximately 0.36π. However, it should be noted that this value exhibits slight variability across different SBG aspect ratios.

$$F_d = \frac{1}{2}\rho U^2(2B)\left[KP_1^*\frac{\dot{p}}{U} + KP_2^*\frac{B\dot{\alpha}}{U} + K^2P_3^*\alpha + K^2P_4^*\frac{p}{B} + KP_5^*\frac{\dot{h}}{U} + K^2P_6^*\frac{h}{B}\right] \tag{41}$$

The angle of attack effect can be determined by defining the motion of the SBG. Suppose that γ is the angle of attack; then, the lateral and vertical motion of the SBG can be expressed as $u_x = -V\cos\gamma + A_x 2\pi f\cos(2\pi ft)$ and $u_y = -V\sin\gamma - A_y 2\pi f\cos(2\pi ft)$. For the sake of brevity, only the aerodynamic coefficients of B1, with an angle of attack of $\pm 3°$, are exhibited in Figure 7. The results show that the drag force still has a second-order effect on the wind flow at the angle of attack of $\pm 3°$, and this phenomenon also appears in the potential flow solution. Additionally, the analytical solution for the potential flow can predict the amplitudes of the lift force and pitching moment in a viscous flow well. As a consequence, the proposed force model solution and the viscous CFD solution showcase remarkable similarities in the magnitude and frequency of the unsteady aerodynamic force. In view of these findings, the expressions in Equations (36) and (39) have the potential to serve as useful references in the development of an aerodynamic force model in a real flow.

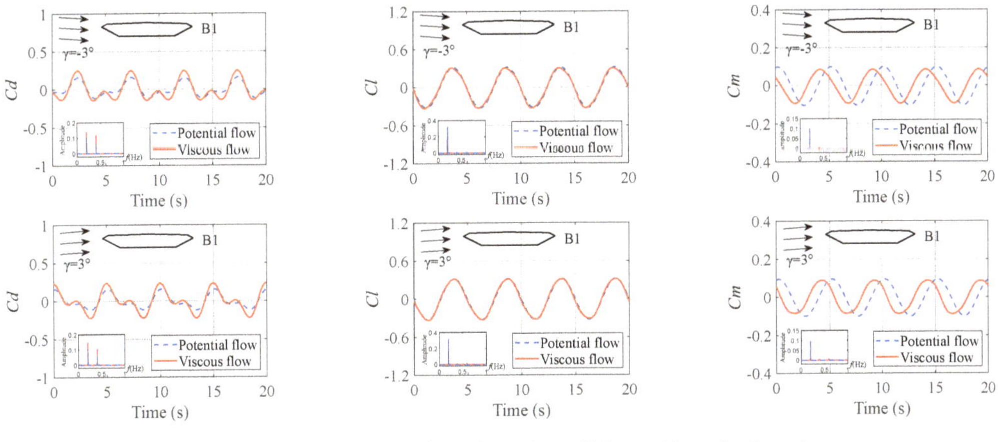

Figure 7. Comparison of aerodynamic coefficients with angle of attack.

The precisions of the amplitude of the aerodynamic forces by the potential theory are displayed in Figure 8, where Cd1 and Cd2 denote the first-order and second-order components of the drag force coefficients, respectively. The precision is defined in Equation (42), where Ap and Av are the amplitudes of the aerodynamic force coefficients in a potential flow and a viscous flow, respectively. It is observed that the precision of the first-order component of the drag force is higher compared to that of the second-order component. However, due to the neglect of viscosity, the drag force in the potential flow is lower than that obtained from the viscous flow. Generally, the precision level of the lift force and pitching moment are better than that of the drag force. Nevertheless, the precision levels of the lift force and pitching moment decrease with an increasing aspect ratio. The average precision levels of the lift force and pitching moment of B1 at the three angles of attack are 98.6% and 95.8%, respectively, indicating that the analytical solution can predict the lift force and pitching moment of an SBG with a small aspect ratio well. The potential flow solution is advantageous in terms of the computation time, which is less than one second when using Equations (36) and (39). Therefore, the analytical solution in a potential flow can be used for a rapid estimation of the aerodynamic forces on an SBG in coupled vibration.

$$prc = \left(1 - \frac{|A_p - A_v|}{A_v}\right) \times 100\% \tag{42}$$

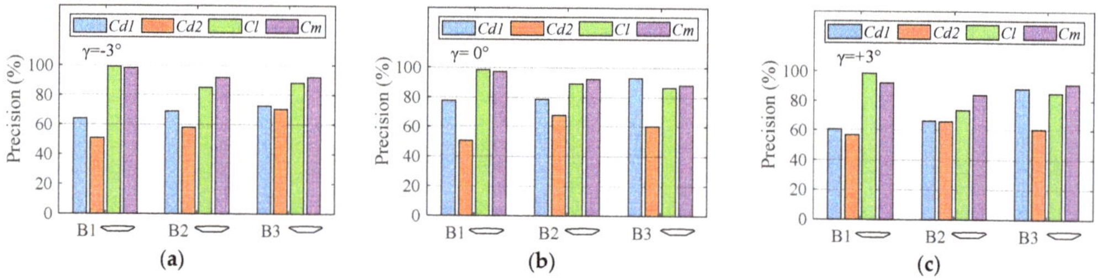

Figure 8. Precision of potential solution for different angles of attack: (**a**) $\gamma = -3°$; (**b**) $\gamma = 0°$; (**c**) $\gamma = 3°$.

5. Conclusions

The aerodynamic force acting on the bridge structure has the potential to cause significant vibrations or even structural damage, thereby endangering the safety and reducing the sustainability of the bridge structure. In the present study, an analytical solution to the unsteady aerodynamic forces of the SBG was investigated based on the potential flow theory. The current study is distinguished from prior research by the comparatively intricate conformal mapping expression for SBGs, which resulted in discernable disparities in the aerodynamic force formulae. The complex potential and pressure functions were derived. The aerodynamic drag force, lift force, and pitching moment were expressed as functions of the motion state of the SBG and the conformal mapping coefficients. Subsequently, a numerical model was developed to simulate the wind field under a viscous flow. Finally, the analytical aerodynamic forces of three SBGs for different angles of attack were compared with the CFD viscous solution. The main conclusions are as follows:

(1) The proposed analytical solution for the unsteady aerodynamic forces of an SBG with coupled vibration in a potential flow is a function of the SBG's shape-related parameters and vibration response, providing a convenient and efficient method for calculating the unsteady aerodynamic forces of SBGs.

(2) The analytical drag force successfully reproduced the second-order component, which mainly results from the multiplication of the vertical vibration and pitching vibration velocity terms and plays a significant role in drag force. On the other hand, the first-order frequency component dominates the lift force and pitching moment.

(3) The proposed analytical solution for the drag force in a potential flow yields lower values than that in a viscous flow. Nonetheless, the analytical solution demonstrates high accuracy in predicting the amplitude of the lift force and pitching moment for an SBG at angles of attack of 0° and ±3°. The explicit formulation and satisfactory precision of the analytical solution enable its effective utilization for the rapid estimation of the aerodynamic forces acting on an SBG with coupled vibration.

(4) It is imperative to underscore that the proposed analytical aerodynamic force model may not exhibit a sufficient degree of accuracy when applied to bluff bodies. It is noteworthy that the potential flow theory, which serves as the foundation for this model, is a linearized theory that may not fully capture the intricate nonlinear unsteady aerodynamic forces that manifest at high angles of attack and velocities or

during large amplitude vibrations. In the future, there will be a greater focus on a more accurate aerodynamic force model for SBGs. To achieve this, further studies shall be conducted to investigate innovative correction methods.

Author Contributions: Conceptualization, L.W. and M.Z.; methodology, L.W.; validation, F.J. and Z.Z.; formal analysis, L.W.; investigation, L.W.; writing—original draft preparation, L.W.; writing—review and editing, M.Z. and Y.L.; visualization, F.J.; supervision, M.Z. and Y.L. All authors have read and agreed to the published version of the manuscript.

Funding: This research received no external funding.

Institutional Review Board Statement: Not applicable.

Informed Consent Statement: Not applicable.

Data Availability Statement: The data presented in this study are available upon request from the corresponding author.

Conflicts of Interest: The authors declare no conflict of interest.

References

1. Liu, Y.; Wang, H.; Xu, Z.; Li, J.; Wu, T.; Mao, J. Modeling Multidimensional Multivariate Turbulent Wind Fields Using a Correlated Turbulence Wave Number–Frequency Spectral Representation Method. *J. Eng. Mech.* **2023**, *149*, 04023010. [CrossRef]
2. Zhang, H.; Wang, H.; Xu, Z.; Zhang, Y.; Tao, T.; Mao, J. Monitoring-Based Analysis of Wind-Induced Vibrations of Ultra-Long Stay Cables during an Exceptional Wind Event. *J. Wind. Eng. Ind. Aerodyn.* **2022**, *221*, 104883. [CrossRef]
3. Amer, T.S.; Bek, M.A.; Abouhmr, M.K. On the Vibrational Analysis for the Motion of a Harmonically Damped Rigid Body Pendulum. *Nonlinear Dyn.* **2018**, *91*, 2485–2502. [CrossRef]
4. Bek, M.A.; Amer, T.S.; Sirwah, M.A.; Awrejcewicz, J.; Arab, A.A. The Vibrational Motion of a Spring Pendulum in a Fluid Flow. *Results Phys.* **2020**, *19*, 103465. [CrossRef]
5. Amer, T.S.; Galal, A.A.; Elnaggar, S. The Vibrational Motion of a Dynamical System Using Homotopy Perturbation Technique. *Appl. Math.* **2020**, *11*, 1081–1099. [CrossRef]
6. Zhang, J.; Zhang, M.; Jiang, X.; Wu, L.; Qin, J.; Li, Y. Pair-Copula-Based Trivariate Joint Probability Model of Wind Speed, Wind Direction and Angle of Attack. *J. Wind Eng. Ind. Aerodyn.* **2022**, *225*, 105010. [CrossRef]
7. Li, Y.; Jiang, F.; Zhang, M.; Dai, Y.; Qin, J.; Zhang, J. Observations of Periodic Thermally-Developed Winds beside a Bridge Region in Mountain Terrain Based on Field Measurement. *J. Wind Eng. Ind. Aerodyn.* **2022**, *225*, 104996. [CrossRef]
8. Diana, G.; Omarini, S. A Non-Linear Method to Compute the Buffeting Response of a Bridge Validation of the Model through Wind Tunnel Tests. *J. Wind Eng. Ind. Aerodyn.* **2020**, *201*, 104163. [CrossRef]
9. Wu, L.; Woody Ju, J.; Zhang, J.; Zhang, M.; Li, Y. Vibration Phase Difference Analysis of Long-Span Suspension Bridge during Flutter. *Eng. Struct.* **2023**, *276*, 115351. [CrossRef]
10. Cui, C.; Ma, R.; Hu, X.; He, W. Vibration Analysis for Pendent Pedestrian Path of a Long-Span Extradosed Bridge. *Sustainability* **2019**, *11*, 4664. [CrossRef]
11. Amman, O.H.; von Kármán, T.; Woodruff, G.B. *The Failure of the Tacoma Narrows Bridge*; Federal Works Agency: Washington, DC, USA, 1941.
12. Chen, A.; He, X.; Xiang, H. Identification of 18 Flutter Derivatives of Bridge Decks. *J. Wind Eng. Ind. Aerodyn.* **2002**, *90*, 2007–2022. [CrossRef]
13. Scanlan, R.H.; Tomko, J.J. Airfoil and Bridge Deck Flutter Derivatives. *J. Eng. Mech.* **1971**, *97*, 1717–1737. [CrossRef]
14. Ma, T.; Zhao, L.; Shen, X.; Ge, Y. Case Study of Three-Dimensional Aeroelastic Effect on Critical Flutter Wind Speed of Long-Span Bridges. *J. Wind Eng. Ind. Aerodyn.* **2021**, *212*, 104614. [CrossRef]
15. Sauder, H.S.; Sarkar, P.P. A 3-DOF Forced Vibration System for Time-Domain Aeroelastic Parameter Identification. *Wind Struct.* **2017**, *24*, 481–500. [CrossRef]
16. Xu, F.; Wu, T.; Ying, X.; Kareem, A. Higher-Order Self-Excited Drag Forces on Bridge Decks. *J. Eng. Mech.* **2016**, *142*, 06015007. [CrossRef]
17. Guo, J.; Tang, H.; Li, Y.; Wu, L.; Wang, Z. Optimization for Vertical Stabilizers on Flutter Stability of Streamlined Box Girders with Mountainous Environment. *Adv. Struct. Eng.* **2020**, *23*, 205–218. [CrossRef]
18. Kusano, I.; Cid Montoya, M.; Baldomir, A.; Nieto, F.; Jurado, J.Á.; Hernández, S. Reliability Based Design Optimization for Bridge Girder Shape and Plate Thicknesses of Long-Span Suspension Bridges Considering Aeroelastic Constraint. *J. Wind Eng. Ind. Aerodyn.* **2020**, *202*, 104176. [CrossRef]
19. Zhou, Z.; Mao, W.; Ding, Q. Experimental and Numerical Studies on Flutter Stability of a Closed Box Girder Accounting for Ground Effects. *J. Fluids Struct.* **2019**, *84*, 1–17. [CrossRef]
20. Zhang, M.; Xu, F.; Zhang, Z.; Ying, X. Energy Budget Analysis and Engineering Modeling of Post-Flutter Limit Cycle Oscillation of a Bridge Deck. *J. Wind. Eng. Ind. Aerodyn.* **2019**, *188*, 410–420. [CrossRef]

21. Chen, Z.Q.; Yu, X.D.; Yang, G.; Spencer, B.F. Wind-Induced Self-Excited Loads on Bridges. *J. Struct. Eng.* **2005**, *131*, 1783–1793. [CrossRef]
22. Siedziako, B.; Øiseth, O. Modeling of Self-Excited Forces during Multimode Flutter: An Experimental Study. *Wind Struct.* **2018**, *27*, 293–309. [CrossRef]
23. Von Karman, T.; Sears, W.R. Airfoil Theory for Non-Uniform Motion. *J. Aeronaut. Sci.* **1938**, *5*, 379–390. [CrossRef]
24. Theodorsen, T. *General Theory of Aerodynamic Instability and the Mechanism of Flup*; Twentieth Annual Report of the National Advisory Committee for Aeronautics; NACA: Washington, DC, USA, 1935; pp. 413–434.
25. Edwards, J.W.; Ashley, H.; Breakwell, J.V. Unsteady Aerodynamic Modeling for Arbitrary Motions. *AIAA J.* **1979**, *17*, 365–374. [CrossRef]
26. Sun, M.; Wang, W. Method for Calculating the Unsteady Flow of an Elliptical Circulation-Control Airfoil. *J. Aircr.* **1989**, *26*, 907–913. [CrossRef]
27. Stangfeld, C.; Rumsey, C.L.; Mueller-Vahl, H.; Greenblatt, D.; Nayeri, C.; Paschereit, C.O. Unsteady Thick Airfoil Aerodynamics: Experiments, Computation, and Theory. In Proceedings of the 45th AIAA Fluid Dynamics Conference, Dallas, TX, USA, 22–26 June 2015.
28. Catlett, M.R.; Anderson, J.M.; Badrya, C.; Baeder, J.D. Unsteady Response of Airfoils Due to Small-Scale Pitching Motion with Considerations for Foil Thickness and Wake Motion. *J. Fluids Struct.* **2020**, *94*, 102889. [CrossRef]
29. Chiu, T.W. A Two-Dimensional Second-Order Vortex Panel Method for the Flow in a Cross-Wind over a Train and Other Two-Dimensional Bluff Bodies. *J. Wind Eng. Ind. Aerodyn.* **1991**, *37*, 43–64. [CrossRef]
30. Chiu, T.W. Prediction of the Aerodynamic Loads on a Railway Train in a Cross-Wind at Large Yaw Angles Using an Integrated Two- and Three-Dimensional Source/Vortex Panel Method. *J. Wind Eng. Ind. Aerodyn.* **1995**, *57*, 19–39. [CrossRef]
31. Valentine, D.T.; Madhi, F. Unsteady Potential Flow Theory and Numerical Analysis of Forces on Cylinders Induced by Nearby Oscillating Disturbances. In Proceedings of the American Society of Mechanical Engineers Digital Collection, Honolulu, HA, USA, 31 May–5 June 2009; pp. 631–639.
32. Henrici, P. Cauchy Integrals on Closed Curves. In *Applied and Computational Complex Analysis, Volume 3: Discrete Fourier Analysis, Cauchy Integrals, Construction of Conformal Maps, Univalent Functions*; John Wiley & Sons: Hoboken, NJ, USA, 1993; pp. 113–119.
33. Wu, L.; Zhou, Z.; Zhang, J.; Zhang, M. A Numerical Method for Conformal Mapping of Closed Box Girder Bridges and Its Application. *Sustainability* **2023**, *15*, 6291. [CrossRef]
34. Wu, L.; Ju, J.W.; Zhang, M.; Li, Y.; Qi, J. Aerostatic Pressure of Streamlined Box Girder Based on Conformal Mapping Method and Its Application. *Wind Struct.* **2022**, *35*, 243–253. [CrossRef]
35. Segletes, S.B.; Walters, W.P. A Note on the Application of the Extended Bernoulli Equation. *Int. J. Impact Eng.* **2002**, *27*, 561–576. [CrossRef]
36. Katz, V.J. The History of Stokes' Theorem. *Math. Mag.* **1979**, *52*, 146–156. [CrossRef]
37. Larsen, A. Aerodynamic Aspects of the Final Design of the 1624 m Suspension Bridge across the Great Belt. *J. Wind Eng. Ind. Aerodyn.* **1993**, *48*, 261–285. [CrossRef]
38. Reinhold, T.A.; Brinch, M.; Damsgaard, A. Wind Tunnel Tests for the Great Belt Link. In Proceedings of the Aerodynamics of Large Bridges: Proceeding of the First International Symposium on Aerodynamics of Large Bridges; Taylor & Francis: Copenhagen, Denmark, 1992; pp. 255–267.
39. Honore Walther, J. *Discrete Vortex Method for Two-Dimensional Flow Past Bodies of Arbitrary Shape Undergoing Prescribed Rotary and Translational Motion*; Technical University of Denmark: Copenhagen, Denmark, 1994.

sustainability

Article

A Numerical Method for Conformal Mapping of Closed Box Girder Bridges and Its Application

Lianhuo Wu [1], Zelin Zhou [2,*], Jinxiang Zhang [1] and Mingjin Zhang [1]

1 Department of Bridge Engineering, Southwest Jiaotong University, Chengdu 610031, China
2 China 19th Metallurgical Corporation, Chengdu 610031, China
* Correspondence: zhouzelin2021@19mcc.com.cn

Abstract: Conformal mapping has achieved many successes in engineering. It can help to solve some complex fluid flow problems. This study proposed a numerical method for conformal mapping of closed box girder bridges and applied it to flutter performance prediction, which is crucial for ensuring the safety and sustainability of bridge structures. The characteristics of conformal mapping coefficients for the closed box were investigated. Thereafter, a numerical method through searching the conformal mapping coefficients was presented. The results show that the proposed numerical method has a smaller error in the existing research. The conformal mapping of six practical bridges agrees well with the closed box girder shapes, indicating the validity of the proposed method. The flutter prediction results by the proposed method are consistent with the wind tunnel test. The conformal mapping and flutter calculations took no more than ten seconds, showing high computing efficiency. This method is easier to understand and implement without complex mathematical derivation, which is helpful for the extensive application of conformal mapping in bridge engineering.

Keywords: numerical method; conformal mapping; closed box girder; sliding searching method; flutter performance

Citation: Wu, L.; Zhou, Z.; Zhang, J.; Zhang, M. A Numerical Method for Conformal Mapping of Closed Box Girder Bridges and Its Application. *Sustainability* **2023**, *15*, 6291. https://doi.org/10.3390/su15076291

Academic Editors: Marc A. Rosen and Chunxu Qu

Received: 6 January 2023
Revised: 21 February 2023
Accepted: 3 April 2023
Published: 6 April 2023

1. Introduction

Bridge structures are a critical component of transportation infrastructure and have a significant impact on the sustainability of society, the economy, and the environment. The growing length of the bridge span makes the wind action gradually become the key factor in bridge design [1,2]. During the entire life cycle, bridges can experience violent vibrations and sustain serious damages due to unsteady wind loads [3,4], compromising their structural safety and undermining their sustainability. At present, wind tunnel tests and computational fluid dynamics simulation are the prevailing methods used to determine the wind load on structures [5–7]. However, both methods have their limitations: wind tunnel tests can be costly, while CFD simulation is time-consuming. The conformal mapping method offers a more efficient approach to analyzing wind force. Therefore, it is of great significance to study the conformal mapping of closed box girders.

Conformal mapping is an outstanding method of analysis with many successful applications in engineering. Many complicated problems, for instance, the fluid flow problem, can be simplified by mapping a complex domain to a regular region. The Joukowski transformation is one of the conformal mappings which helps obtain the flows around airfoils [8,9]. Poozesh and Mirzaei [10] applied the Joukowski transform to quickly generate a mesh grid for flow simulations around airfoils. Malonek and De Almeida [11] proposed a generalized Joukowski transformation mathematical model with high dimensions. In addition, conformal mapping was employed to analyze the stress distributions of solid structures. Jia et al. [12] proposed an analytical function expressed as a power series to compute the gravity-induced underground stress distribution of the elastic half-plane with slope. The results produced by the analytical model are consistent with the numerical solution. Kuliyev [13] presented a conformal mapping function to determine the stress–strain state of the bending beam. The majority of conventional conformal mappings concentrate on the mappings on a planar domain. Gu et al. [14] introduced conformal mapping on three-dimensional objects. In addition, conformal mapping also achieves great success in hydrodynamics [15], heat transfer [16], and electromagnetics [17] fields.

There are many studies that focus on the conformal mapping of closed-form shape bodies. Natarajan et al. [18] proposed a numerical conformal mapping method on arbitrary polygonal inner domains, which does not need a two-level isoparametric mapping. Wang et al. [19] presented a mathematical method based on the Schwarz–Christoffel transform [20] and successfully mapped the strip polygon to a rectangular region. The shape of a closed box girder in bridge engineering commonly takes the form of a symmetrical convex hexagon. It is one of the two main girder types of long-span bridges. Wu et al. [21] derived a direct iteration expression for conformal mapping of the closed box girder, which does not need to mutually interpolate between even and odd points. The initial iteration points were defined through an ellipse transformation. Nevertheless, there are many complicated mathematical derivations in the existing conformal mapping methods. Furthermore, in order to minimize mapping errors, a larger number of series is often required.

To ensure the sustainability of bridges throughout their lifecycle, it is crucial to determine the structural wind loads. This study presents a numerical method for calculating the conformal mapping of closed box girders, which is simple to understand and implement due to the absence of complex derivation. Using the numerical method, the flutter performance of closed box girders was analyzed. The paper is organized as follows. In Section 2, the characteristics of the conformal mapping coefficients were analyzed. Afterward, the numerical method was proposed and verified in Section 3. Section 4 investigated the application of the numerical method in flutter performance estimation. Finally, some main conclusions of this study were summarized.

2. Characteristics of the Conformal Mapping Coefficients

Conformal mapping uses functions of complex variables to map one region of the complex plane onto another region that is easier to analyze. The conformal mapping function from a unit circle to a closed box girder is shown in Figure 1, which can be expressed as the Laurent series [21]:

$$z = \sum_{k=1}^{n} c_k \zeta^{2-k}, \tag{1}$$

where z and ζ are points on the closed box and unit circle, respectively; $c_k = a_k + ib_k$ (k = 1,2, ... ,n) is the conformal mapping coefficient; n is the series number.

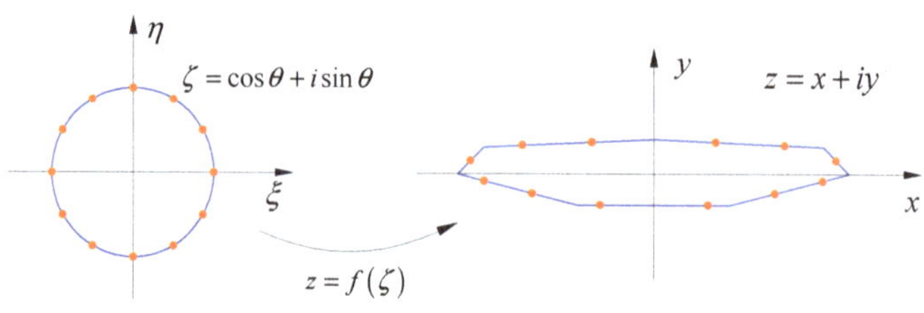

Figure 1. Conformal mapping from a unit circle to a closed box.

The points on the unit circle can be expressed as $\zeta = \cos\theta + i\sin\theta$, where θ denotes the phase angle. Substituting it into Equation (1), then we obtain:

$$x = \sum_{k=1}^{n} (a_k \cos(2-k)\theta - b_k \sin(2-k)\theta), \tag{2}$$

$$y = \sum_{k=1}^{n} (a_k \sin(2-k)\theta + b_k \cos(2-k)\theta). \tag{3}$$

For a long-span bridge, the shape of the closed box is usually symmetrical with respect to the y-axis. Thus, we can always find a point $z_2 = x_2 + iy_2$, which is the symmetry of the other point $z_1 = x_1 + iy_1$ about the y-axis. That is, we have $x_1 = -x_2$ and $y_1 = y_2$. Assume that the arguments of their mapping points on the unit circle are θ_1 and θ_2. Substituting $\theta_2 = \pi - \theta_1$ into Equation (2), we have:

$$x_1 = \sum_{k=1}^{n} (a_k \cos(2-k)\theta_1 - b_k \sin(2-k)\theta_1), \tag{4}$$

$$y_1 = \sum_{k=1}^{n} (a_k \sin(2-k)\theta_1 + b_k \cos(2-k)\theta_1), \tag{5}$$

$$x_2 = \sum_{k=1}^{n} (\cos(k\pi)(a_k \cos(2-k)\theta_1 + b_k \sin(2-k)\theta_1)), \tag{6}$$

$$y_2 = \sum_{k=1}^{n} (\cos(k\pi)(-a_k \sin(2-k)\theta_1 + b_k \cos(2-k)\theta_1)). \tag{7}$$

To investigate the characteristics of conformal mapping coefficients a_k and b_k, add x_1 and x_2, then we have:

$$\begin{aligned} x_1 + x_2 &= \sum_{k=1}^{n} (a_k \cos(2-k)\theta_1 - b_k \sin(2-k)\theta_1) \\ &+ \sum_{k=1}^{n} (a_k \cos(2-k)(\pi - \theta_1) - b_k \sin(2-k)(\pi - \theta_1)) \\ &= (1 + \cos(2-k)\pi) \sum_{k=1}^{n} a_k \cos(2-k)\theta_1 - (1 - \cos(2-k)\pi) \sum_{k=1}^{n} b_k \sin(2-k)\theta_1 \end{aligned} \tag{8}$$

When k is an even number, $x_1 + x_2 = 2\sum_{k=1}^{n} a_k \cos(2-k)\theta_1 = 0$ is true for any k only if $a_k = 0$. On the contrary, when k is an odd number, $x_1 + x_2 = -2\sum_{k=1}^{n} b_k \sin(2-k)\theta_1 = 0$ is true for any k only if $b_k = 0$. Therefore, $c_k = a_k$ is true when k is an odd number, or $c_k = ib_k$ is true when k is an even number. The same conclusion can also be obtained through the calculation of $y_1 - y_2 = 0$.

The Great Belt East Bridge (GBEB) [22], a long-span suspension bridge located in Danmark with a main span length of 1624 m, is one of the most famous bridges in the world. The shape of the closed box girder of GBEB, which has a width of 31.0 m and a height of 4.4 m, respectively, was selected to verify the above finding. The conformal mapping coefficients when n = 10 are listed in Table 1 based on the Direct Iteration Conformal Mapping method (DICM) by Wu et al. [21]. Upon analyzing the results, it was observed that the values of a_k converge to zero when k is an even number, and the values of b_k tend towards zero when k is an odd number. These results provide strong evidence that the conclusion drawn previously is indeed valid.

Table 1. The results of coefficients a_k and b_k (n = 10).

k	1	2	3	4	5	6	7	8	9	10
a_k	8.7621	-1.5×10^{-4}	6.4213	4.4×10^{-5}	−0.0841	-4.6×10^{-6}	0.1502	-2.9×10^{-5}	0.0771	-4.6×10^{-5}
b_k	5.3×10^{-4}	−0.5759	-4.2×10^{-4}	0.4411	2.6×10^{-5}	0.1616	-6.1×10^{-5}	−0.0881	-3.1×10^{-5}	−0.0450

Through Table 1, we also found that the absolute values of a_k and b_k generally decrease with the increase of k. Meanwhile, the coefficients a_1 and a_3 are close to (B + H)/4 and

(B − H)/4, respectively, where B and H are the width and height of the closed box. To study the range of a_k and b_k, 493 thousand box girders are generated to calculate Equation (1). The sizes of these box girders were $B \in [15, 60]$, $H \in [2.5, 6.0]$, $B_u \in [0.75B, B]$, $B_d \in [0.2B_u, B_u]$, and $H_d \in [0.5H, H - 0.5B_u \tan(0.02)]$, where B_u is the width of the upper deck of the closed box, B_d is the width of the lower deck of the closed box, and H_d is the height from the triangular fairing to the lower deck of the closed box, as displayed in Figure 2. The shapes of these box girders almost include all possible closed box girder designs.

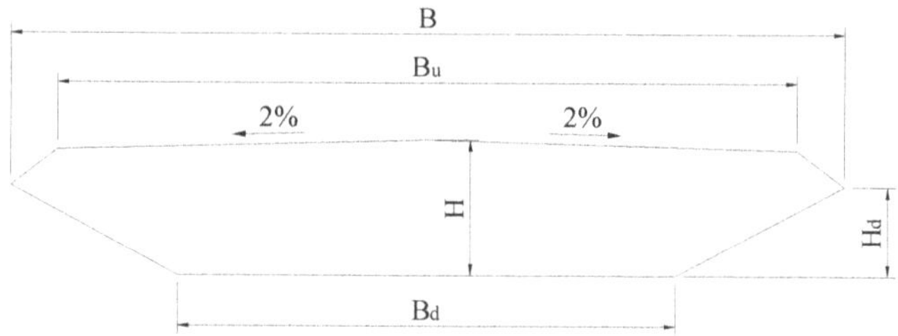

Figure 2. Parameters of the bridge box girder.

The results of a_k and b_k ($k \leq 50$) are rendered in Figure 3. However, since the values of coefficients a_1 and a_3 exceed those of the other coefficients, they are not included in Figure 3. Instead, they will be analyzed in greater detail below. It is noted that a_k range between −1 and 1 when $k \geq 5$, while b_k ranges from −2 and 2 when $k \geq 2$. The range of a_k and b_k are limited and substantially diminishing with the increase of k. This can be attributed to the decreasing index of the Laurent series in Equation (1), which causes ζ^{2-k} to grow exponentially. As a result, a_k and b_k must be reduced in order to counteract this effect and maintain equilibrium.

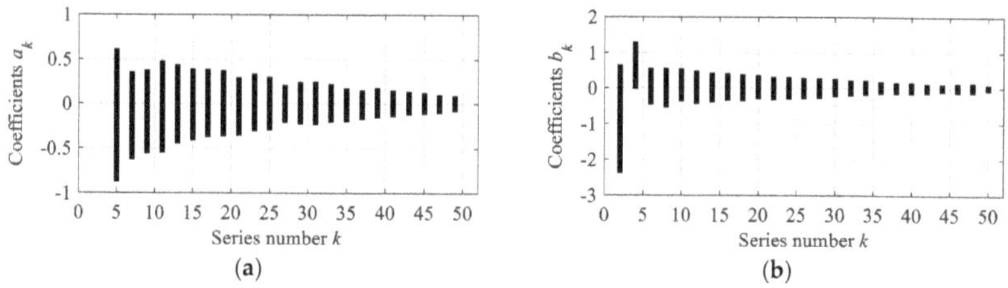

Figure 3. The range of conformal mapping coefficients a_k and b_k; (**a**) Coefficients a_k; (**b**) Coefficients b_k.

To study the characteristics of a_1 and a_3, let $a_1^* = \frac{W+H}{4}$, $a_3^* = \frac{W-H}{4}$. Divide a_1 and a_3 by a_1^* and a_3^*, respectively. The percentage error of a_1^* and a_3^* are shown in Figure 4. The results indicate that, in most cases, the error of a_1^* and a_3^* remains within ±5%, especially when the aspect ratio B/H is larger than 10, implying a high level of accurate prediction. Moreover, as the aspect ratio B/H increases, the distribution range of the error for a_1^* and a_3^* gradually becomes narrower. The envelope of the error distribution can be expressed as Equation (9). To facilitate further analysis, the range of conformal mapping coefficients is listed in Table 2.

$$f_1 = 0.0729e^{-0.1453B/H} + 0.0308e^{-0.0296B/H}, \tag{9}$$

$$f_2 = -0.1397e^{-0.0962B/H} + 0.0502e^{-0.2258B/H}, \tag{10}$$

$$f_3 = 0.8339e^{-0.8028B/H} + 0.0683e^{-0.0911B/H}, \tag{11}$$

$$f_4 = -0.2996e^{-0.6509B/H} - 0.0398e^{-0.0644B/H}. \tag{12}$$

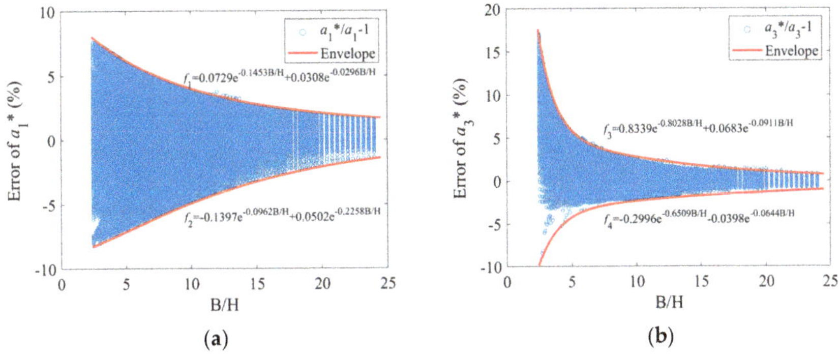

Figure 4. The percentage error of a_1^* and a_3^*; (**a**) Error of a_1^*; (**b**) Error of a_3^*.

Table 2. The range of conformal mapping coefficients.

j	1	2	3	4	5
a_{2j-1}	$(f_2\sim f_1)\ a_1^*$	$(f_4\sim f_3)\ a_3^*$	−0.8770~0.6097	−0.6294~0.3481	−0.5646~0.3740
b_{2j}	−2.3874~0.6518	−0.0434~1.2909	−0.4762~0.5408	−0.5597~0.5474	−0.3782~0.5328
j	6	7	8	9	10
a_{2j-1}	−0.5535~0.4761	−0.4523~0.4350	−0.4178~0.3847	−0.3804~0.3798	−0.3688~0.3697
b_{2j}	−0.4501~0.4792	−0.4060~0.4203	−0.3597~0.4136	−0.3711~0.3768	−0.3180~0.3512
j	11	12	13	14	15
a_{2j-1}	−0.3596~0.2913	−0.3104~0.3328	−0.3018~0.2942	−0.2155~0.2098	−0.2327~0.2414
b_{2j}	−0.3369~0.3004	−0.3136~0.3040	−0.3038~0.2854	−0.2887~0.2736	−0.2467~0.2545
j	16	17	18	19	20
a_{2j-1}	−0.2364~0.2441	−0.2077~0.2158	−0.2057~0.1724	−0.1789~0.1484	−0.1593~0.1736
b_{2j}	−0.2195~0.2164	−0.1991~0.2090	−0.1667~0.1845	−0.1609~0.1605	−0.1621~0.1557
j	21	22	23	24	25
a_{2j-1}	−0.1441~0.1503	−0.1280~0.1342	−0.1164~0.1223	−0.1068~0.0982	−0.0873~0.0886
b_{2j}	−0.1557~0.1277	−0.1138~0.1116	−0.1438~0.1371	−0.1458~0.1411	−0.0861~0.0832

3. Numerical Method for Conformal Mapping Coefficients

3.1. Dichotomy Method

On the basis of the Riemann Mapping Theorem [23], there exists a unique conformal mapping between the unit circle and the closed box girder. Therefore, it is possible to find the optimal combinations of a_k and b_k for Equation (1) by searching coefficients within the range in Table 2.

According to the formula of distance from a point to a line segment, we have:

$$\frac{\partial \varepsilon_m}{\partial c_k} = \frac{\left| \sum_{k=1}^{n} (A\cos(2-k)\theta_m + B\sin(2-k)\theta_m) \right|}{\sqrt{A^2+B^2}}, \quad (13)$$

where ε_m denotes the distance from a mapping point to the boundary of the closed box, and A and B denote the coefficients of the line segment equation. For a specific mapping point, $\frac{\partial \varepsilon_m}{\partial c_k}$ is a constant, which means that the position of this point is linearly related to coefficients a_k and b_k. Hence, there exists a unique point with the minimum distance from the boundary of the bridge box girder when we search a particular coefficient of a_k and

b_k. To find this point, the Dichotomy Method (DM) is a good choice. However, due to different phase angles θ_m, the maximum ε_m of all points is no longer linearly related to the coefficient a_k and b_k. The DM may miss the final optimal solution in the process of iteration, as shown in Figure 5. The blue points denote the endpoints of the search interval. The orange points indicate the optimal solution of a_k and b_k in the current search interval. The green points represent the final optimal solution for $n = 50$ in Equation (1) by DICM [21]. In the l iteration, the combination of a_k and b_k with the minimum ε_{max} is chosen, which are marked by red points as they are closer to the orange points. However, the orange and green points may have large errors, resulting in the computation of an incorrect interval in the next iteration, such as c_{j+1}.

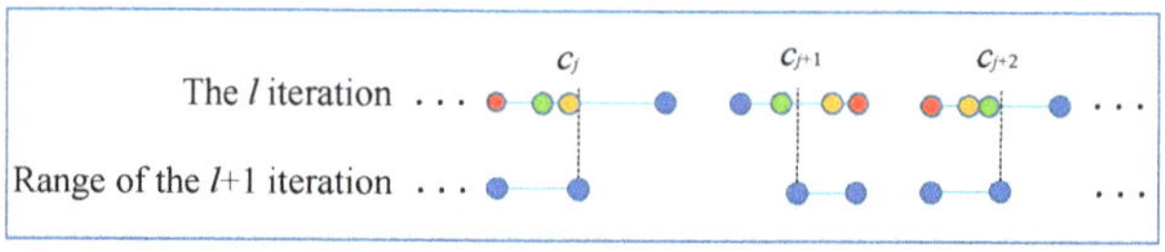

Figure 5. The search process by the DM.

The search histories of the coefficients a_7 to b_{10} by the DM are revealed in Figure 6. After conducting no more than five iterations, it has been observed that the search intervals of various coefficients tend to deviate from the final optimal solution and cannot be corrected back. This phenomenon is a result of a flaw in the DM process, as shown in the c_{j+1} process illustrated in Figure 5. The incorrect coefficients can have an interactive impact on the iterations of other coefficients, resulting in a cascading effect of solution errors. This ultimately leads to an increase in the number of coefficients containing inaccuracies, which can have detrimental effects on the accuracy of conformal mapping.

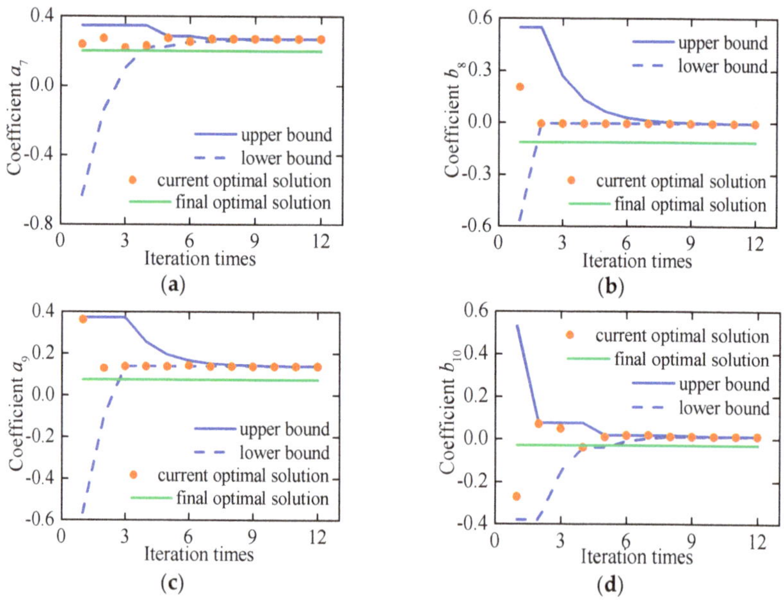

Figure 6. The search history of the coefficients by the DM. (**a**) Coefficient a_7; (**b**) Coefficient b_8; (**c**) Coefficient a_9; (**d**) Coefficient b_{10}.

3.2. Improved Dichotomy Method

In view of the above problems, this paper proposes an Improved Dichotomy Method (IDM). As shown in Figure 7, the meaning of colors is the same as that in Figure 5. To begin with, we select the endpoints and midpoints of the interval of coefficients for our search. In each iteration, we evaluate the midpoint of the interval and determine whether it represents the best combination. If it does, we take this midpoint as the center and divide the interval in half, resulting in a reduced interval for the next iteration, such as c_j. If the midpoint is not the best combination, keep the interval length unchanged and choose the endpoint that is closer to the orange point as the new midpoint for the next iteration, denoted as c_{j+1}.

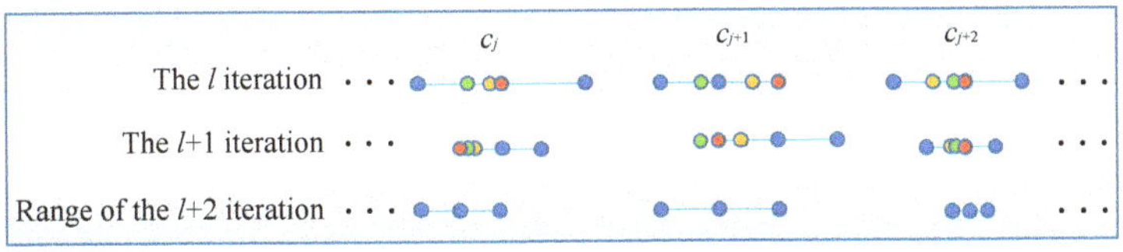

Figure 7. The search process by the IDM.

As before, the search processes of the coefficients a_7 to b_{10} are displayed in Figure 8. It is of particular interest that the IDM adjusts the search interval many times during the iteration depending on the position of the currently selected coefficient. This adaptive behavior allows the IDM to effectively explore the search space and converge towards the optimal solution in a more efficient manner. Compared with the DM in Figure 6, the results searched by the IDM are much closer to the final optimal solutions. The IDM's ability to dynamically adjust the search interval offers a distinct advantage over the DM, enabling it to find optimal solutions with improved accuracy.

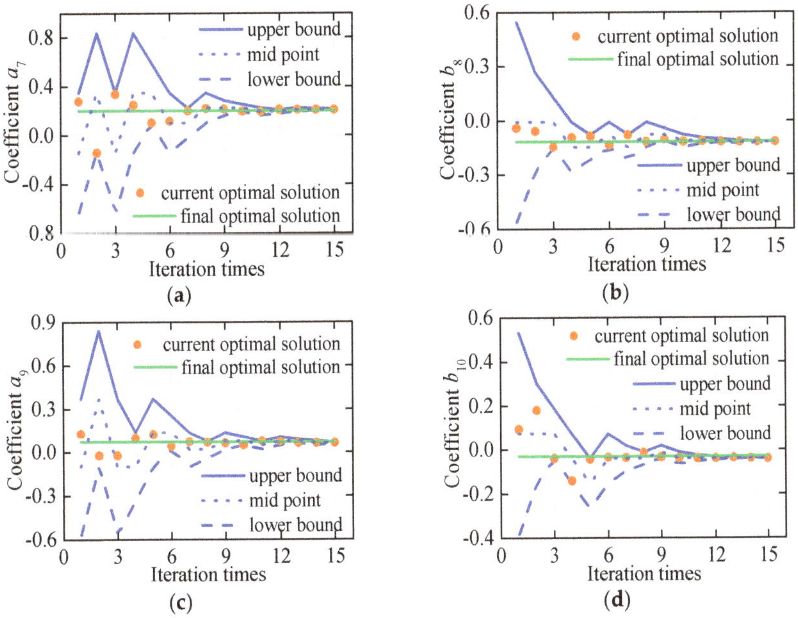

Figure 8. The search history of the coefficients by the IDM; (**a**) Coefficient a_7; (**b**) Coefficient b_8; (**c**) Coefficient a_9; (**d**) Coefficient b_{10}.

The maximum errors ε_{max} of different methods are presented in Figure 9. The results indicate that the DM exhibits the poorest performance in error controlling. Notably, for $n \geq 14$, as a result of the errors of coefficients, the maximum error of the DM tends to remain stable rather than decrease. On the other hand, the IDM proves to be the most effective in controlling errors. Nevertheless, due to the huge amount of calculation, the computation of the IDM requires a substantial amount of time, exceeding 7000 s when $n > 12$, making it excessively time-consuming.

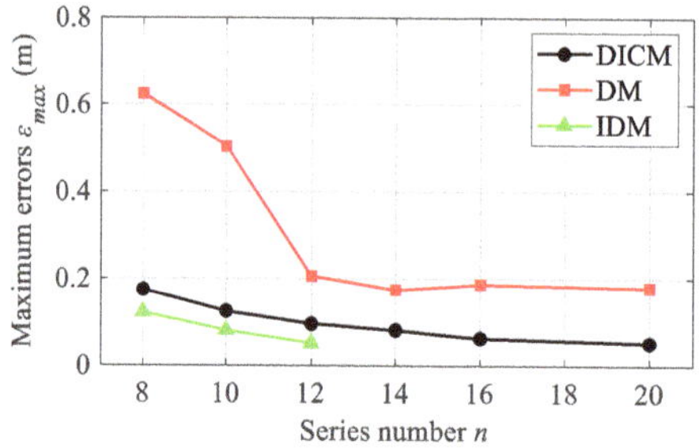

Figure 9. Maximum errors of different methods.

3.3. *Sliding Searching Method*

To speed up the solution, here we propose a Sliding Searching Method (SSM). As depicted in Figure 10, this method involves three main steps. The first step entails the computation of several foundation coefficients through the use of the IDM process. The foundation coefficients are the basis of subsequent calculations. Secondly, we select a specific number of coefficients to quantify the length of the sliding window. To increase the serial number, start with a coefficient, leave the coefficients in front of the window unchanged, and calculate coefficients within the window using the IDM. Subsequently, proceed to displace the window by a predetermined step to a new coefficient, followed by a recalculation of the coefficients contained within the updated window. Repeat the above procedures until the serial number meets the requirements. Finally, for the sake of higher precision, we introduce a new sliding window and recalculate coefficients within it. After several times of sliding and circular computations, the precision of coefficients can be significantly improved.

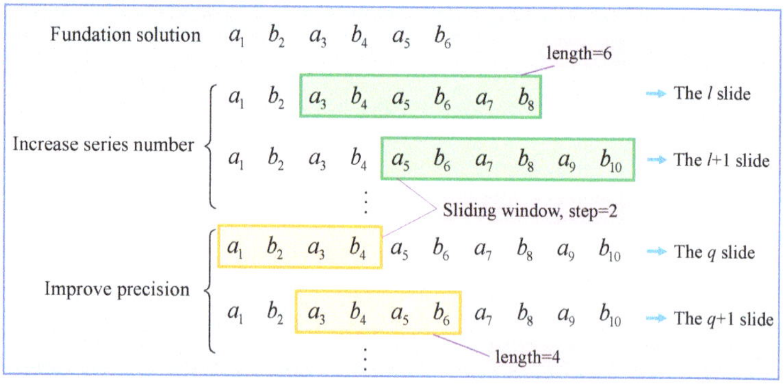

Figure 10. The process of the SSM.

Take the GBEG as an instance, the length and step of the first sliding window are six and two, respectively. The series number increasing process starts from a_5. The length and step of the second sliding window are four and two, starting from a_1 and repeating the sliding three times. For $n = 12$, it only takes 5.85 s in a personal computer (CPU: AMD-R9−5950X @3.4GHz) to finish the computation, which is much less than that of IDM. The maximum errors ε_{max} and time consumption of different methods are shown in Figure 11. The results reveal that the time consumption of DICM [21] is so short that it can be neglected. The time consumption of the SSM is higher than that of the DICM and linearly increases with respect to n. Nevertheless, the calculation time for a closed box usually does not exceed 20 s. On the other hand, the SSM can reduce the maximum error, making the increase in the acceptable calculation time worthwhile. In other words, with the same $\varepsilon_{\max}$, the SSM needs less series number n, bringing a briefer Equation (1).

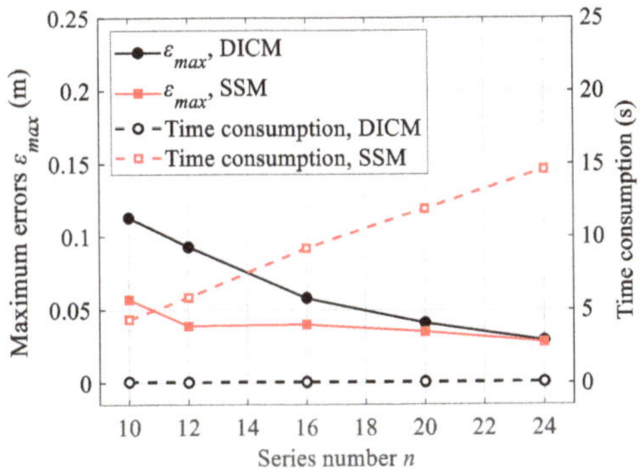

Figure 11. Maximum errors and time consumption of different methods.

To further verify the validity of SSM, the conformal mapping points using DICM and SSM are compared in six closed box girder bridges, as displayed in Figure 12. The six closed boxes have large differences in shape, such as width, height, and the inclination of the lower web. Therefore, they are highly representative. The series number n is set to 24 because the maximum error ε_{max} of the two methods is very close, as presented in Figure 11. The results clearly demonstrate a high level of consistency between the mapping points calculated by DICM and SSM and the original shape. The mapping points by the two methods agree very well. There are no abnormal phenomena, such as disorderly distribution and accumulation of mapping points, implying that SSM is an effective approach for performing conformal mapping calculations.

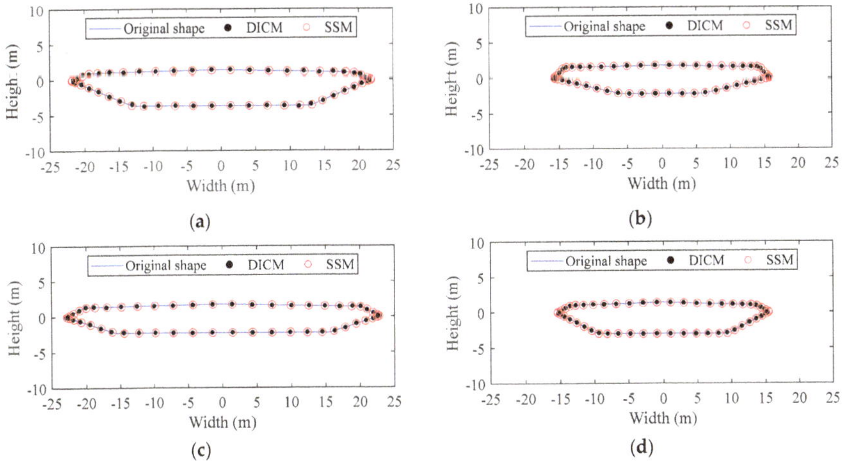

Figure 12. *Cont.*

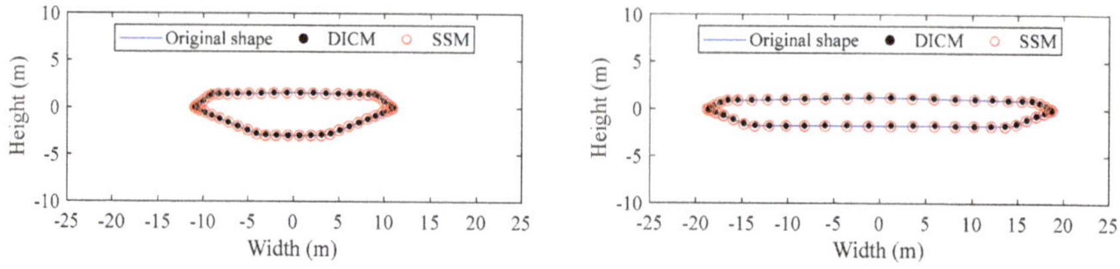

(e) (f)

Figure 12. Comparison of conformal mapping points using different methods; (**a**) Akashi Strait Bridge (Box scheme) [24]; (**b**) Xianxin Road Bridge [21]; (**c**) Second Humen Bridge [25]; (**d**) Great Belt East Bridge [22]; (**e**) Humber Bridge [26]; (**f**) Yangluo Yangtze River Bridge [27].

4. Application

Bridge flutter is a phenomenon that occurs when a bridge's structure vibrates in response to strong wind. This can cause the bridge to sway or oscillate violently, which can cause damage and even collapse of bridge structures. With the increase of the span length of the bridge, the flutter performance has gradually become one of the control factors of bridge design [28–31]. A bridge that is able to withstand wind load is less likely to require frequent repairs or replacements, which can be expensive and have negative environmental impacts. Hence, the design of a bridge that can prevent flutter vibration throughout its entire life cycle is crucial for promoting sustainability.

The flutter critical wind speed of a bridge is strongly linked to the shape of a closed box girder. In this study, SSM was utilized to assess the flutter performance of four closed-box girders. First, the conformal mapping coefficients of the closed boxes are calculated using SSM. Subsequently, the flutter derivatives of the closed boxes are computed. Finally, the flutter critical wind speeds are determined. The shapes of the closed box girders, method for acquiring flutter derivatives, dynamic parameters for flutter analysis, and wind tunnel test data remain consistent with those presented in reference [21], in which the wind tunnel test was conducted at the XNJD−2 wind tunnel of Southwest Jiaotong University. Additionally, according to Figure 11, the mapping error calculated using SSM remains relatively stable beyond an n value of 12. From the perspective of computational efficiency, the series number n is set to 12 in this study.

The flutter prediction results obtained from SSM are exhibited in Figure 13, in which the flutter wind speeds are normalized using the results of Box 4. It is demonstrated that, among the four closed boxes, Box 4 outperforms the other three closed boxes, while Box 3 exhibits the weakest flutter performance. Moreover, the flutter critical wind speed calculated by SSM are in agreement with those obtained from the wind tunnel test, confirming the effectiveness of SSM in flutter performance estimation of closed box girder bridges. The SSM method exhibits exceptional efficiency, with a computation time of fewer than 10 s for calculating the flutter wind speed of a closed box girder. Given its satisfactory accuracy and efficiency levels, the proposed SSM is well suited for preliminary girder shape selection in the design of closed-box girder bridges. Furthermore, there is no complex derivation for SSM. Thus, it is uncomplicated to understand and implement SSM, which can help the application of conformal mapping in bridge engineering.

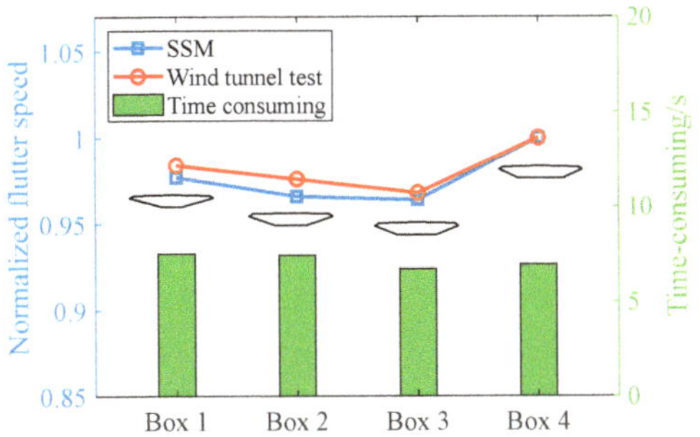

Figure 13. Flutter performance prediction.

5. Conclusions

The investigation of wind load on structures holds paramount significance for ensuring the safety and sustainability of bridges. This study investigates the characteristics of the conformal mapping coefficients from a unit circle to closed box girders, which is essential for analyzing wind load. Based on the distribution range of the conformal mapping coefficients, a numerical method, the Sliding Searching Method (SSM), was proposed to calculate the coefficients. Finally, the application of SSM in flutter prediction was studied. The main conclusions are summarized below:

(1) For conformal mapping of the closed box girder, the coefficients $a_k \to 0$ when series number k is an even number and $b_k \to 0$ when k is an odd number. This will help to reduce the computation time of the solving of the conformal mapping coefficient.

(2) The range of coefficients a_k and b_k are limited and substantially diminishing with the increase of the series number. Especially the coefficients a_1 and a_3 are closely related to the width and height of the closed box.

(3) The SSM can calculate the coefficient solution with high accuracy when the number of series is small. The mapping points by SSM are very consistent with six practical box-girder shapes, indicating the effectiveness of SSM. The SSM manifests good simplicity and easy comprehensibility without convoluted mathematical operations and deductions.

(4) The estimation of flutter performances by SSM agrees well with wind tunnel tests and only takes several seconds. The satisfactory accuracy and efficiency make the SSM a good method for girder shape selection in the preliminary design stage of a closed-box girder bridge.

(5) The limitation of this study is that SSM can take tens of seconds when the number of series is large due to the increase in the calculation. In future research, we will delve deeper into the relationship between the shape of the closed-box girder and the mapping coefficient. This can contribute to decreasing the number of iterations required by the SSM and enhance its overall efficiency.

Author Contributions: Conceptualization, L.W.; methodology, L.W.; validation, L.W. and J.Z.; investigation, J.Z.; writing—Original draft preparation, L.W.; writing—Review and editing, Z.Z. and M.Z.; supervision, Z.Z. and M.Z. All authors have read and agreed to the published version of the manuscript.

Funding: This research received no external funding.

Institutional Review Board Statement: Not applicable.

Informed Consent Statement: Not applicable.

Data Availability Statement: The data presented in this study are available on request from the corresponding author.

Conflicts of Interest: The authors declare no conflict of interest.

References

1. Li, Y.; Jiang, F.; Zhang, M.; Dai, Y.; Qin, J.; Zhang, J. Observations of Periodic Thermally-Developed Winds beside a Bridge Region in Mountain Terrain Based on Field Measurement. *J. Wind Eng. Ind. Aerodyn.* **2022**, *225*, 104996. [CrossRef]
2. Zhang, J.; Zhang, M.; Jiang, X.; Wu, L.; Qin, J.; Li, Y. Pair-Copula-Based Trivariate Joint Probability Model of Wind Speed, Wind Direction and Angle of Attack. *J. Wind Eng. Ind. Aerodyn.* **2022**, *225*, 105010. [CrossRef]
3. Amman, O.H.; von Kármán, T.; Woodruff, G.B. *The Failure of the Tacoma Narrows Bridge*; Federal Works Agency: Washington, DC, USA, 1941.
4. Cui, C.; Ma, R.; Hu, X.; He, W. Vibration Analysis for Pendent Pedestrian Path of a Long-Span Extradosed Bridge. *Sustainability* **2019**, *11*, 4664. [CrossRef]
5. Li, Y.; Chen, X.; Yu, C.; Togbenou, K.; Wang, B.; Zhu, L. Effects of Wind Fairing Angle on Aerodynamic Characteristics and Dynamic Responses of a Streamlined Trapezoidal Box Girder. *J. Wind Eng. Ind. Aerodyn.* **2018**, *177*, 69–78. [CrossRef]
6. Wang, Z.; Tang, H.; Li, Y.; Guo, J.; Liu, Z. Windproof Ability of Aerodynamic Measures to Improve the Wind Environment above a Truss Girder. *Wind Struct.* **2021**, *32*, 423–437.
7. Ke, Y.; Shen, G.; Yang, X.; Xie, J. Effects of Surface-Attached Vertical Ribs on Wind Loads and Wind-Induced Responses of High-Rise Buildings. *Sustainability* **2022**, *14*, 11394. [CrossRef]
8. Selig, M.S.; Maughmer, M.D. Multipoint Inverse Airfoil Design Method Based on Conformal Mapping. *AIAA J.* **1992**, *30*, 1162–1170. [CrossRef]
9. Chen, H.; Jaworski, J.W. Aeroelastic Interactions and Trajectory Selection of Vortex Gusts Impinging upon Joukowski Airfoils. *J. Fluids Struct.* **2020**, *96*, 103026. [CrossRef]
10. Poozesh, A.; Mirzaei, M. Flow Simulation Around Cambered Airfoil by Using Conformal Mapping and Intermediate Domain in Lattice Boltzmann Method. *J. Stat. Phys.* **2017**, *166*, 354–367. [CrossRef]
11. Malonek, H.R.; De Almeida, R. A Note on a Generalized Joukowski Transformation. *Appl. Math. Lett.* **2010**, *23*, 1174–1178. [CrossRef]
12. Jia, X.; Lu, A.; Cai, H.; Ma, Y. An Analytical Method for Solving Gravity-Induced Stresses in Slope. *Appl. Math. Model.* **2021**, *98*, 665–679. [CrossRef]
13. Kuliyev, S.A. Conformal Mapping Function of a Complex Domain and Its Application. *Arch. Appl. Mech.* **2020**, *90*, 993–1003. [CrossRef]
14. Gu, X.D.; Zeng, W.; Luo, F.; Yau, S.-T. Numerical Computation of Surface Conformal Mappings. *Comput. Methods Funct. Theory* **2012**, *11*, 747–787. [CrossRef]
15. Dyachenko, A.I.; Lushnikov, P.M.; Zakharov, V.E. Non-Canonical Hamiltonian Structure and Poisson Bracket for Two-Dimensional Hydrodynamics with Free Surface. *J. Fluids Struct.* **2019**, *869*, 526–552. [CrossRef]
16. Song, K.; Yin, D.; Dai, M.; Schiavone, P. Design of Non-Circular Nanoinhomogeneities with Uniform Heat Flux in Two-Dimensional Heat Conduction. *Int. J. Heat Mass Transf.* **2021**, *166*, 120789. [CrossRef]
17. Chui, S.T.; Chen, X.; Liu, M.; Lin, Z.; Zi, J. Scattering of Electromagnetic Waves from a Cone with Conformal Mapping: Application to Scanning near-Field Optical Microscope. *Phys. Rev. B* **2018**, *97*, 081406. [CrossRef]
18. Natarajan, S.; Bordas, S.; Mahapatra, D.R. Numerical Integration over Arbitrary Polygonal Domains Based on Schwarz–Christoffel Conformal Mapping. *Int. J. Numer. Meth. Eng.* **2009**, *80*, 103–134. [CrossRef]
19. Wang, Y.; Feng, B.; Duan, Q.; Cao, F. Numerical Calculation Method for the Boundary of Closed Polygon and Rectangle Region. In Proceedings of the 2020 International Conference on Computer Information and Big Data Applications (CIBDA), Guiyang, China, 17–19 April 2020; pp. 421–424.
20. Driscoll, T.A.; Trefethen, L.N. *Schwarz-Christoffel Mapping*; Cambridge University Press: Cambridge, UK, 2002; Volume 8, ISBN 978-1-139-43392-1.
21. Wu, L.; Ju, J.W.; Zhang, M.; Li, Y.; Qi, J. Aerostatic Pressure of Streamlined Box Girder Based on Conformal Mapping Method and Its Application. *Wind Struct.* **2022**, *35*, 243–253. [CrossRef]
22. Kuroda, S. Numerical Simulation of Flow around a Box Girder of a Long Span Suspension Bridge. *J. Wind Eng. Ind. Aerodyn.* **1997**, *67–68*, 239–252. [CrossRef]
23. Walsh, J.L. History of the Riemann Mapping Theorem. *Am. Math. Month.* **1973**, *80*, 270–276. [CrossRef]
24. Fujino, Y.; Iwamoto, M.; Ito, M.; Hikami, Y. Wind Tunnel Experiments Using 3D Models and Response Prediction for a Long-Span Suspension Bridge. *J. Wind Eng. Ind. Aerodyn.* **1992**, *42*, 1333–1344. [CrossRef]
25. Wang, X.; Fei, P.; Zhang, X.; Dong, Y.; Melchy, O.D. Design and Construction of the Second Humen Bridge, China. *Proc. Inst. Civ. Eng.-Civ. Eng.* **2019**, *172*, 161–166. [CrossRef]
26. Adanur, S.; Günaydin, M.; Altunişik, A.C.; Sevim, B. Construction Stage Analysis of Humber Suspension Bridge. *Appl. Math. Model.* **2012**, *36*, 5492–5505. [CrossRef]
27. Zhu, Y.-F.; Ren, W.-X.; Wang, Y.-F. Structural Health Monitoring on Yangluo Yangtze River Bridge: Implementation and Demonstration. *Adv. Struct. Eng.* **2022**, *25*, 1431–1448. [CrossRef]
28. Wu, L.; Woody Ju, J.; Zhang, J.; Zhang, M.; Li, Y. Vibration Phase Difference Analysis of Long-Span Suspension Bridge during Flutter. *Eng. Struct.* **2023**, *276*, 115351. [CrossRef]
29. Tang, H.; Zhang, H.; Mo, W.; Li, Y. Flutter Performance of Box Girders with Different Wind Fairings at Large Angles of Attack. *Wind Struct.* **2021**, *32*, 509–520. [CrossRef]

30. Hu, P.; Han, Y.; Cai, C.S.; Cheng, W. Wind Characteristics and Flutter Performance of a Long-Span Suspension Bridge Located in a Deep-Cutting Gorge. *Eng. Struct.* **2021**, *233*, 111841. [CrossRef]
31. Zhang, M.; Xu, F. Tuned Mass Damper for Self-Excited Vibration Control: Optimization Involving Nonlinear Aeroelastic Effect. *J. Wind Eng. Ind. Aerodyn.* **2022**, *220*, 104836. [CrossRef]

sustainability

MDPI

Article

Study on Seismic Performance Optimization of Assembly Concrete-Filled Steel Tubular (CFST)-Laced Piers

Liang Chen [1,2], Rui Zuo [1], Yingao Zhang [1], Dahai Yang [2], Jianluan Li [2], Zhigang Wu [2,*] and Xuekai Ji [2]

1 College of Civil Engineering, Hefei University of Technology, Hefei 230009, China
2 Anhui Transport Consulting and Design Institute Co., Ltd., Hefei 230088, China
* Correspondence: 2021170685@mail.hfut.edu.cn

Abstract: This study aims to investigate the seismic behavior of concrete-filled steel tubular (CFST)-laced piers; after verifying the model through engineering tests, the simplified finite element models (S-FEM) and refined ones (R-FEM) with CFST-laced piers are developed in this manuscript, respectively. Through comparison, it is found that the S-FEM can effectively improve analyzing efficiency while meeting the requirements of engineering analysis accuracy. In addition, the seismic response of assembled flange-connected CFST-laced piers bridge was studied based on the S-FEM, and different structural parameters, including pier height, axial compression ratios, steel ratios of CFST columns, steel lacing tube arrangement, and longitudinal slope, are considered to optimize the bridge design scheme. Results indicate that the parameters of 0.1 axial pressure ratios and 1:30 longitudinal slope show superior seismic performance. Meanwhile, the peak axial force and peak bending moment of CFST column limbs occur at the pier bottom, and the flanges, which are subject to larger bending moments, are generally located at the two connection positions above the pier bottom.

Keywords: concrete-filled steel tubular (CFST); laced piers; modeling method; seismic performance; design optimization

Citation: Chen, L.; Zuo, R.; Zhang, Y.; Yang, D.; Li, J.; Wu, Z.; Ji, X. Study on Seismic Performance Optimization of Assembly Concrete-Filled Steel Tubular (CFST)-Laced Piers. *Sustainability* **2023**, *15*, 8318. https://doi.org/10.3390/su15108318

Academic Editor: Syed Minhaj Saleem Kazmi

Received: 20 April 2023
Revised: 12 May 2023
Accepted: 14 May 2023
Published: 19 May 2023

1. Introduction

Concrete-filled steel tubular (CFST) four-limb-laced columns (piers) use four circular CFST column limbs as chord tubes and hollow steel tubes as lacing tubes (Figure 1). Due to many advantages such as high bearing capacity, high structural stiffness, superior seismic performance, simple construction, material saving, and environmental protection, etc., assembled CFST-laced piers began to be widely used in mountainous areas.

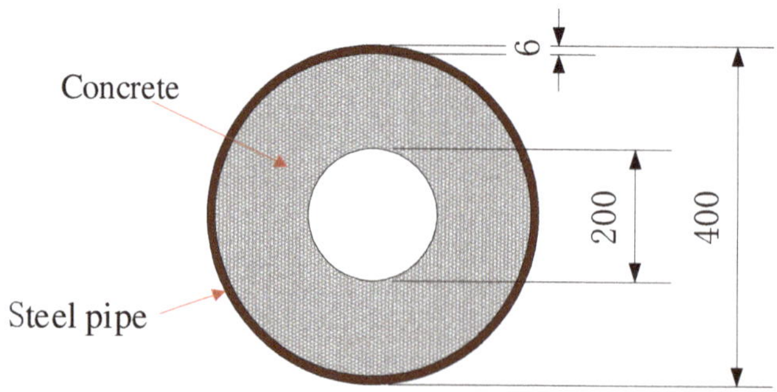

Figure 1. Cross-sectional view of the SC1 specimen.

CFST has been deeply studied by scholars in the past decades. For static behavior of CFST-laced columns, Han et al. [1] investigated the fire performance of CFST-laced columns by a test, and the test results indicate that CFST triple-limb-laced columns have good integrity because of the steel lacing tubes.

To study the failure mechanism of CFST columns at ultimate loads, Ou Z et al. [2] conducted a series of tests, and the experimental results were used to validate several analytical models. Huang et al. [3] studied the ultimate load of laced columns with initial stress by axial loading test, and found that the initial stress has little effect on the mechanical properties of lacing tubes.

To date, studies on the overall seismic performance of bridges with CFST-laced piers are still insufficient. Previous studies on seismic performance usually focused on the component level. Scholars [4–8] carried out the hysteresis performance tests and finite element analysis to investigate the seismic performance of CFST-laced piers; the results showed that the CFST columns have good ductility and energy dissipation capacity. In more detail, to improve the rapid repair capacity of bridges after earthquakes, Zhang et al. [8] proposed a replaceable connection component for laced piers. Hajjar and Gourley [9,10] put forward two theoretical models to calculate the hysteretic curve of concrete-filled square steel tubular structures. Moreover, Aval et al. [11] used the fiber model method to analyze the load–displacement hysteresis curves of CFST under cyclic load, which takes the steel tube combined effect into account with core concrete and bond–slip effect. Additionally, similar studies also can be seen in the research of [2,12].

It should be noted that the application research of CFST-laced columns in bridge engineering is still in its infancy. In particular, the current study on CFST-laced piers in segmental prefabricated assemblies is limited and lacks engineering examples to promote its development and application. The new type of assembled flange-connected CFST-laced columns proposed in this paper has a novel structure. It adopts a new form of hollow CFST column limbs and internal flange connection (see Figure 1), which effectively realizes the construction concept of standardized design, factory prefabrication, and assembled construction, and also greatly reduces the construction difficulty of mountainous bridges. However, due to the innovative structure of this new lace pier, there is still a lack of relevant design codes and standards, which poses many difficulties in the design process, such as the mechanical properties of key components, seismic performance, and identification of key design parameters. These difficulties will have an important impact on the safety, durability, and economy of the CFST-laced piers [13,14]. Moreover, due to the low stiffness of the overall structural system of this bridge, the mass and stiffness distribution of the superstructure may have some influence on the seismic response of the structure, which needs efficient seismic analysis models for the bridge with CFST-laced piers.

Due to the above considerations, to further investigate the seismic performance of a real bridge with hollow CFST-laced columns, this manuscript used the finite element analysis software SAP2000 to verify the experimental steel–concrete model, and developed two finite element models: a simplified model using an equivalent section of superstructure replaced, and a refined model taking into account the mass and stiffness distribution of the superstructure [15]. The applicability of the two models was evaluated by analyzing the static properties, dynamic properties, and seismic response of the structure. In addition, the seismic response mechanisms of assembled flange-connected CFST-laced piers were studied, and their seismic design was optimized with five optimized parameters, including pier height, axial compression ratios, steel ratios of CFST columns, and longitudinal slope. Generally, this study aims to obtain a simple and effective seismic analysis model, and will provide useful information for the design of CFST-laced columns in composite bridges.

2. Materials and Methods

To validate the proposed modeling method for CFST in this paper, finite element simulations were carried out based on the bending test of CFST conducted by Reference [16]. The correctness of the proposed modeling method can be validated by comparing the numerical results with the experimental data. This section takes the SC1 specimen of Reference [16] as an analysis object, which has an outer diameter of 400 mm, an inner diameter of 200 mm, and a steel tube wall thickness of 6 mm made of Q235B steel, and is filled with C80 high-strength concrete with a wall thickness of 94 mm. The loading device adopts a hydraulic jack for two-point symmetrical loading, with a loading span of 3.6 m and a pure bending beam length of 1 m (see Figures 1 and 2). The material properties of the specimen are shown in Table 1.

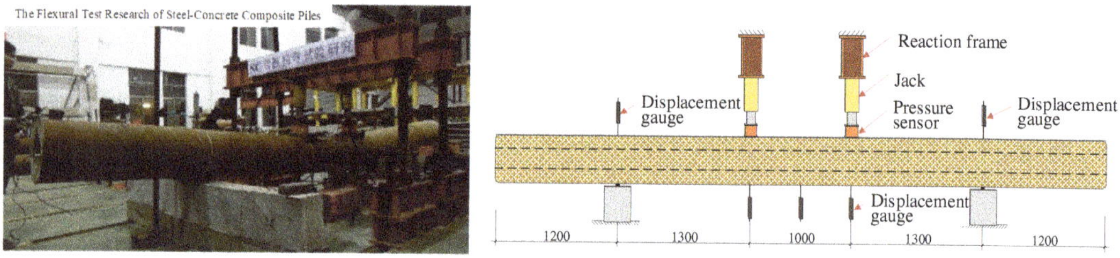

(**a**) (**b**)

Figure 2. (**a**) In situ photos of bending test [16]; (**b**) bending test apparatus. Bending performance test of the SC1 specimen.

Table 1. SC1 test specimen material properties.

Component	Compressive Strength (Mpa)	Yield Strength (Mpa)	Tensile Strength (Mpa)
Grout-filled concrete (C80)	50.2		
Steel pipe (Q235B)		243.5	512.3

2.1. Numerical Modeling

As illustrated in Figures 3 and 4, the model was established based on a 1:1 scale of the experimental setup, and the entire model was built using beam elements. The stress–strain relationship of concrete used in the numerical model was based on the envelope curve recommended by Mander [17], where f'_{co} represents the compressive strength of the cylinder, E_c represents the elastic modulus, and ε_{co} represents the strain corresponding to the peak stress of the unconfined concrete, which is generally taken as 0.002. The steel tube was modeled using the bilinear model for its constitutive behavior, where E_s represents the elastic modulus and is measured in 2.06×10^5 MPa.

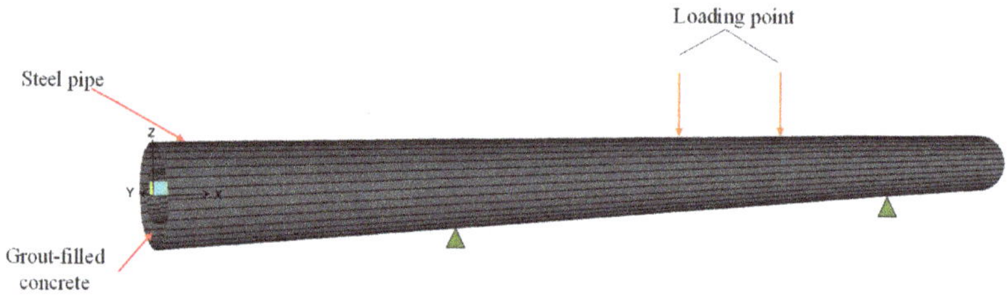

Figure 3. FEM of test specimens.

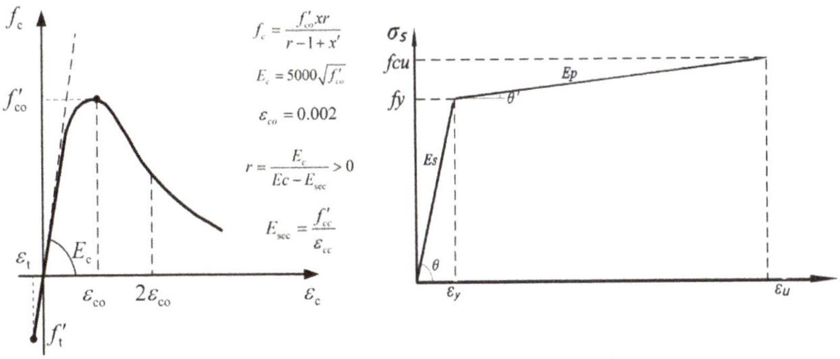

Figure 4. Material constitutive models in SAP2000. (**a**) Mander constitutive model for infilled concrete; (**b**) steel tube bilinear model.

2.2. Model Verification

According to the experimental loading scheme, a force-controlled step loading is first used until reaching yield, followed by displacement-controlled loading with each step of 2 mm, up to a total displacement of 110 mm. The moment–displacement curve at mid-span of the structure is obtained, and the comparison of numerical simulation and experimental data is shown in Figure 5. Apparently, by comparing the FEM results with the experimental results, FEM results correspond well with experimental results, which validates the effectiveness of the proposed numerical models of CFST column.

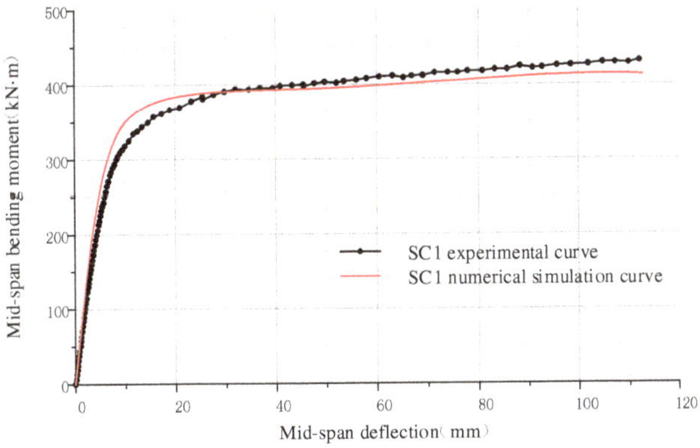

Figure 5. Comparison of experimental and simulated bending moment–deflection data.

2.3. Finite Element Model (FEM)

2.3.1. Bridg Overview

To investigate the seismic performance of concrete-filled steel tubular (CFST) laced columns, this study takes a real bridge as the object of analysis. The bridge is a 4 × 35 m four-span continuous composite bridge, the substructure is assembled flange connected hollow CFST-laced piers, and the superstructure is made of steel–concrete composite girder; the bridge layout is shown in Figure 6.

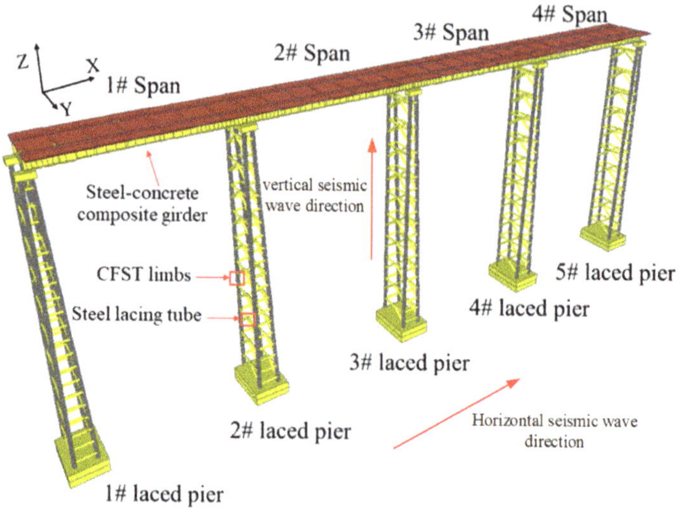

Figure 6. General layout of bridge with CFST-laced piers.

For the steel–concrete composite girders of the bridge superstructure, which consist of I-shaped steel girders and concrete deck slabs. The concrete deck slab is 12.9 m wide and takes C40 concrete (i.e., the cubic compressive strength is 40 MPa). The steel girder consists of two I-shaped beams 1.7 m high; more detailed information about the superstructures is illustrated in Figure 7. The bridge substructures are 50 m high and consist of 5 sections of four-limb CFST-laced columns connected by flanges. Additionally, CFSTs are 700 mm in diameter, and their total wall thickness is 110 mm, of which the wall thickness of steel tube is 12 mm. Steel pipes of 406 mm diameter and wall thickness of 12 mm are adopted in steel lacing tube, and arrangement type of steel lacing tube is K-shaped layout. Q345B steel is used for the Lacing tube, steel pipes, and flanges in CFST-laced columns. C80 high-strength concrete was used for the core concrete in CFTC, C40 concrete was used for cover beam, and C35 underwater concrete was used for the piles. Detailed information about the substructures and the material properties are shown in Figure 8 and Table 2.

The bridge is oriented in the longitudinal direction as the X-axis, the transverse direction as the Y-axis, and the vertical direction as the Z-axis, and FEM of CFST-laced columns bridges was conducted based on SAP2000 (2014). The design parameters of the original bridge scheme are shown in Table 3 and Figures 6–8.

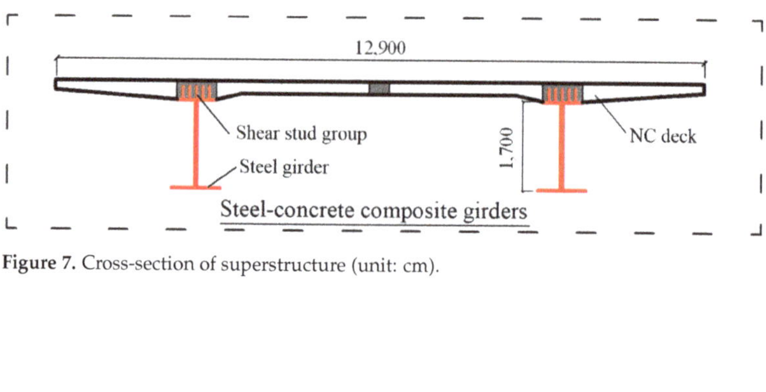

Figure 7. Cross-section of superstructure (unit: cm).

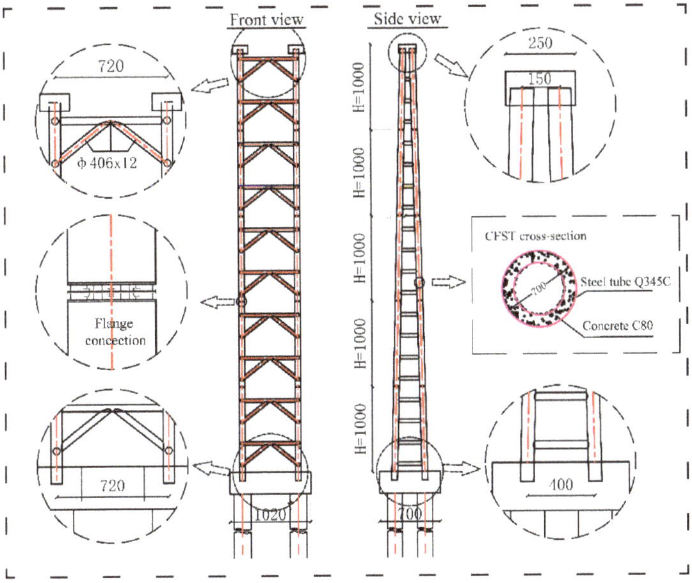

Figure 8. General layout of CFST-laced piers (unit: cm).

Table 2. Material properties.

Component	Compressive Strength (Mpa)	Yield Strength (Mpa)	Tensile Strength (Mpa)
C35	23.4		
C40	26.8		
C80	50.2		
Q235B		243.5	512.3
Q345B		356.4	623.8

Table 3. The design parameters of the original bridge scheme.

Parameters	Values	Parameters	Values
Pier height	50 m	Single span	35 m
Longitudinal slope	1:40	Steel ratio of CFST (wall thickness of steel tube)	12.7% (12 mm)
Axial compression ratio	0.1	Strength of steel	Q345
Arrangement type of lacing tube	K-shaped layout	Concrete strength of CFST	C80

2.3.2. Steel–Concrete Composite Girders (SCCG) of the Bridge Superstructure

In general, the bridge superstructure is in an elastic state under seismic action [18–24]. Due to the small stiffness and mass of the steel–concrete composite girders in this bridge, a reasonable simulation of the actual distribution of stiffness and mass may have some influence on the seismic response of the bridge. Thus, to investigate the effect of the simulation method of steel–concrete composite girders on the seismic response, the frame model (simplified model) and refined model were established in this section. For the frame model, the elastomeric frame element (frame) was chosen to simulate the steel–concrete composite girders without considering the effect of its actual distribution of stiffness and mass on the seismic response. For the refined model, shell element was used to simulate concrete deck slabs, steel girders were simulated by elastomeric frame element, and the shear nail connections between girders and slabs were simulated by Link units, and shear nail stiffnesses were considered in this model. The refined model is able to take into account not only the effect of the actual distribution of stiffness and mass of the composite girder, but also the effect of the mechanical properties of the shear nail connection on the seismic response. However, the refined model is more computationally expensive, especially when performing nonlinear dynamic time history analysis.

2.3.3. Support Installment

Spherical bearings were used in this bridge, which was simulated by master–slave constraint + Link unit for the dynamic characteristic analysis and elastic dynamic response spectrum analysis. In the nonlinear dynamic time history analysis, the bearings were simulated by the Plastic–Wen elements, and the bearing arrangement of the bridge is shown in Figure 9. This bridge is located in a mountainous area, which has good geological conditions; the soil layer of foundation is hard and can be treated as a rigid foundation; thus, pile–soil interaction is directly simulated by the consolidation method.

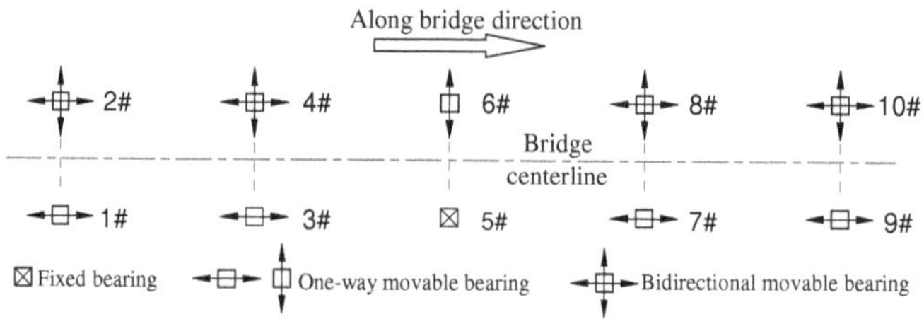

Figure 9. The layout of the support.

2.3.4. Concrete-Filled Steel Tubular (CFST) Laced Columns

Frame elements (frame) were adopted to simulate CFST-laced columns in this study, where the column limbs are steel–concrete combined sections (see Figure 8), and there are two methods of FEM for column limbs.

a. The combined section of CFST;
b. Equivalent section replaced CFST combined section, i.e., simplifying the cross-section based on the principle of equivalent bending stiffness and axial stiffness, and the specific calculation formula is as follows.

Equivalent section flexural stiffness,

$$EI = E_c I_c + E_s I_s \tag{1}$$

Equivalent cross-sectional axial stiffness,

$$EA = E_c A_c + E_s A_s \tag{2}$$

In which E_c and E_s are elastic moduli of concrete and steel, respectively. A_c and A_s are the cross-sectional areas of concrete and steel, respectively. I_c and I_s are the area moment of inertia of concrete and steel, respectively.

2.3.5. Seismic Wave Selection

According to the Specification for Seismic Design of Highway Bridges (JTG/T2231-01-2020) [25], this bridge is located in a Class II site, which is a Class B bridge with a seismic intensity of VII degrees. Therefore, the transverse and vertical design acceleration response spectra corresponding to E2 seismic level are generated, as shown in Figure 10. The Reference [26] presents a methodology for synthesizing artificial seismic waves, enabling the generation of seven seismic waves that correspond to the horizontal and vertical design acceleration response spectra of E2. These seven seismic waves are then employed for comprehensive time history analysis, facilitating a thorough investigation of the seismic response.

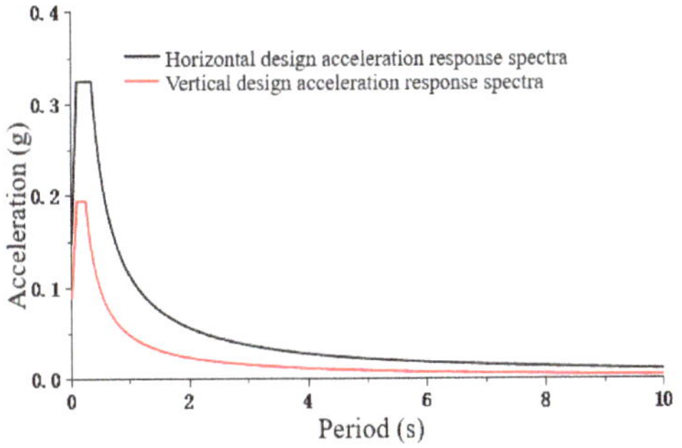

Figure 10. E2 acceleration response spectrum.

3. Results and Discussions

3.1. Comparative Analysis of Different Modeling Methods

To investigate the effect of the simulation method of this bridge on the seismic response, the refined model and simplified model were established based on the validated FEM of the CFST column (see Section 2). The differences between the FEM results of those models are compared by static and dynamic analysis and seismic response analysis. The differences between the above two models are shown in Table 4, and FEM is shown in Figure 11.

Table 4. The difference between the two finite element models.

	Superstructures	**Substructures (Laced Columns)**	**Others**
Simplified model	Frame element	Equivalent section	Identical
Refined model	shell element, frame element, and Link units	Combined section of CFST	

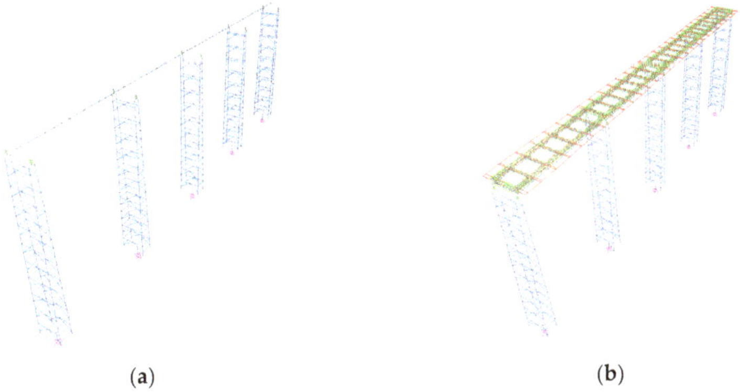

Figure 11. (**a**) S-model; (**b**) R-model. Diagrams of FEM.

3.1.1. Static Characterization Analysis

To compare the differences between the above two models for the bridge under self-weight and secondary dead load, static analysis was carried out; the comparison factors were mid-span deflection, pier bottom axial force, and support reaction force, and the analysis results are shown in Figure 12. From Figure 12, it can be seen that the changing pattern of deflection, pier bottom axial force, and vertical reaction force of bearing in each span of the two models are consistent, and the difference in the results is very limited. For the vertical displacement of each span, the error of span 1# and span 4# is the largest, at

5.1%; for the axial force at the bottom of each pier, the error of pier 1# and pier 5# is the largest, at 1.36%. Similarly, the error of bearing reaction force in 1#, 2#, 9#, and 10# is the largest, at 2.44%.

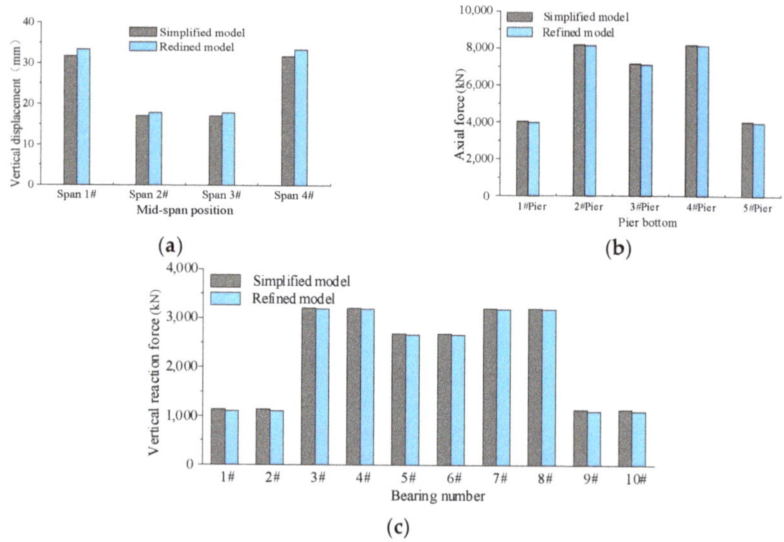

Figure 12. (**a**) Comparison of mid-span deflection; (**b**) comparison of pier bottom axial force; and (**c**) vertical reaction force of bearing. Comparison of static performance of two models.

3.1.2. Dynamic Characterization Analysis

To compare the differences in the dynamic characteristics of the simplified model and the refined model, the two models were also analyzed based on SAP2000, and the natural periods, mass participation factors, and shapes of the first 10 orders of vibration are listed, which are shown in Tables 5 and 6. Meanwhile, the first three modes of the simplified and refined model are listed in Figure 8. From Tables 5 and 6 and Figure 13, the differences in structural natural periods between the two models are small, but the mode shapes are slightly different, because the refined model can take into account the actual distribution of the superstructure stiffness and mass; for example, in the third order vibration shape, the refined model can show the torsional vibration pattern of the superstructure, while the rod system model cannot.

Table 5. The first ten cycles comparison table.

Mode Order	Simplified Model		Refined Model		Relative Errors
	Natural Periods (s)	Mass Participation Factors	Natural Periods (s)	Mass Participation Factors	
1	6.225	0.597	6.237	0.597	0.19%
2	1.530	0	1.581	0	3.23%
3	1.256	0	1.318	0	4.70%
4	0.897	3.753×10^{-10}	0.895	6.773×10^{-16}	0.22%
5	0.897	0.0176	0.895	0.0177	0.22%
6	0.897	2.041×10^{-5}	0.895	2.844×10^{-13}	0.22%
7	0.897	3.25×10^{-10}	0.892	1.002×10^{-6}	0.56%
8	0.494	0.0451	0.514	0.04521	3.89%
9	0.475	0	0.476	0	0.21%
10	0.395	2.692×10^{-6}	0.439	1.421×10^{-6}	9.94%

Note: The relative error in the table is the ratio of the absolute value of the difference between S-model and R-model. The mass participation factors in the table vary along the longitudinal direction of the bridge.

Table 6. The first ten vibration characteristics.

Mode Order	Mode Shapes	
	Simplified Model	**Refined Model**
1	Longitudinal vibration of superstructure	
2	Symmetrical lateral bending of superstructure	
3	Antisymmetrical lateral bending of superstructure	Transverse antisymmetric bending and twisting of superstructure
4	5# Pier longitudinal bending	
5	1# Pier longitudinal bending	
6	4# Pier longitudinal bending	
7	3# Pier longitudinal bending	
8	Symmetrical lateral bending of superstructure, 1#, 5# pier floating horizontally	Symmetrical lateral bending of superstructure
9	3# Pier longitudinal bending	
10	Antisymmetric vertical bending of superstructure	

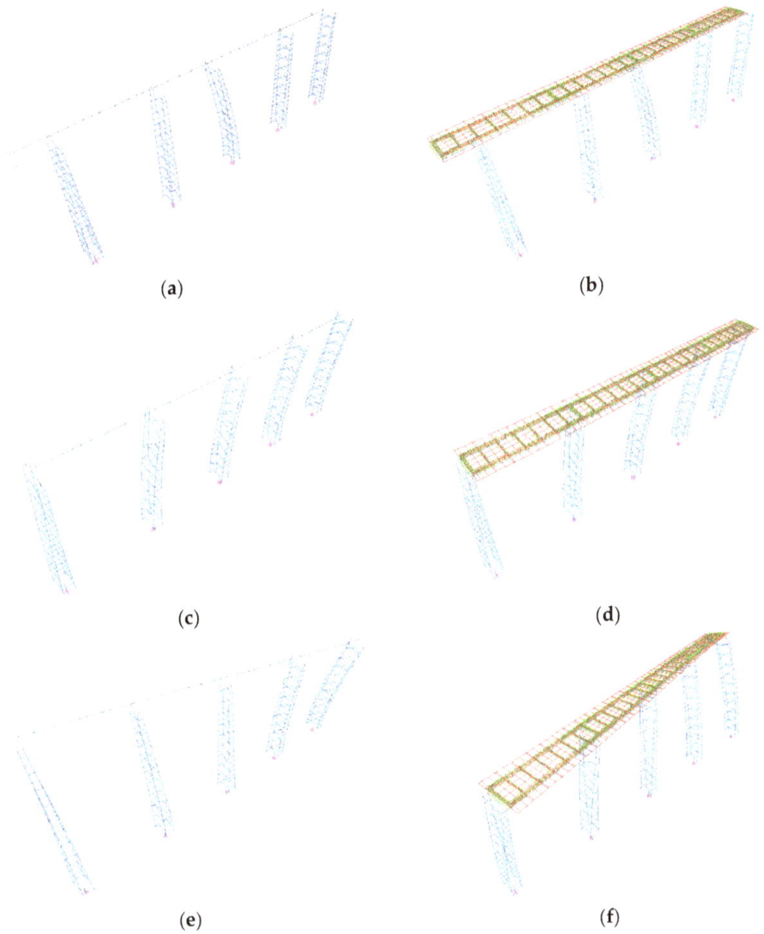

Figure 13. Comparison of the first three modes of the two models. (**a**) Simplified model—the first mode; (**b**) refined model—the first mode; (**c**) simplified model—the second mode; (**d**) refined model—the second mode; (**e**) simplified model—the third mode; and (**f**) refined model—the third mode.

3.1.3. Time History Analysis

According to the combination of seven horizontal and seven vertical artificial waves generated in Section 3.1.2, nonlinear dynamic time history analysis was performed on the simplified and refined models, with the input directions in the longitudinal + vertical and lateral + vertical directions, respectively. Taking the top displacement of the pier and superstructure displacement as the seismic response parameters, half of the structure was chosen for analysis according to the symmetry of the bridge, and the displacement response of the bridge under seismic action is obtained, as shown in Figures 14 and 15. It is clear that the two models are basically the same for the seismic response in the lateral and longitudinal directions, including the seismic response time curve and the response peak.

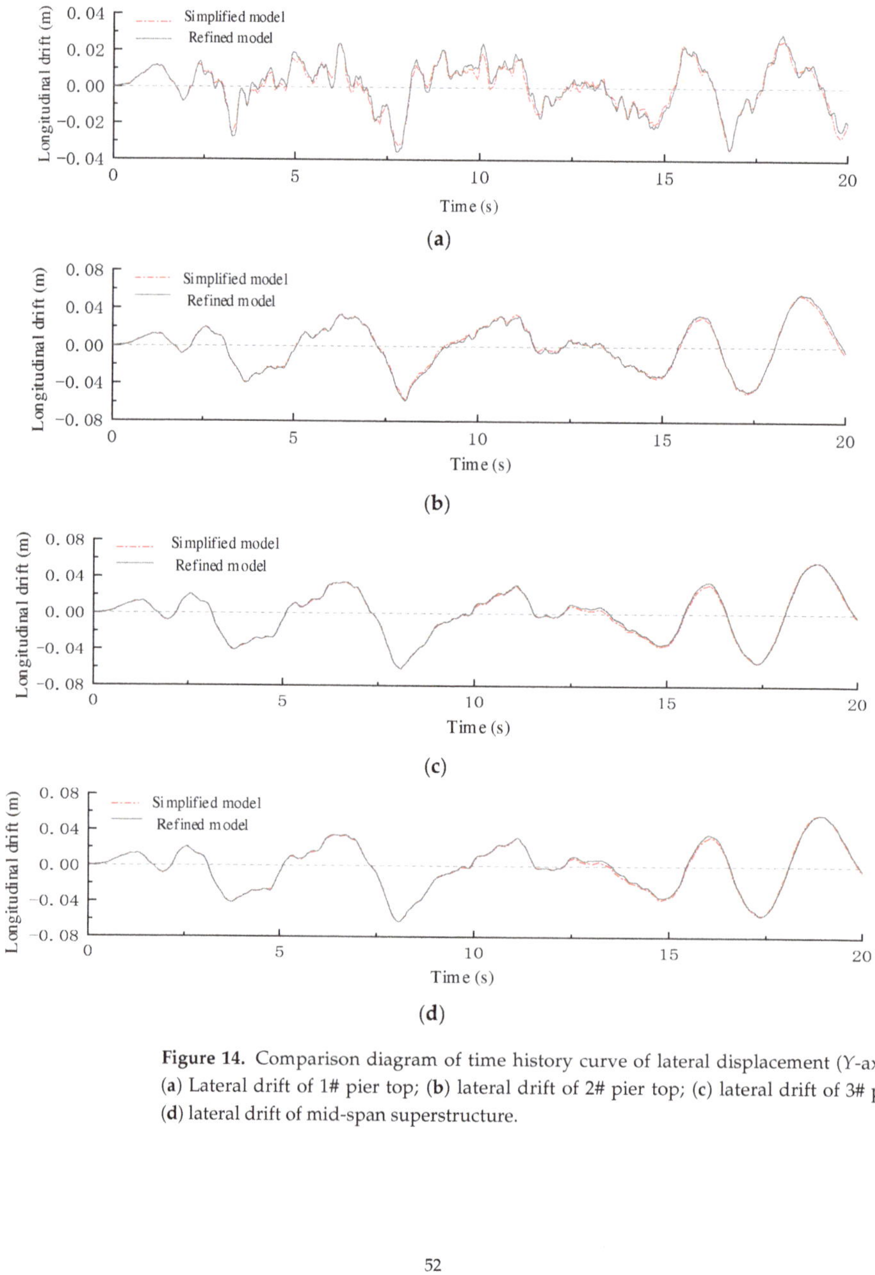

Figure 14. Comparison diagram of time history curve of lateral displacement (Y-axis direction). (**a**) Lateral drift of 1# pier top; (**b**) lateral drift of 2# pier top; (**c**) lateral drift of 3# pier top; and (**d**) lateral drift of mid-span superstructure.

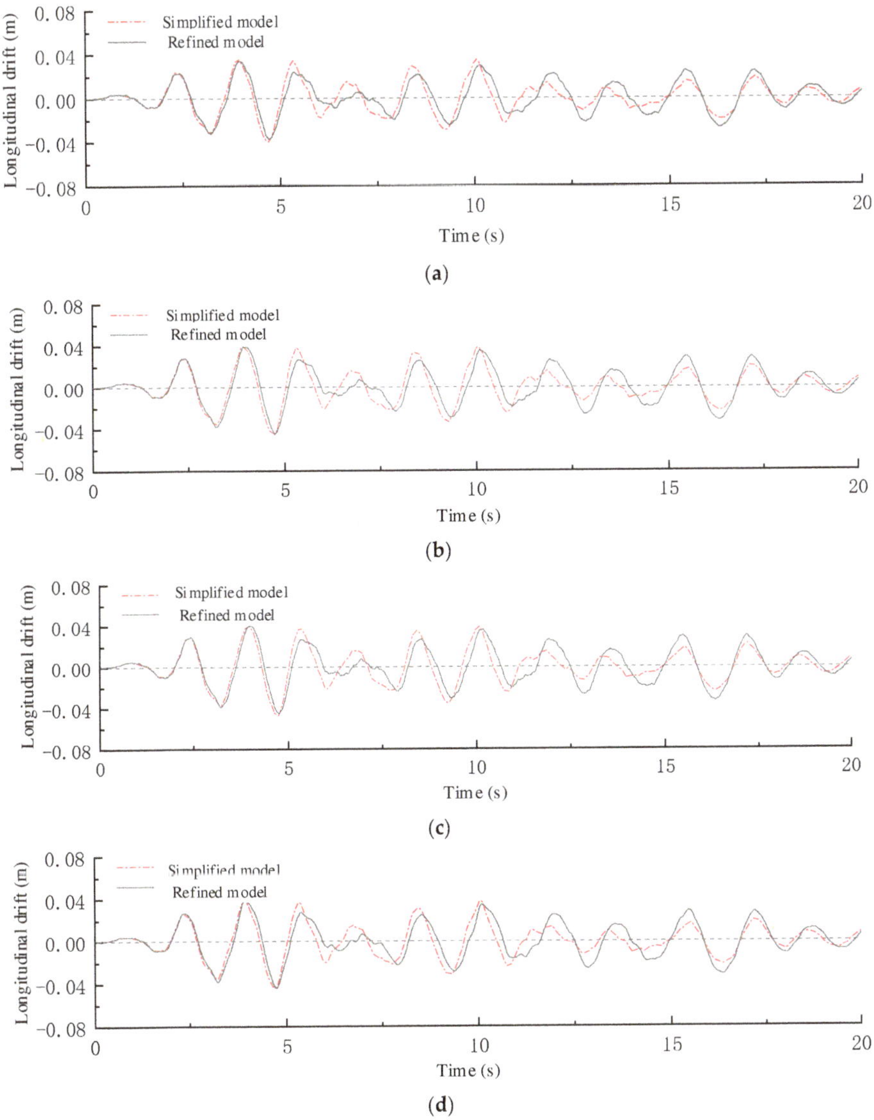

Figure 15. Comparison diagram of time history curve of longitudinal displacement (*X*-axis direction). (**a**) Longitudinal drift of 1# pier top; (**b**) longitudinal drift of 2# pier top; (**c**) longitudinal drift of 3# pier top; (**d**) and longitudinal drift of mid-span superstructure.

Based on the above discussion, There is little difference between the simplified model and the refined model for the static and dynamic characterization and seismic response, while the refined model for the seismic response analysis is significantly less efficient than the simplified model due to its more complex model. Thus, the seismic response of the bridge with a laced pier will be analyzed by the simplified model in the subsequent study.

3.1.4. Seismic Responses

In this section, the S-model and the seismic waves selected in Section 2.3.5 are used to calculate the seismic responses of the bridge. The metrics that reflect the seismic responses include structural displacements (displacement of the girder and columns), axial forces, and bending moments (steel pipe concrete column limbs at the pier bottom and flange connections). According to the symmetry, analysis of half of the bridge is permitted, which is listed in Figures 16–19.

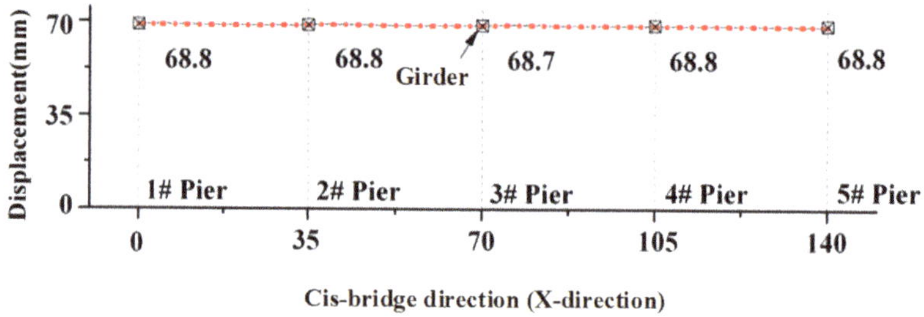

Figure 16. Peak longitudinal displacement of the girder.

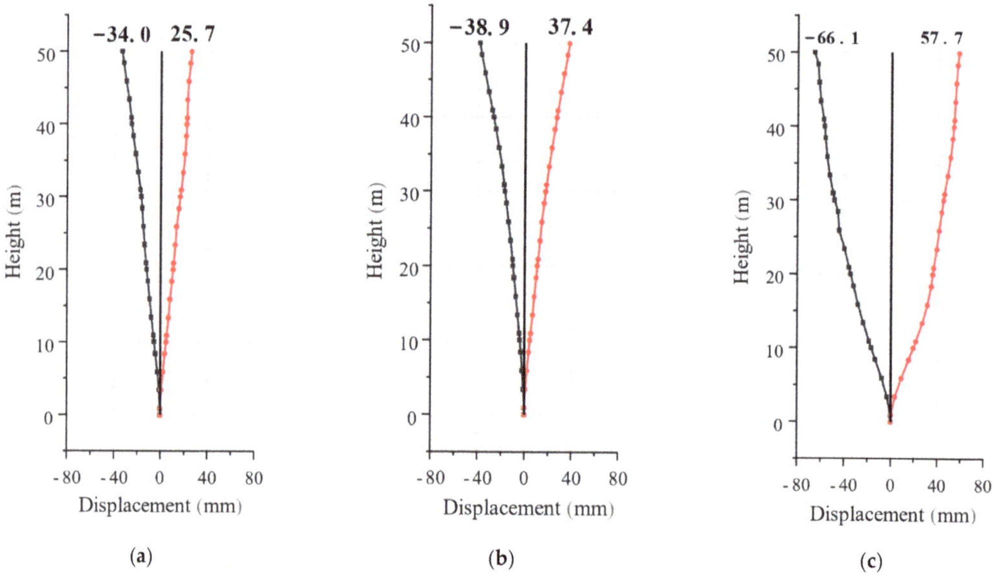

Figure 17. Horizontal displacement envelope diagram of each pier. (**a**) 1#Pier; (**b**) 2#Pier; (**c**) 3#Pier.

Figures 16 and 17 show that the longitudinal displacement of each node of the girder is basically the same under the ground motion, and the horizontal displacement of each point of the bridge pier develops with the increase in pier height.

From Figures 18 and 19, the displacement and force of the 3#pier are significantly larger than the movable pier; the maximum axial force and bending moment occur at the bottom of the pier, the maximum axial force and bending moment of the flange connection also occurs near the bottom of the pier, and the two flange connections near the bottom of the pier may be subjected to large loads.

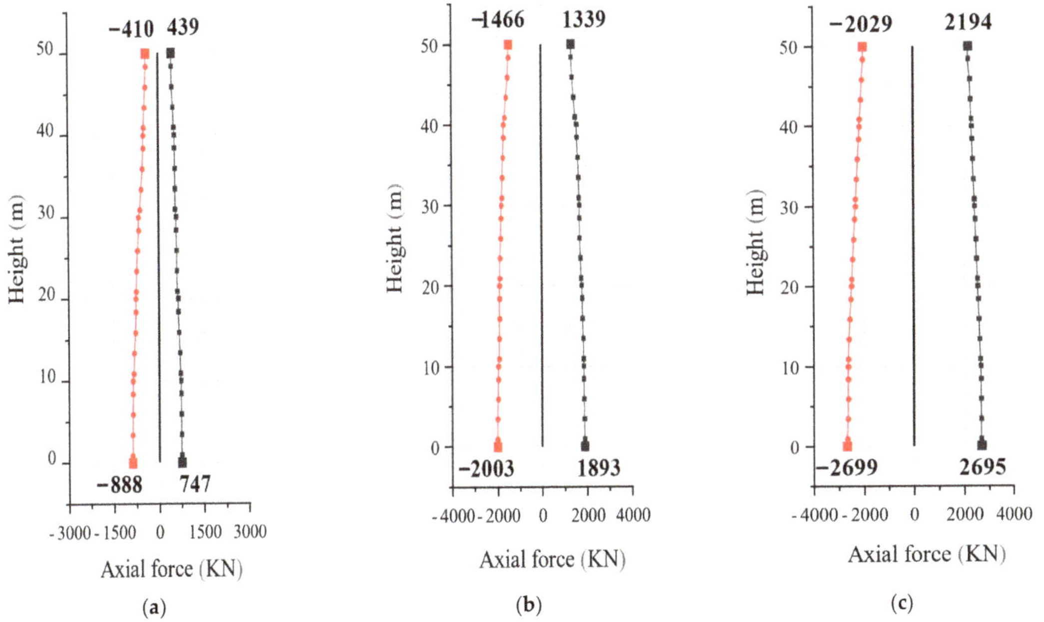

Figure 18. Axial force envelope diagram of each pier. (**a**) 1#Pier; (**b**) 2#Pier; (**c**) 3#Pier.

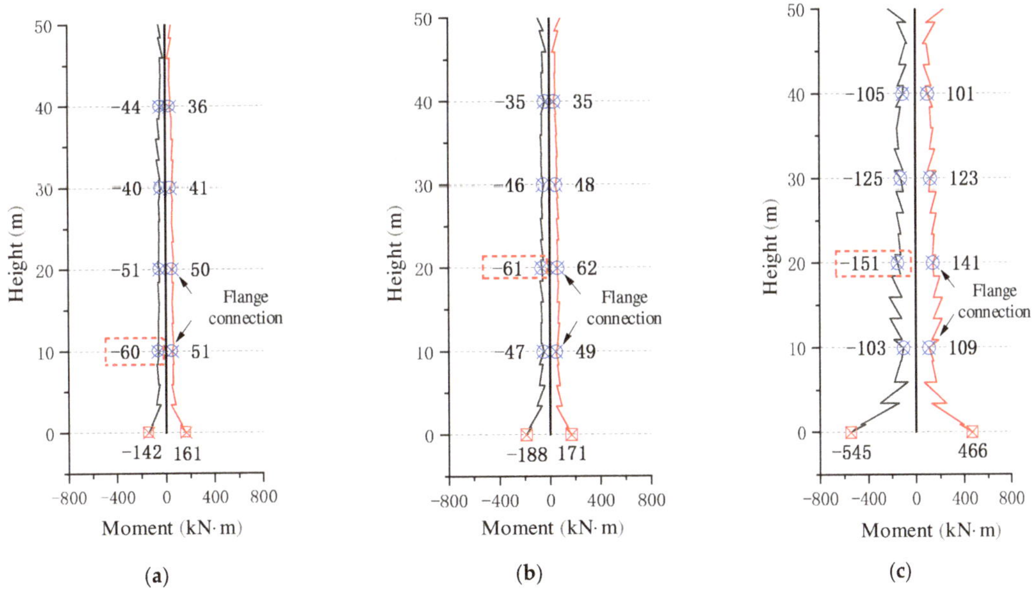

Figure 19. Envelope diagram of bending moment of each pier. (**a**) 1#Pier; (**b**) 2#Pier; (**c**) 3#Pier.

3.2. Sensitivity Analysis of Key Design Parameters for Seismic Resistance of Laced Columns

To further explore the seismic response law of laced piers, this section investigates the seismic performance of laced piers by considering the effects of parameters, for instance, pier height, axial compression ratio, steel ratio of CFST column, arrangement type of lacing tube, and longitudinal slope of column limbs, on their seismic performance. The seismic input direction is x and z direction, and the maximum displacement of the pier top, the maximum axial force of the pier bottom, the maximum bending moment of the pier bottom, the maximum bending moment at the flange connection of 3# pier's maximum concrete stress of pier bottom, and maximum steel tube stress of pier bottom are discussed.

3.2.1. Pier Height

On the basis of the original design scheme, a total of nine models are formed by changing the pier height, with the pier height of 0 m, 40 m, 50 m, 60 m, 70 m, 80 m, 90 m, 100 m, and 110 m in order. The structural response under the seismic action in the cis-bridge direction (X-direction) is shown in Figure 20 and the first-mode period of different structures is shown in Table 7. From Figure 20, the horizontal drift of the pier top gradually increases with the pier height, while the axial force and bending moment of the flange connection and stress at the bottom section of the pier gradually decreases, but the degree of decrease is inapparent.

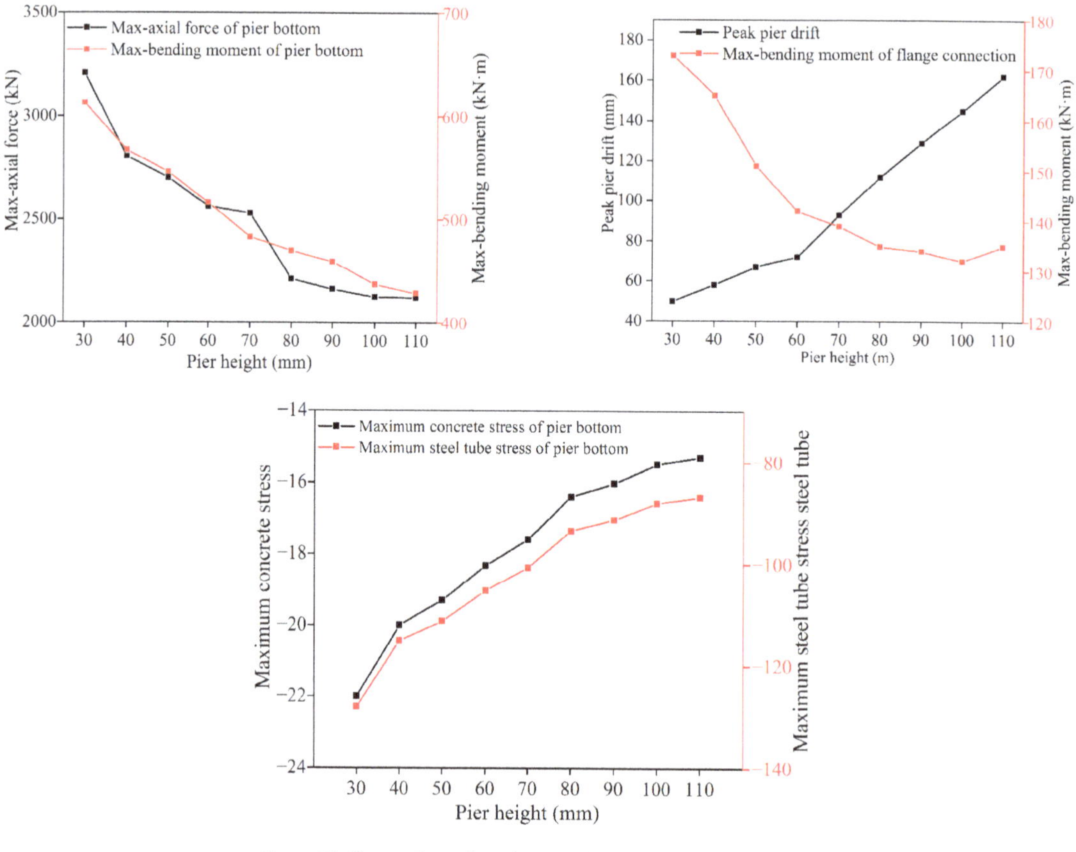

Figure 20. Comparison of maximum seismic response under different pier heights.

Table 7. Comparison of first mode period under different pier heights.

Pier Height	First Mode Period (s)
30 m	4.648
40 m	5.423
50 m	6.225
60 m	6.973
70 m	7.938
80 m	8.522
90 m	9.591
100 m	10.072
110 m	11.175

3.2.2. Axial Compression Ratios

By changing the axial compression ratios of the bridge pier in this section, a total of five models were formed, with the axial compression ratios of 0.1, 0.15, 0.2, 0.25, and 0.3 in order. The FEM results are shown in Figure 21. From Figure 21, increasing the axial compression ratios will enhance the top displacement, the bottom axial force of the pier, and the stress at the bottom section of the pier, but the impact on the bottom bending moment of the pier and the bending moment of the flange connection is inapparent. So, it is recommended that the axial compression ratios of the pier should be about 0.1 in such bridges.

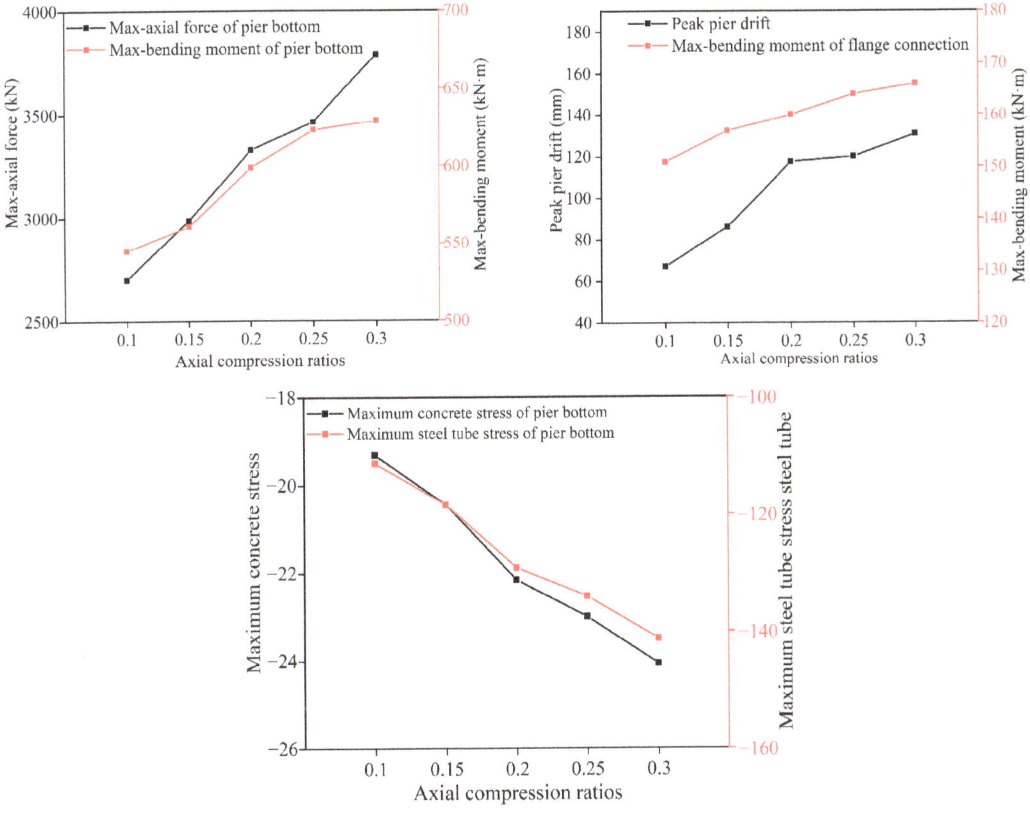

Figure 21. Comparison of response peaks under different axial compression ratios.

3.2.3. Steel Ratios of CFST Columns

To investigate the steel rations of CFST columns on the seismic performance of the bridge, five models are carried out with 4.3%, 8.5%, 12.7%, 16.9%, and 21.0% of steel ratios of CFST by changing the wall thickness of the CFST columns to 4 mm, 8 mm, 12 mm, 16 mm, and 20 mm, respectively. The structural response under the seismic action in the cis-bridge direction is shown in Figure 22 and the first-mode period of different structures is shown in Table 8. From Figure 22, the effect of increasing the steel ratios of CFST column limbs on the displacement of the top of the pier is not obvious, but it will enhance the force of the column limbs and flange connection. At the same time, increasing the wall thickness of CFST can effectively improve the flexural stiffness of the section, reduce the stress in the bottom section of the pier, and thus effectively enhance the seismic performance of the CFST pier. Considering that increasing the steel ratios of CFST columns may significantly increase the cost of the project, a laced column with a variable cross-section can be considered according to the force characteristics of the laced pier.

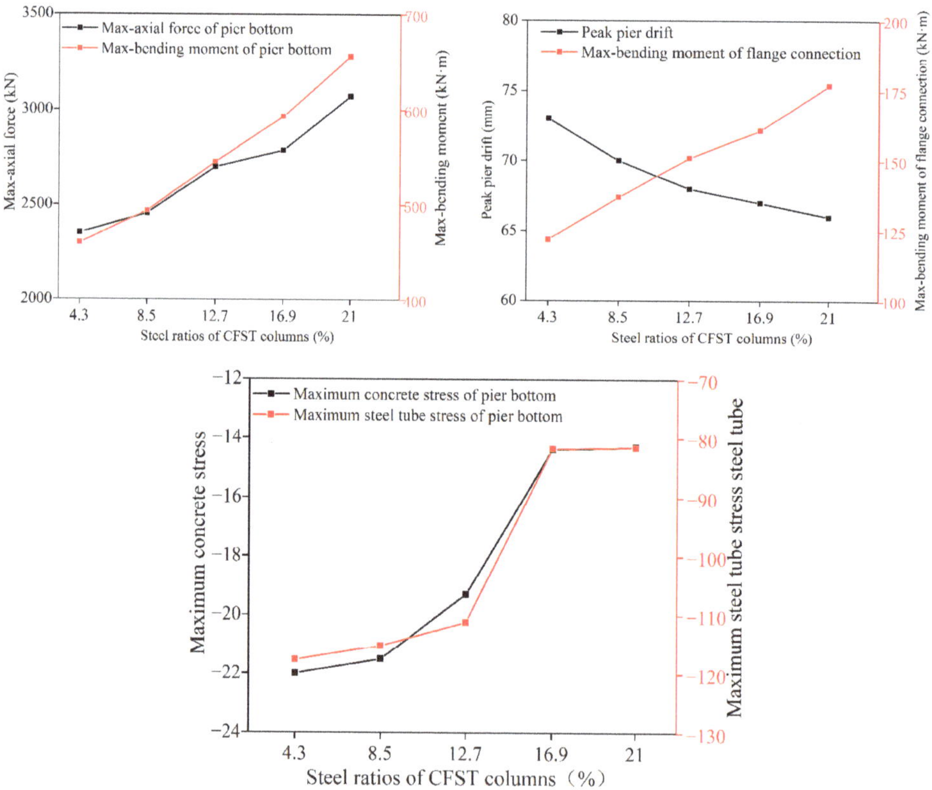

Figure 22. Comparison of maximum seismic response under different column steel ratios.

Table 8. Comparison of first mode period under different column steel ratios.

Column Steel Ratios	First Mode Period (s)
4.3	6.840
8.5	6.496
12.7	6.225
16.9	6.005
21	5.712

3.2.4. Arrangement Type of Steel Lacing Tube

To investigate the arrangement type of lacing tube on the seismic performance of the bridge, a total of six kinds of laced piers were designed, as shown in Figure 23. The seismic response curves for each seismic action in the X-direction are shown in Figure 24 and the first-mode period of different structures is shown in Table 9. To some extent, the use of C-type, D-type, and E-type arrangement forms can reduce the displacement of the top of the pier and column limbs and the force of flange connection, but considering the comprehensive economy and construction difficulties, it is recommended to give priority to the A-type, C-type, and D-type forms in this study.

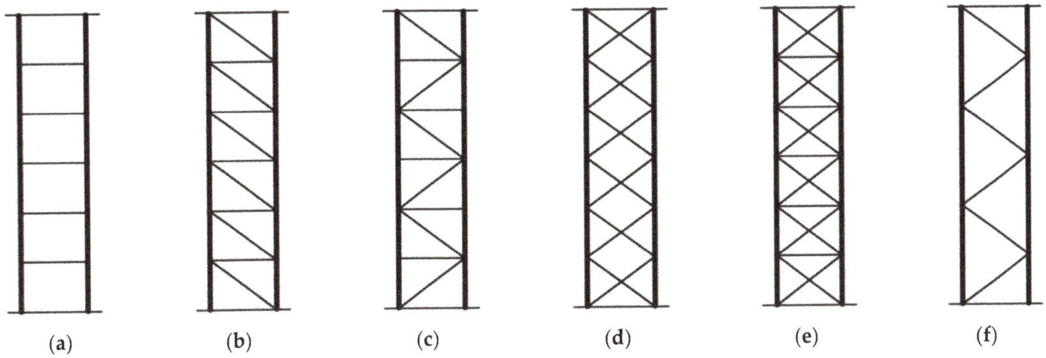

Figure 23. Arrangement type of steel lacing tube. (**a**) A-type; (**b**) B-type; (**c**) C-type; (**d**) D-type; (**e**) E-type; and (**f**) F-type.

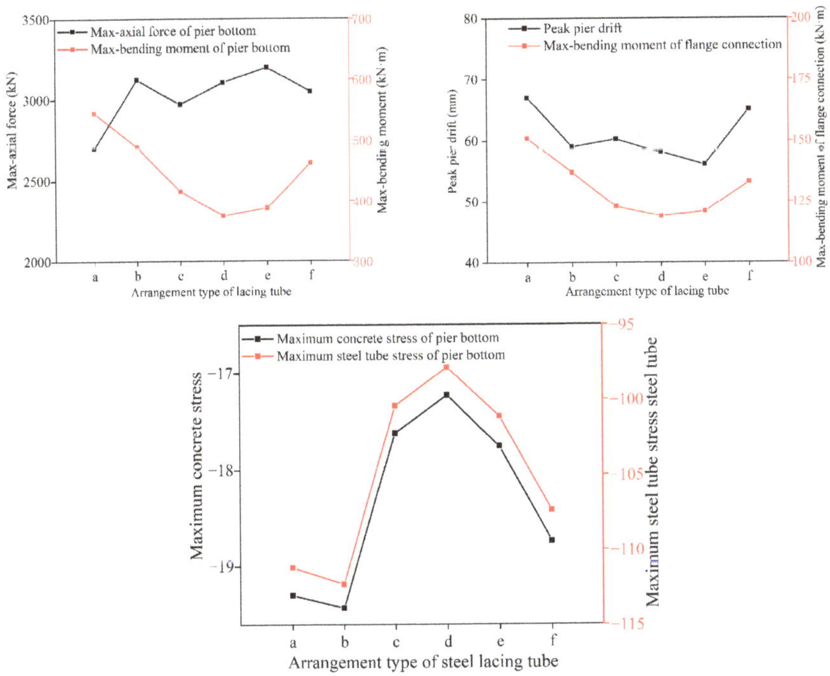

Figure 24. Comparison of the maximum seismic response under different arrangement types of steel lacing tube.

Table 9. Comparison of first mode period under different arrangement types of steel lacing tube.

Arrangement Type of Steel Lacing Tube	First Mode Period (s)
A	9.290
B	6.039
C	6.030
D	6.048
E	6.076
F	8.906

3.2.5. Longitudinal Slope

In order to study the effect of the longitudinal slope of the column limbs on the seismic performance of the bridge, five conditions are designed in this section, which consider the longitudinal slope as no slope, 1:100, 1:50, 1:40, and 1:30, respectively. The structural response under the seismic action in the cis-bridge direction is shown in Figure 25 and the first-mode period of different structures is shown in Table 10. Increasing the longitudinal slope has no significant effect on the displacement of the top of the pier, the force of the column limb, and the flange connection. When the slope exceeds 1:50, the stress on the pier bottom section decreases significantly, reducing seismic response to some extent. Therefore, it is recommended to adopt a slope of 1:50 to 1:30.

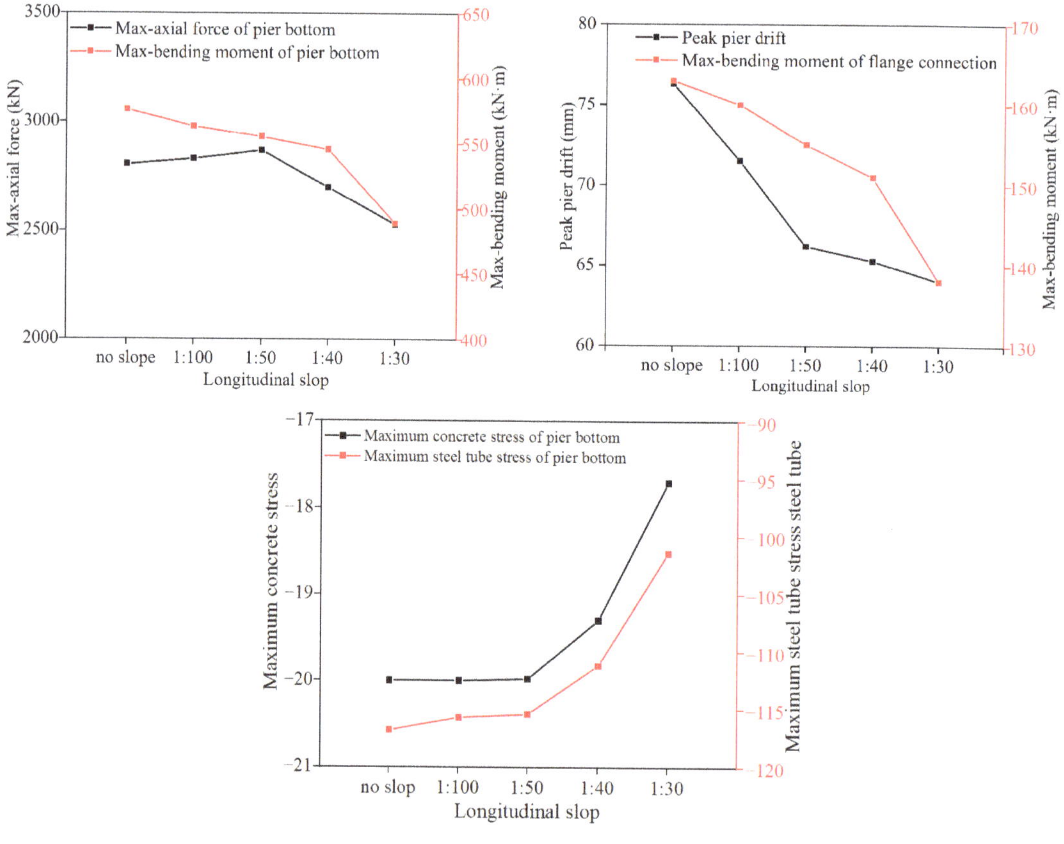

Figure 25. Comparison of maximum seismic response under different column slopes.

Table 10. Comparison of first mode period under different Longitudinal Slope.

Column Slopes	First Mode Period (s)
No slope	7.367
1:100	7.034
1:50	6.641
1:40	6.225
1:30	5.167

Based on the above discussion, the effect of each parameter on the seismic response of the bridge is derived, as shown in Table 11. It should be noted that a positive correlation indicates that the seismic response value increases with the increase in the design parameter value; a negative correlation indicates that the seismic response value decreases with the increase in the design parameter value. Moreover, insignificant (insign), less important (less im), important (im), and more important (more im) means the impact range is 0~10%, 10~30%, 30~50%, and more than 10%, respectively.

Table 11. Influence law of different design parameters of lattice pier.

Parameter	Displacement of Top of Pier	Axial Force of Bottom of Pier	Bending Moment of Bottom of Pier	Bending Moment of Flange	Importance	Recommended Value
Pier height	positive correlation	positive correlation	negative correlation	negative correlation	Im	——
Axial compression ratios	positive correlation	positive correlation	positive correlation	positive correlation	Im	About 0.1
Steel ratios of CFST columns	negative correlation	positive correlation	positive correlation	positive correlation	More im	variable cross-section
Arrangement type of lacing tube	It mainly affects the bending moment of pier bottom and flange				Less im	A, C, D-type
Longitudinal slope	negative correlation	negative correlation	negative correlation	negative correlation	Insign	1:50~1:30

3.2.6. Seismic Performance Optimization of CFST-laced Piers

To obtain an optimal design of laced piers, it is necessary to consider the seismic performance of the structure, as well as to take into account various factors such as economy, aesthetics, and construction ease. Based on the analysis of Section 3.2, the original design scheme of the laced piers was optimized, in which the pier height was kept unchanged, and the rest of the design parameters were selected according to the recommended values in Table 12. The comparison of the parameters before and after optimization is shown in Table 8. The optimized model was analyzed, and a comparison of the seismic response results between the original model and the optimized model was obtained, which is illustrated in Figure 23. In Table 8, variable cross-section means that the wall thickness of Section 1 is 14 mm, the wall thickness of Sections 2 to 3 is 12 mm, and the wall thickness of Sections 4 to 5 of the CFST pier is 10 mm with a height of 50 m.

Table 12. Comparison of design parameters of laced piers before and after optimization.

Parameters	Original Design Scheme	Optimized Design Scheme
Axial compression ratios	0.1	0.1
Steel ratios of CFST	12.7%	variable cross-section of steel tube
Arrangement type of lacing tube	A-type	D-type
Longitudinal slope	1:40	1:30

Figure 26 shows that the differences between the peak displacement of the pier top and the axial force of the pier bottom are small between the original and optimized models. This is because the bridge superstructure, input seismic waves, and bearing arrangement system are unchanged. However, the differences in the maximum bending moment of the pier bottom and the bending moment of the flange connection are large; therefore, the seismic response of the laced piers can be effectively reduced by using the optimized design scheme, and the seismic resistance of the key sections (such as the pier bottom) and members (such as column limbs and flanges) of the laced piers are not reduced. Therefore, the seismic response of the laced piers can be effectively reduced by using the optimized scheme.

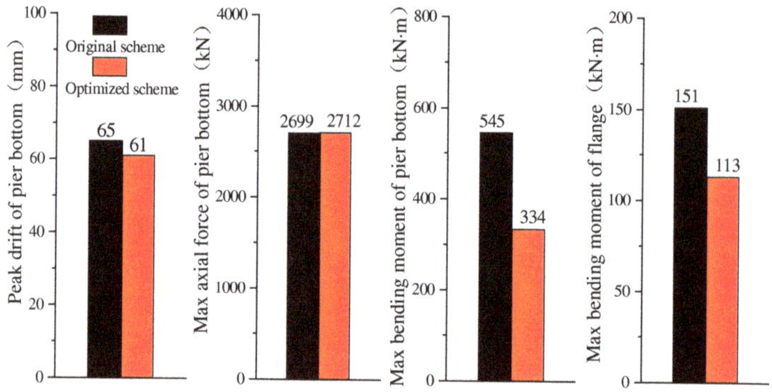

Figure 26. Comparison of seismic response peaks before and after optimization.

4. Conclusions

Two kinds of FEM were developed for the bridge with hollow CFST-laced piers in this study, namely, a simplified model for simple engineering analysis considerations, and a refined model considering the mass and stiffness distribution of the superstructure. The applicability of the two models was evaluated through the analysis of structural static and dynamic properties and seismic response. Moreover, the seismic response law of assembled flange-connected hollow CFST lattice piers was studied, and their seismic design was optimized with five optimized parameters, such as pier height, axial compression ratios, steel ratios of CFST columns, arrangement type of steel lacing tube and longitudinal slope. The main conclusions obtained were as follows:

(1) By comparison, the differences between the simplified model and the refined model for the static and dynamic characterization and seismic response are very small. Thus, this study suggests the use of a simplified FEM for the seismic analysis of a bridge with hollow CFST-laced piers;

(2) FEM results show that the maximum axial force and bending moment of CFST column limbs occur at the bottom of the pier, and the flanges that are subject to large bending moments are generally located at the two connections above the bottom of the pier. In addition, the pier height, axial compression ratio, steel ratio of CFST column, and the arrangement type of lacing tube have an influence on the seismic performance of this laced pier, but the influence of the longitudinal slope of the column limbs is inapparent;

(3) Based on the results of the parametric analysis, the optimization of the design scheme for the CFST-laced pier is carried out by changing the longitudinal slope, steel ratios of CFST, and arrangement type of the lacing tube, and the results indicate that 0.1 axial pressure ratios and 1:30 longitudinal slope of column limbs with variable cross-section CFST have good seismic performance.

Author Contributions: Conceptualization, L.C., Z.W. and X.J.; Methodology, L.C. and Z.W.; Writing—original draft, L.C. and R.Z.; Writing—review & editing, Y.Z., D.Y., J.L. and X.J. All authors have read and agreed to the published version of the manuscript.

Funding: This research received no external funding.

Informed Consent Statement: Not applicable.

Data Availability Statement: Not applicable.

Acknowledgments: This work was supported by the National Natural Science Foundation of China (Grant No: 52178135) and the Anhui Province Natural Science Foundation of China (Grant No: 2208085ME151).

Conflicts of Interest: The authors declare no conflict of interest.

References

1. Han, L.H.; Song, T.Y.; Zhou, K.; Cui, Z.Q. Fire Performance of CFST Triple-Limb Laced Columns. *J. Struct. Eng.* **2018**, *144*, 400–412. [CrossRef]
2. Ou, Z.; Chen, B.; Hsieh, K.H.; Halling, M.W.; Barr, P.J. Experimental and Analytical Investigation of Concrete Filled Steel Tubular Columns. *J. Struct. Eng.* **2011**, *137*, 635–645. [CrossRef]
3. Huang, F.; Chen, B.C.; Li, J.Z.; Yu, G. Experimental study on influence of initial stress on concrete filled steel tubular lattice columns subjected to axial load. *J. Build. Struct.* **2013**, *34*, 109–115.
4. Cheng, B.; Xiang, S.; Zuo, W.; Teng, N. Behaviors of partially concrete-filled welded integral T-joints in steel truss bridges. *Eng. Struct.* **2018**, *166*, 16–30. [CrossRef]
5. Tang, Y.C.; Li, L.J.; Feng, W.X.; Liu, F.; Zhu, M. Study of seismic behavior of recycled aggregate concrete-filled steel tubular columns. *J. Constr. Steel Res.* **2018**, *148*, 1–15. [CrossRef]
6. Han, L.H.; Hou, C.C.; Wu, X. Seismic performance of concrete-encased column base for hexagonal concrete-filled steel tube: Numerical study. *J. Constr. Steel Res.* **2018**, *149*, 225–238. [CrossRef]
7. Aguirre, D.; Kowalsky, M.; Nau, J.; Gabr, M.; Lucier, G. Seismic performance of reinforced concrete filled steel tube drilled shafts with inground plastic hinges. *Eng. Struct.* **2018**, *165*, 106–119. [CrossRef]
8. Zhang, X.; Xia, X.; Han, S. Quasi-static test of the lattice-type railway bridge pier with replaceable connection component. *Structures* **2022**, *43*, 635–644. [CrossRef]
9. Hajjar, J.F.; Gourley, B.C. Representation of Concrete-Filled Steel Tube Cross-Section Strength. *J. Struct. Eng.* **1996**, *122*, 1327–1336. [CrossRef]
10. Hajjar, J.F.; Gourley, B.C. A cyclic nonlinear model for concrete-filled tubes. I: Formulation. *J. Struct. Eng.* **1997**, *123*, 736–744. [CrossRef]
11. Aval, S.B.B.; Saadeghvaziri, M.A.; Golafshani, A.A. Comprehensive Composite Inelastic Fiber Element for Cyclic Analysis of Concrete-Filled Steel Tube Columns. *J. Eng. Mech.* **2002**, *128*, 428–437. [CrossRef]
12. Ou, Z.; Lin, J.; Chen, S.; Lin, W. Experimental Research on Seismic Performance of Four-Element Variable Cross-Sectional Concrete Filled Steel Tubular Laced Columns. *IOP Conf. Ser. Mater. Sci. Eng.* **2017**, *250*, 012039. [CrossRef]
13. Hu, Z.; Wei, B.; Jiang, L.; Li, S.; Yu, Y.; Xiao, C. Assessment of optimal ground motion intensity measure for high-speed railway girder bridge (HRGB) based on spectral acceleration. *Eng. Struct.* **2021**, *252*, 113728. [CrossRef]
14. Wei, B.; Hu, Z.; He, X.; Jiang, L. Evaluation of optimal ground motion intensity measures and seismic fragility analysis of a multi-pylon cable-stayed bridge with super-high piers in Mountainous Areas. *Soil Dyn. Earthq. Eng.* **2020**, *129*, 105945. [CrossRef]
15. CSI. *SAP2000 User's Manual*; Computers and Structures, Inc.: Walnut Creek, CA, USA, 2016.
16. Mao, Y.; Ye, F.; Tang, Q.; Lu, T. Experimental study on flexural bearing capacity of SC pile. *Eng. Constr.* **2015**, *29*, 24–29.
17. Mander, J.B.; Priestley MJ, N.; Park, R. Theoretical stress-strain model for confined concrete. *J. Struct. Eng.* **1988**, *1114*, 1804–1826. [CrossRef]
18. Yang, T.; Yuan, X.; Zhong, J.; Yuan, W. Near-fault pulse seismic ductility spectra for bridge columns based on machine learning. *Soil Dyn. Earthq. Eng.* **2023**, *164*, 107582. [CrossRef]
19. Xiang, N.; Chen, X.; Alam, M.S. Probabilistic seismic fragility and loss analysis of concrete bridge piers with superelastic shape memory alloy-steel coupled reinforcing bars. *Eng. Struct.* **2020**, *207*, 110229. [CrossRef]
20. Chen, X.; Xiang, N.; Guan, Z.; Li, J. Seismic vulnerability assessment of tall pier bridges under mainshock-aftershock-like earthquake sequences using vector-valued intensity measure. *Eng. Struct.* **2022**, *253*, 113732. [CrossRef]
21. Chen, L.; Yan, J.; Xiang, N.; Zhong, J. Shear performance of ultra-high performance concrete multi-keyed epoxy joints in precast segmental bridges. *Structures* **2022**, *46*, 1696–1708. [CrossRef]
22. Wei, B.; Hu, Z.; He, X.; Jiang, L. Probabilistic Seismic Evaluation and Experimental Tests of Multi-Direction Damping System on a Super-Long Column-Pylon Cable-Stayed Bridge. *J. Earthq. Eng.* **2023**, 1–24. [CrossRef]
23. Zhong, J.; Zhu, Y.; Han, Q. Impact of vertical ground motion on the statistical analysis of seismic demand for frictional isolated bridge in near-fault regions. *Eng. Struct.* **2023**, *278*, 115512. [CrossRef]

24. Yang, T.; Wei, Y.; Zhong, J. Potential bias of conventional structural seismic fragility for bridge structures under pulse-like ground motions: Bias evaluation and strategy improvement. *Soil Dyn. Earthq. Eng.* **2023**, *166*, 197787. [CrossRef]
25. *JTG/T2231-01-2020*; Specification for Seismic Design of Highway Bridges. Ministry of Transport of the People's Republic of China: Beijing, China, 2020.
26. Liang, Y.; Xiao, C.; Wu, T.; Xu, Z.; Fan, J.G. Synthesis and Application of Artificial Seismic Waves Based on Building Code Response Spectra. *J. Sichuan Univ. Sci. Eng.* **2016**, *29*, 83–87.

sustainability MDPI

Article

Experimental Study on the Seismic Performance of a Steel Slag CFDST T-Joint

Wuhua Zeng [1,2,*], Mingliang Fu [1], Yuezong Lian [1,2], Hai Zhong [1,2] and Wei Wang [3]

[1] School of Civil Engineering, Sanming University, Sanming 365004, China
[2] Key Laboratory of Engineering Material & Structure Reinforcement in Fujian Province College, Sanming University, Sanming 365004, China
[3] College of Civil Engineering, Tongji University, Shanghai 200092, China
* Correspondence: zwhua@fjsmu.edu.cn

Abstract: In this paper, a kind of steel slag, concrete-filled double-skin steel tube (CFDST) T-joint is proposed to promote sustainable structural development. In order to examine the seismic performance of the steel slag CFDST T-joint, a series of hysteresis experiments were carried out on 4 CFDST T-joints with the main pipe under axial compression load and the brace pipe subjected to cyclic axial loading. The seismic performance of the CFDST T-joint was experimentally investigated in terms of the failure mode, load–displacement hysteresis relationship, stiffness degradation, energy dissipation and ductility. The effects of the hollow ratio and the steel slag concrete of the CFDST main pipe on the seismic performance of CFDST T-joint specimens were compared and analyzed. The experimental results show that the failure modes of the CFDST T-joint mainly included two kinds of failure, with those being main pipe fracture and joint area compression-bending failure. The seismic performance of the joints could be improved with a 12% ultimate bearing capacity and 54% ultimate deformation capacity with the hollow ratio of the CFDST main pipe increasing from 0 to 0.5. The seismic performance of the joints could be improved with a 54% ultimate deformation capacity by filling the steel slag concrete with a 0.5 hollow ratio.

Keywords: steel slag; concrete-filled double-skin steel tube (CFDST); T-joint; seismic performance

Citation: Zeng, W.; Fu, M.; Lian, Y.; Zhong, H.; Wang, W. Experimental Study on the Seismic Performance of a Steel Slag CFDST T-Joint. *Sustainability* **2023**, *15*, 7991. https://doi.org/10.3390/su15107991

Academic Editors: Constantin Chalioris and Paulo Santos

Received: 22 March 2023
Revised: 30 April 2023
Accepted: 10 May 2023
Published: 13 May 2023

1. Introduction

Steel slag is a kind of industrial waste with more than 10% being produced in steel production [1,2]. According to incomplete statistics, the amount of steel slag is estimated to be more than 180 million tons globally; however, in China, the utilization rate is only 22% [3]. Steel slag causes environmental pollution through disposal in landfills without proper processing; moreover, it has resulted in a considerable economic burden for steel companies [4]. Therefore, it is urgent for increased recycling practices and comprehensive utilization of steel slag to achieve sustainable development of the steel industry. With the development of steel slag as a construction material, scholars have conducted many studies on the application of steel slag in concrete [5–7]. The results show that the use of steel slag powder in concrete could improve the workability and viscosity of concrete, while also reducing the self-shrinkage of the concrete. Therefore, steel slag powder can be used to configure self-compacting concrete [8]. In order to compensate for the void effect of the concrete-filled steel tube, the application of steel slag powder in the concrete-filled steel tube structure needs to be discussed.

The CFDST has the advantages of good mechanical performance and convenient construction performance. In addition, it is widely used in high-rise buildings, bridges and other practical projects [9–12]. Due to the complicated structure area of the CFDST joint, many studies have focused on experimental research on the mechanical behavior of the CFDST joint. Wang (2017) conducted an experimental study on the T-shaped joints of high-strength CFDST. The results showed that under the tension condition of the branch pipe, the final failure mode of the member was the cracking of the outer steel pipe and the separation of the steel and concrete at the joints; the initial position of the fracture was mostly at the crown point [13]. Gao (2018) conducted experimental research and a finite element analysis on CFDST K-shaped joints. The results showed that when the hollow ratio is 0.529, the seismic performance of K-shaped joints is the best [14]. Ma (2018) studied the static properties of CFDST T-joints under horizontal axial compression and vertical axial compression. The results showed that there were mainly four failure modes of CFDST T-joints [15]. Shi (2019) conducted experimental and theoretical investigations on the behavior and mechanism of circular CFDST T-joints. The results show that the failure mode of CFDSTs is similar to that of concrete-filled steel tubular T-joints, and the failure mode of the member is mainly controlled by the outside diameter ratio of the abdominal chord [16]. More attention to the static performance of CFDST T-joints in the literature is necessary [17]. In the case of an earthquake, the CFDST T-joints are the most integral feature as the failure of the joint will lead to total structural collapse. It can be seen that it is necessary to study the failure mechanism and seismic performance of the joint experiencing an earthquake.

Compared with uniaxial compressive strength, the triaxial compressive strength of the steel slag concrete was greater, and the influence of the stress ratio on the triaxial compressive strength was obvious. At present, some researchers have carried out experiments and theoretical studies on the mechanical behavior of steel slag CFST columns. Fang et al. (2020) studied the axial compression performance of steel slag concrete columns with circular steel tubes, and the analysis results showed that, compared with ordinary concrete-filled steel tubes, the restraint effect of steel slag concrete columns with circular steel tubes was more obvious. However, the whole process of loading was similar when they were subjected to axial pressure [18]. Shen (2021) studied the axial compression performance of steel tubular slag concrete short columns with elliptical sections, and the results showed that the failure modes of steel tubular slag concrete short columns with elliptical sections mainly included local buckling of the steel tube, local crushing and shear failure of the concrete. As the replacement rate of the steel slag increased, the axial compression strength and stiffness of the components increased but the ductility decreased [19]. Yu et al. (2020; 2022) studied the interfacial bonding and sliding behavior of steel tubular slag concrete and analyzed the influence of parameters such as diameter to thickness ratio and expansion rate on the bonding and sliding behavior. The results showed that shear failure at the bonding interface was the main failure mode [20,21]. At present, there are few studies on the application of steel slag concrete-filled double-skin steel tubes. There is even less research on the seismic performance of steel slag CFDST joints.

This paper aimed to make full use of the advantages of steel tube and steel slag concrete, promote the reuse of solid waste and reduce project costs. In this paper, a kind of steel slag concrete-filled double-skin steel tube (CFDST) main pipe and circular hollow section (CHS) brace pipe composite T-joint was proposed and 4 specimens of CFDST T-joints were made to investigate their seismic behavior.

2. Experimental Program

2.1. Specimen Design

In this paper, a total of 4 specimens (1 commercial Portland CFDST T-joint and 3 steel slag CFDST T-joints) were designed (Table 1). The influences of hollow ratio and steel slag concrete of the CFDST main pipe on hysteretic behavior were compared and studied, and the design diagram of the specimens are exhibited in Figure 1. Table 1 lists the design parameters of the CFDST T-joint specimens. The length of the CFDST main pipes was 900 mm. The outer diameters of the CFDST main pipes were 50 mm, 80 mm and 168 mm, respectively. The thickness of the CFDST main pipe outer tubes was 4 mm. The thickness of the CFDST main pipe inner tubes was 3 mm. The outer diameter, length and thickness of the CHS brace pipe were 114 mm, 400 mm and 4 mm, respectively. The brace-to-main pipe diameter ratio of the CFDST was 0.68. The axial compression ratio of the CFDST main pipe was 0.1.

Table 1. Parameters of the CFDST T-joint specimens.

Number	$D \times L \times T$ (mm × mm × mm)	$D_n \times L \times T_n$ (mm × mm × mm)	$D_0 \times L_0 \times t$ (mm × mm × mm)	χ (D_n/D)	β (D_0/D)	n
ST0	168 × 900 × 4	0 × 900 × 3	114 × 400 × 4	0.0	0.68	0.1
ST3	168 × 900 × 4	50 × 900 × 3	114 × 400 × 4	0.3	0.68	0.1
ST5	168 × 900 × 4	80 × 900 × 3	114 × 400 × 4	0.5	0.68	0.1
CT5	168 × 900 × 4	80 × 900 × 3	114 × 400 × 4	0.5	0.68	0.1

Note: ST is the steel slag concrete of the CFDST T-joint, CT represents the commercial Portland concrete CFDST T-joint; D is the outer diameter of the main pipe outer tubes; D_n is the outer diameter of the main pipe inner tubes; D_0 is the outer diameter of the branch pipe; L is the length of the main pipe outer tubes; L_n is the length of the main pipe inner tubes; L_0 is the length of the branch pipe; T is the thickness of the main pipe outer tubes; T_n is the thickness of the main pipe inner tubes; t is the thickness of the branch pipe; χ is the hollow ratio of the CFDST main pipe; β is the brace-to-main pipe diameter ratio of the CFDST T-joint; n is the axial compression ratio of the CFDST chord T-joint.

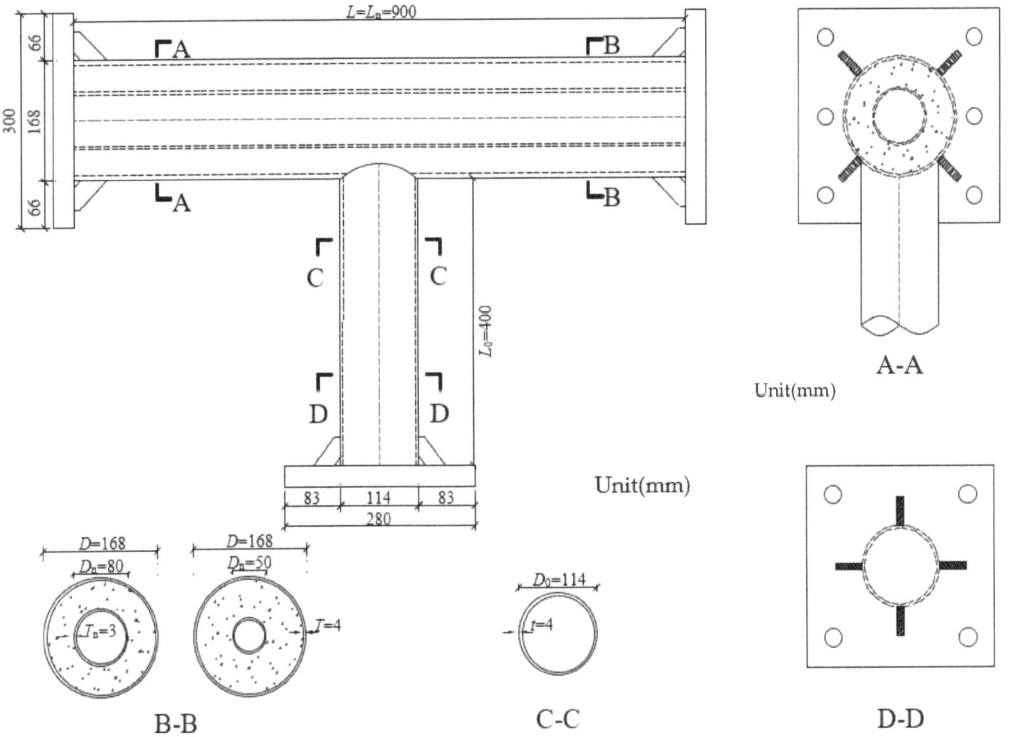

Figure 1. Design diagram of the CFDST T-joint specimens.

2.2. Mechanical Properties of Materials

Q235 steel tubes with straight-welded main pipes and seamless welded branch pipes were used in this experiment. Size specifications of the material specimens are shown in Figure 2. The mechanical properties of the steel were tested according to the Chinese metallic materials—tensile testing code GB/T228.1-2021 [22]. The yield and ultimate strength of the 3-mm and 4-mm thick steel are shown in Table 2, respectively, by tensile experiment.

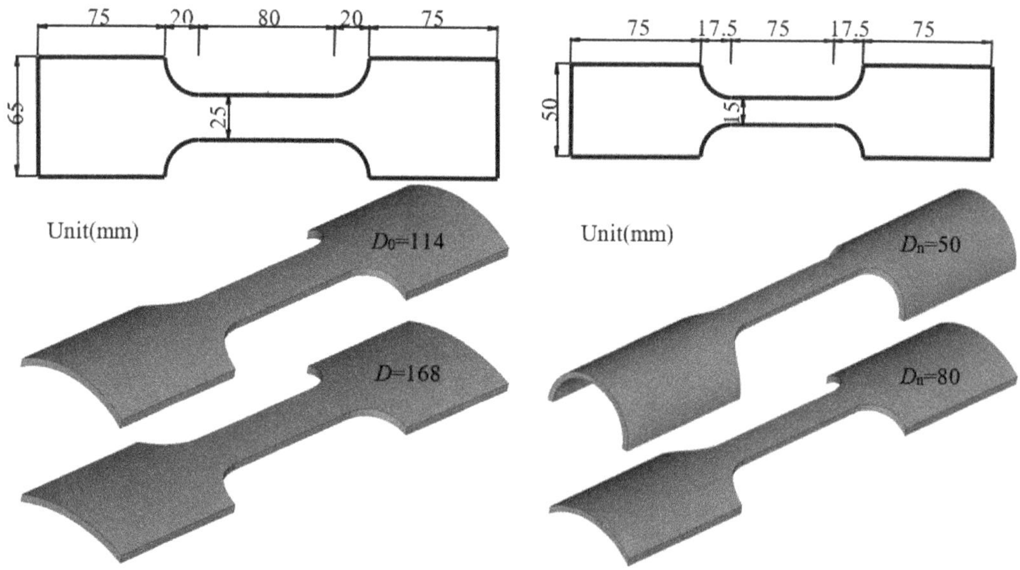

Figure 2. Size specifications of material specimens.

Table 2. Material properties of steel tubes.

Dimension of Steel Tube $D \times t$ (mm × mm)	Yield Strength *fy* (MPa)	Ultimate Strength *fu* (MPa)
168 × 4	280	355
114 × 4	328	494
80 × 3	283	386
50 × 3	372	507

The steel slag concrete was used to filled the CFDST main pipe. The concrete-filled double-skin steel tube was prepared according to Chinese specifications for mix proportions expressed in the design of ordinary concrete code JGJ55-2011 [23]. The ordinary Portland cement with a strength grade of 42.5 was used. The steel slag was obtained from the local enterprise of Fujian Sangang (Group) Co., Ltd. (Sanming, China) [3]. Table 3 shows the chemical compositions of the commercial Portland cement and steel slag powder. Table 4 lists the mix proportions of the commercial Portland concrete and steel slag concrete. The steel slag powder was added to replace part of the ash and cement.

Table 3. Chemical compositions of commercial Portland cement and steel slag powder.

Type	CaO	SiO_2	Al_2O_3	Fe_2O_3	MgO	MnO	P_2O_5	SO_3	Na_2O_{eq}	Loss
Commercial Portland cement	56.11	23.46	7.93	3.46	3.15	-	-	3.49	0.72	2.31
Steel slag powder	44.57	18.11	7.26	19.01	4.17	2.55	1.87	-	0.29	1.38

Note: $Na_2O_{eq} = Na_2O + 0.685K_2O$.

Table 4. Mix proportion of concrete.

Type	Water (kg/m^3)	Cement (kg/m^3)	Steel Slag (kg/m^3)	Flyash (kg/m^3)	Sand (kg/m^3)	Stone (kg/m^3)	Superplasticizer (kg/m^3)
Commercial Portland concrete	181	450	0	170	815	815	6.3
Steel slag concrete	181	405	63	153	815	815	6.3

The mechanical properties of the commercial Portland concrete and steel slag concrete were experimentally measured according to the specifications for the standard test methods for the physical and mechanical properties of concrete, expressed in Chinese code GB/T 50081-2019 [24]. Table 5 lists the concrete's working behavior and the measured compressive strength. It can be seen from the result listed in Table 5 that the elastic modulus and cube compressive strength of the steel slag concrete were increased within 10% of the commercial Portland concrete.

Table 5. Mechanical properties of concrete.

Number	Slump (mm)	Spread (mm)	Ec (GPa)	fcu (MPa)
ST0	280	600–610	31.1	55.7
ST3	280	600–610	31.1	53.7
ST5	280	600–610	31.1	55.4
CT5	285	600–610	30.6	51.2

Note: fcu is the cube compressive strength of the concrete, 150 mm × 150 mm × 150 mm.

2.3. Experiment Setup and Loading Scheme

All CFDST T-joint specimens were completed through the self-balancing reaction rack experiment system in which the horizontal and vertical hydraulic actuator could provide a force of 1000 kN and 2000 kN, respectively. The experiment equipment is shown in Figure 3. During the experiment, the hollow sandwich steel tubular of the CFDST T-joint main pipe was placed horizontally to apply vertical cycle load. The experiment boundary condition of CFDST T-joint specimens were hinged. The end plates at both ends of the string rod were connected to the flat hinge, with one side of the flat hinge connected to the support seat and the other side of the support seat connected to the side actuator to facilitate the transmission of axial load. The other side of the plate is hinged with the lateral connection support, which can transfer the force to the reaction frame and improve the lateral boundary stiffness. During the experiment, the actuator was controlled by the system to keep the axial pressure constant. Cyclic loading was transferred to the belly rod by the vertical actuator through the connector.

To study the seismic performance of the CFDST T-joint, cyclic loading with varying amplitude vertical loading schemes was applied to the joints. The vertical loading experiment adopts a load–displacement dual control loading system. The loading scheme of the CFDST T-joint was generated according to the specification for the seismic testing of buildings code JGJ/T 101-2015 [25], which can be seen in Figure 4. Before the specimen reaches the yield strength, its load–displacement curve shows a roughly linear relationship. Therefore, in this stage, the loading is controlled by force with an increment of 0.25 Pmax, where Pmax is the bearing capacity calculated by finite element analysis. The loading rate is 1 KN/s during this stage. When the specimen reaches the plastic stage, the loading is controlled by the displacement, with the increment of yield displacement and each displacement being cycled two times. The loading rate was 1 mm/min during this stage. The loading is terminated if one of the following conditions is met: the steel tube is fractured, the steel tube is loaded to the point where the load drops beyond the 15% peak load or the joint is subjected to significant plastic deformation failure.

The real-time curve of the load–displacement relationship can be collected by the data acquisition system of the servo actuator. The longitudinal displacement in the direction of the cyclic load of the branch pipe can be measured by the displacement meter and verified with the data collected by the system. In order to measure the vertical displacement of the CFDST main pipe, a total of 11 displacement sensors were installed in the device. The measuring No. 1, No. 2 and No. 8 displacement meters were installed in the middle of the main span, the left of the middle of the span and the end plate of the branch pipe. The No. 3 and No. 4 displacement meters were installed at the bottom of the left end plate of the main pipe and the left hinge center of the plate. The No. 5 and No. 6 displacement meters were installed at the bottom of the right end plate of the main pipe and the right hinge center of the plate. The No. 7 displacement measurement was used to measure the relative displacement of the branch main pipe, and to measure whether the change in the lower main pipe displacement was consistent with the branch pipe displacement. The No. 9 displacement meter was used to measure whether out-of-plane displacement occurs, and the No. 10 and No. 11 displacement meters were used to measure the vertical displacement of the support. Strain gauges were installed on the main pipe and branch pipe to measure the strain distribution, as shown in Figure 5.

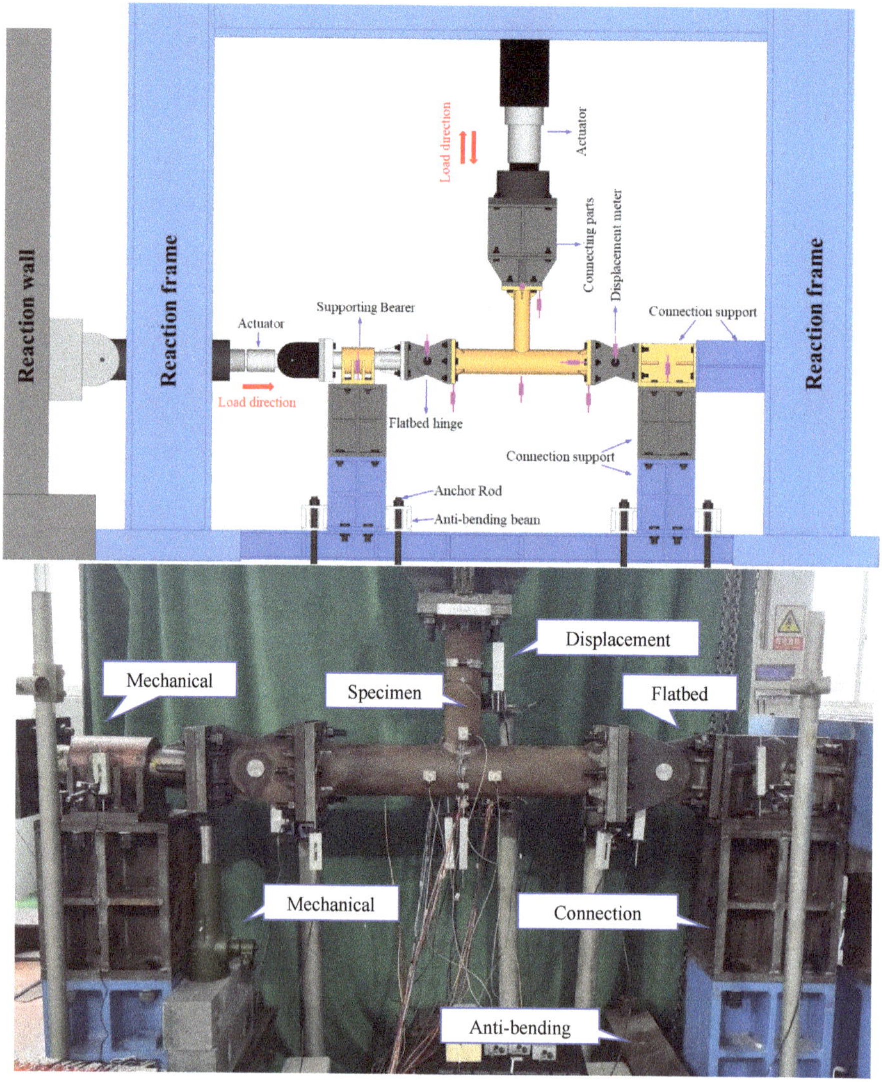

Figure 3. Loading devices.

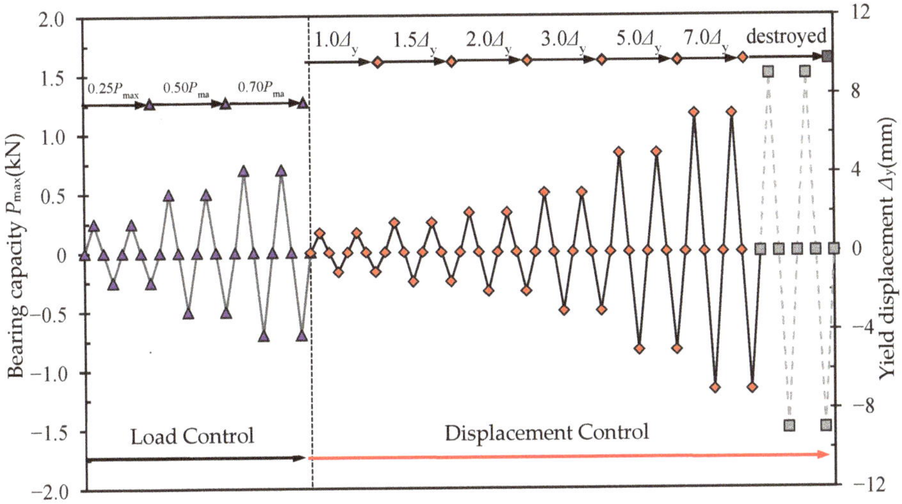

Figure 4. Loading scheme.

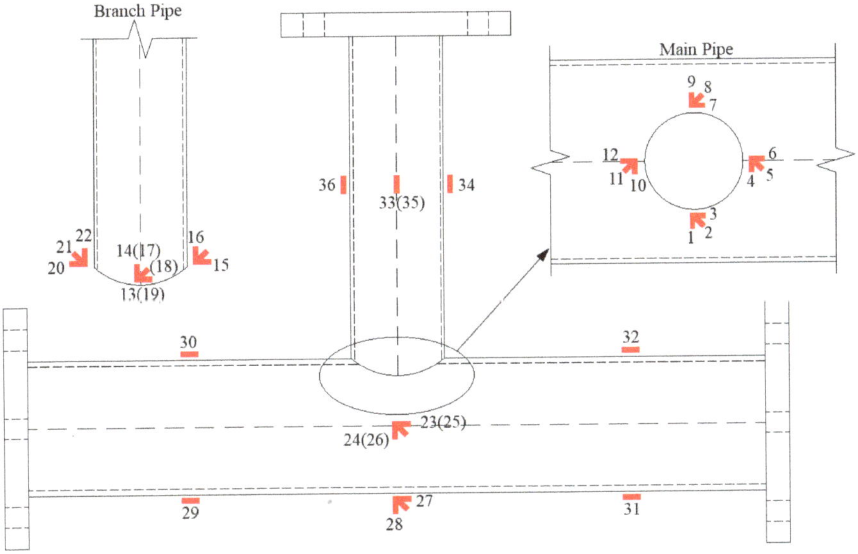

Figure 5. Measuring points arrangement.

3. Results

3.1. Failure Modes of Specimens

In order to study the failure process and failure mode influence law of CFDST T-joints under cyclic loading, the whole process of experimental observation was conducted on each specimen. It was found that none of the specimens had obvious appearance changes before the yield stage. With further increases of the loading displacement at the branch end, the components eventually failed and there were two failure modes: the regional head of the joint fracture and compression-bending failure in the joint area (see Figure 6).

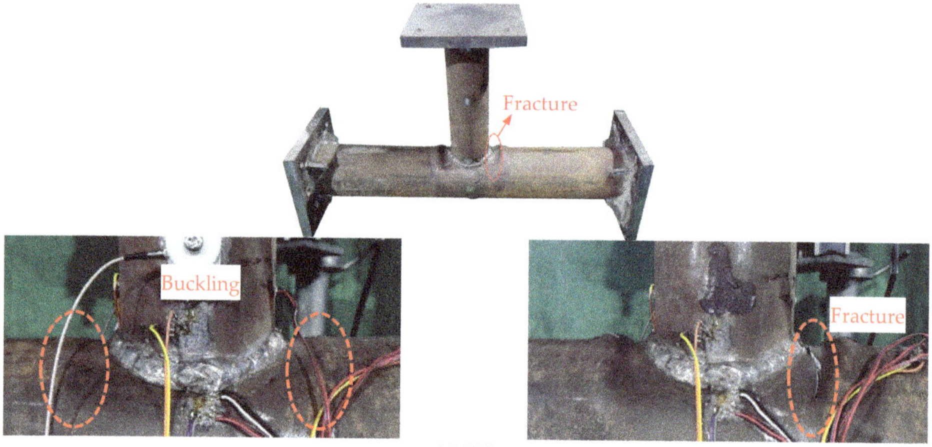

(**a**) ST0

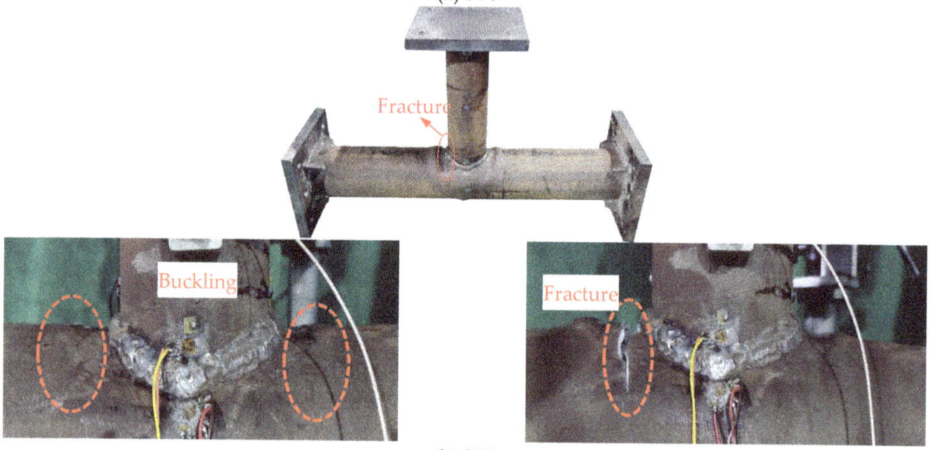

(**b**) ST3

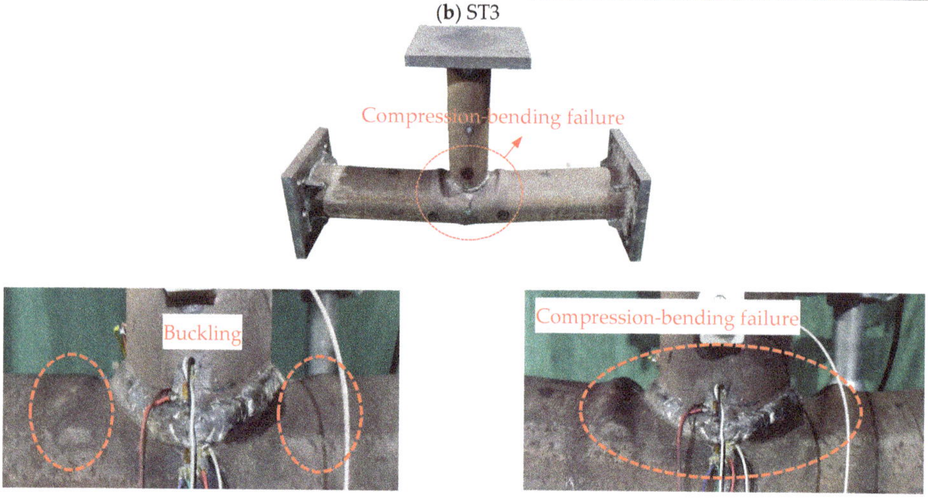

(**c**) ST5

Figure 6. *Cont.*

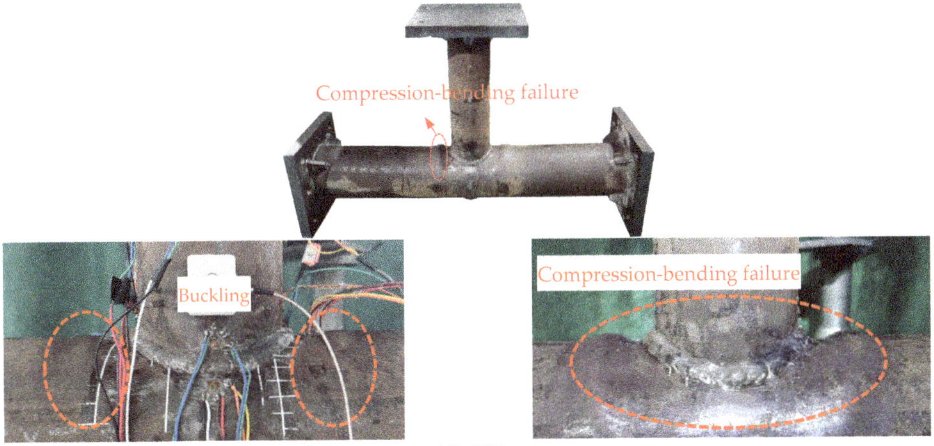

(**d**) CT5

Figure 6. Failure modes of specimens: (**a**) Fracture failure of ST0 specimen; (**b**) Fracture failure of ST3 specimen; (**c**) Compression-bending failure of ST5 specimen; (**d**) Compression-bending failure of CT5 specimen.

3.2. Load–Displacement Curves

Figure 7 shows the measured load–displacement curves of each experimental specimen. It can be seen that the hysteretic curves of each component are fusiform, full and symmetrical in the positive direction, without obvious pinching phenomena, indicating that the CFDST T-joint specimens have good energy dissipation capacity.

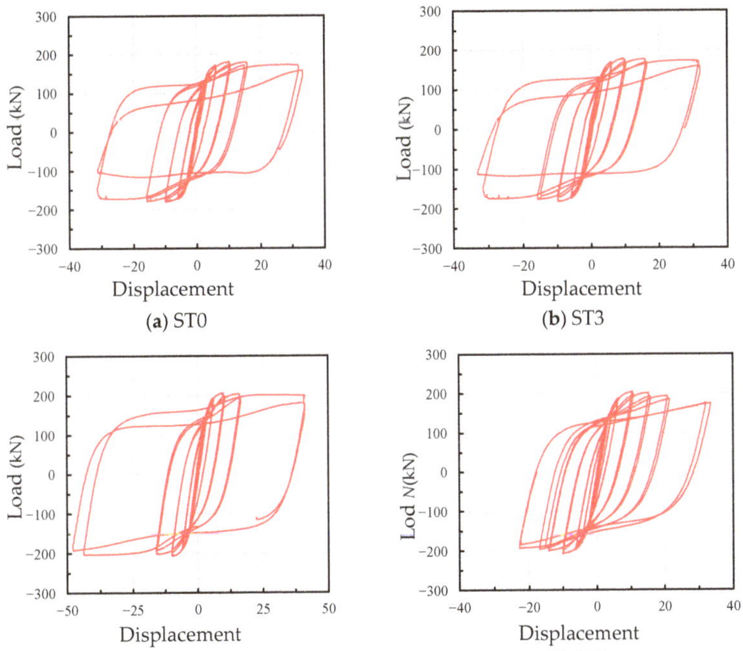

(**a**) ST0 (**b**) ST3 (**c**) ST5 (**d**) CT5

Figure 7. Load–displacement curves of specimens.

The skeleton curves of the CFDST T-joint specimens were formed by connecting the peak load point of the first circle of each stage of the hysteresis curve [25]. The skeleton curve in this paper is an "S" shape, indicating that the curve has experienced yield, maximum bearing capacity, maximum displacement and other characteristic points, according to which the ductility, stiffness and strength of the joint can be analyzed. Figure 8 shows the characteristic points of the skeleton curve. The experiment results of the specimens at the characteristic points are shown in Table 6.

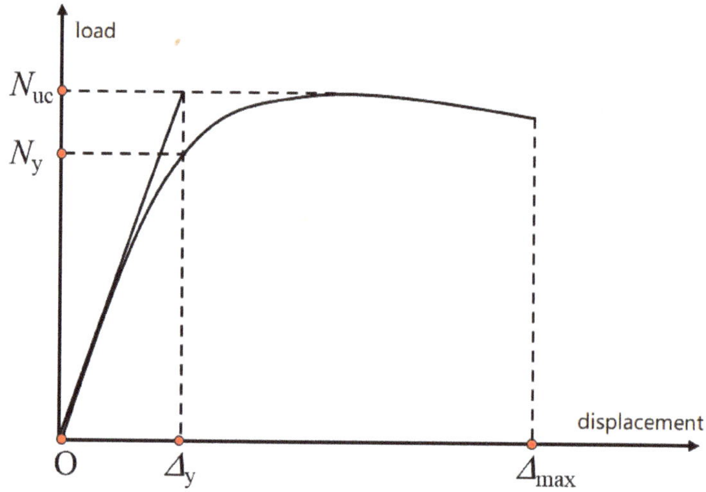

Figure 8. Characteristic points of the skeleton curve.

Table 6. The experimental results of the specimens.

Number	Yield Load		Yield Displacement		Ultimate Load		Ultimate Displacement	
	N_y^+ (kN)	N_y^- (kN)	Δ_y^+ (mm)	Δ_y^- (mm)	N_u^+ (kN)	N_u^- (kN)	Δ_u^+ (mm)	Δ_u^- (mm)
ST0	119.2	−124.7	2.57	−2.75	181.2	−179.8	28.3	−26.8
ST3	120.7	−120.1	2.74	−2.38	178.1	−182.2	30.7	−28.0
ST5	144.9	−144.1	3.27	−3.46	206.2	−206.6	40.9	−43.7
CT5	151.2	−152	3.73	−3.69	204.3	−206.7	32.1	−22.8

Note: N_y^+, N_y^-, N_u^+ and N_u^- are the loads at the yield point and failure point, respectively. Corresponding to N_y^+, N_y^-, N_u^+ and N_u^-, the displacement is Δ_y^+, Δ_y^-, Δ_u^+ and Δ_u^-, respectively.

The influence of various parameters on the characteristic points of the specimen was quantitatively compared. Taking CT5 as the benchmark, the standardization coefficient of characteristic points were calculated by average values of benchmark. Table 7 lists the standardization experiment results of specimens during the cycle loading. It can be seen from Table 5 that the ultimate capacity of the specimens could be improved with 12% load and 54% displacement and the hollow ratio of the CFDST main pipe increasing from 0 to 0.5. The reason for this is that the steel slag powder, with its properties of slight expansion, can strengthen the inter-reaction between the concrete and steel tube. The ultimate capacity of the specimens could be improved with 54% displacement by filling the steel slag concrete with the 0.5 hollow ratio of the CFDST main pipe. The filling of the steel slag concrete in the CFDST T-joint main pipe has no effect on its ultimate bearing capacity, but results in a larger ultimate displacement value and a fuller hysteresis curve.

Table 7. The standardization coefficient of characteristic points.

Number	Yield Load	Yield Displacement	Ultimate Load	Ultimate Displacement
ST0	80%	72%	88%	100%
ST3	79%	69%	88%	107%
ST5	95%	91%	100%	154%
CT5	100%	100%	100%	100%

The load–displacement curve's effect on the hollow ratio of the CFDST T-joint specimens are shown in Figure 9. It can be seen from the Figure 9 that changing the hollow ratio on the CFDST T joint main pipe has little influence on the elastic stiffness. The ultimate load and deformation all increased when the hollow ratio increased from 0 to 0.5. The load–displacement curve's effect on the CFDST specimen with steel slag or commercial Portland concrete is shown in Figure 10. The stiffness of the elastoplastic stage and ductility of the CFDST specimen with steel slag (ST5) all increased compared with the commercial Portland concrete (CT5).

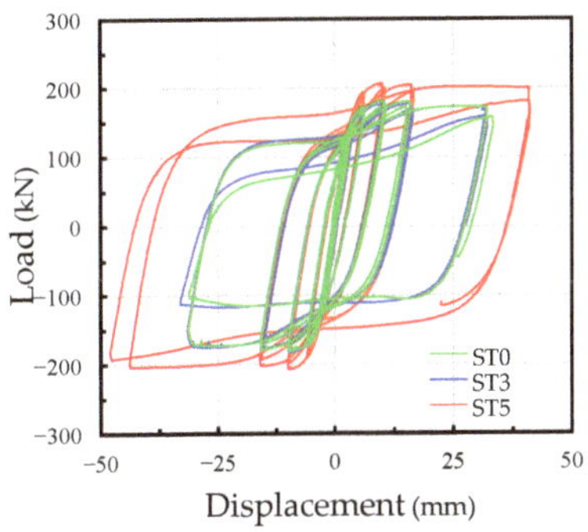

(**a**) Load–displacement hysteresis curves

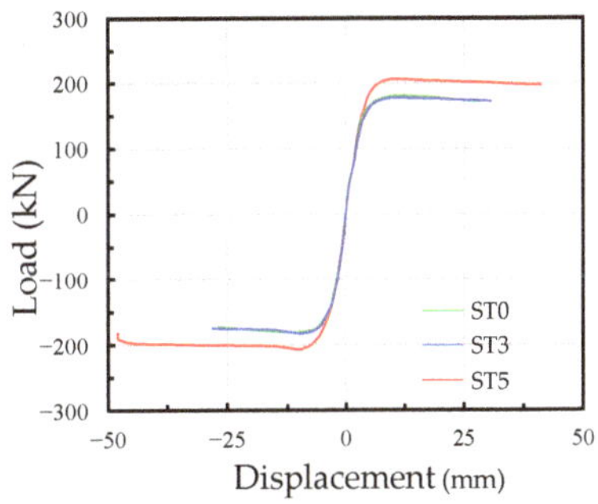

(**b**) Load–displacement skeleton curves

Figure 9. Effect of the hollow ratio on the CFDST T-joint main pipe.

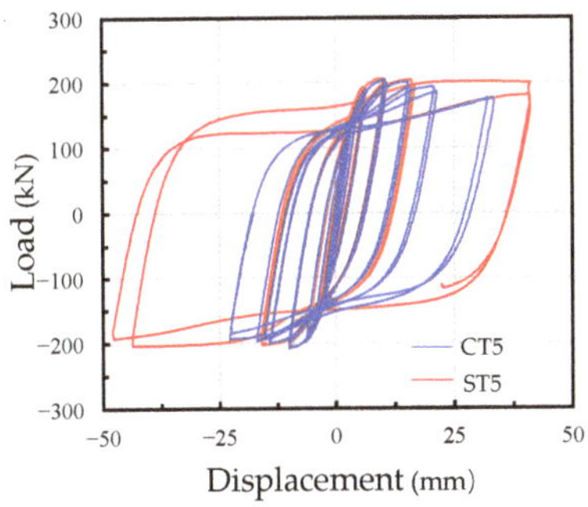

(**a**) Load–displacement hysteresis curves

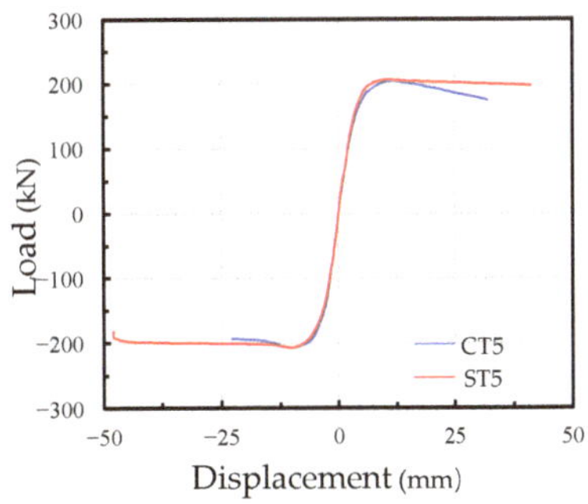

(**b**) Load–displacement skeleton curves

Figure 10. Effect of the CFDST specimen with steel slag or commercial Portland concrete.

3.3. *Stiffness Degradation*

In this paper, the stiffness can be expressed by the secant stiffness of the peak point of the hysteresis ring. The average secant stiffness of each loading stage is calculated by the following Formula (1) [25].

$$K_i = \frac{\sum_{i=1}^{j} N_j^i}{\sum_{i=1}^{j} \Delta_j^i} \tag{1}$$

where K_i is the secant stiffness under the *i-th* cycle and $N_j{}^i$, $\Delta_j{}^i$ is the *i-th* cycle peak load and peak displacement at *j*-level loading.

The stiffness degradation curve of each specimen is shown in Figure 11. It can be seen from Figure 11 that the trend and shape of the stiffness degradation curve of each specimen is almost identical and the stiffness degradation curve is relatively symmetrical. The stiffness degradation curves show that the stiffness decreases with the increase of loading displacement. The stiffness degradation is fast in the early stage and slower in the late loading stage. The interaction between the steel tube and concrete makes the failure of the concrete inside steel tube change from brittle failure to plastic failure; therefore, the stiffness degradation is slower in the late loading stage.

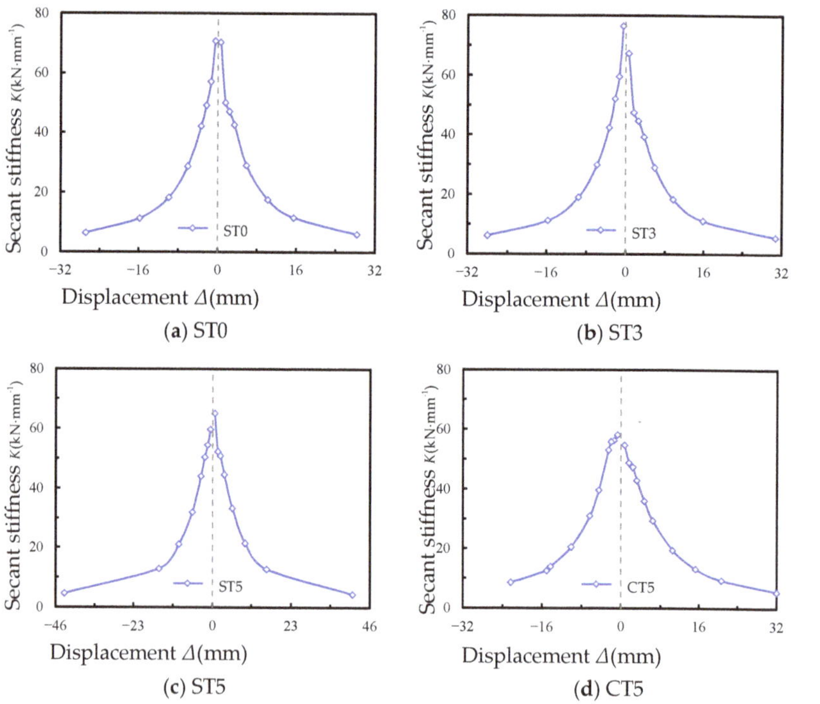

Figure 11. Degradation of secant stiffness.

At the beginning of loading, the stiffness degradation curves almost coincide (see Figure 11). Furthermore, the hollow rate increases to 0.5 and the stiffness degradation curve of the early stage coincides; however, the stiffness degrades more slowly (see Figure 12a). It can be seen from Figure 12b that the steel slag concrete has a greater stiffness in the plastic stage and wider stiffness degradation range than ordinary concrete.

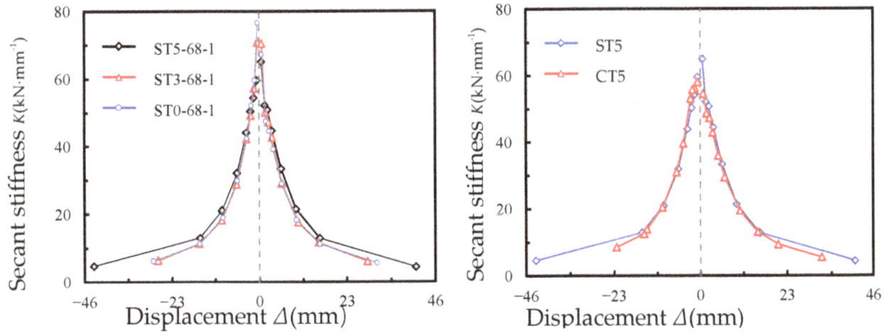

(**a**) Effect of the hollow ratio (**b**) Effect of the steel slag

Figure 12. The stiffness degradation analysis of influencing factors.

3.4. Energy Dissipation

The cumulative energy dissipation coefficient E is a key index of structural energy dissipation capacity, and it can be calculated by the following Formula (2) [25].

$$E = \frac{S_{(ABC+CDA)}}{S_{(OBE+ODF)}} \tag{2}$$

where S_{ABCD} represents the hysteresis loop area, as shown in Figure 13, and S_{OBE} and S_{ODF} represent the areas of the triangles Δ_{OBE} and Δ_{ODF}.

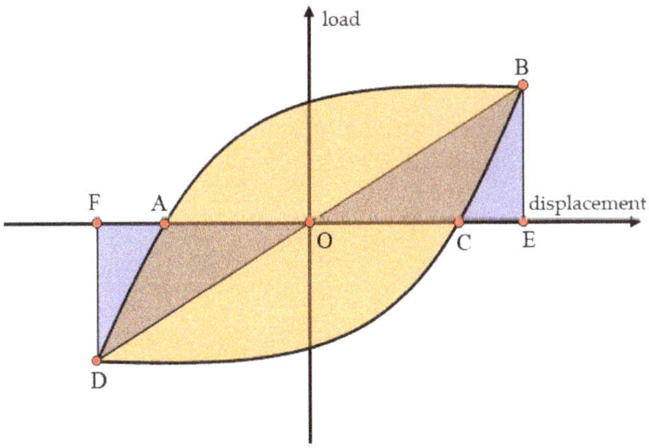

Figure 13. Load–displacement curve.

In this experiment, the quantity dissipation coefficient (E)-cycle (n) curve is used to describe the energy dissipation characteristics of the specimens. Figure 14 shows the quantity dissipation coefficient of the specimens during loading. It can be seen from Figure 14 that the trend and shape of the quantity dissipation coefficient of each specimen are almost identical. It decreases slightly in the early stage and rises rapidly in the late stage. Using the same loading level, the quantity dissipation coefficient obtained by the second cycle loading calculation is less than that of the first loading calculation. The joints have a good energy dissipation capacity with about a 3.0 quantity dissipation coefficient.

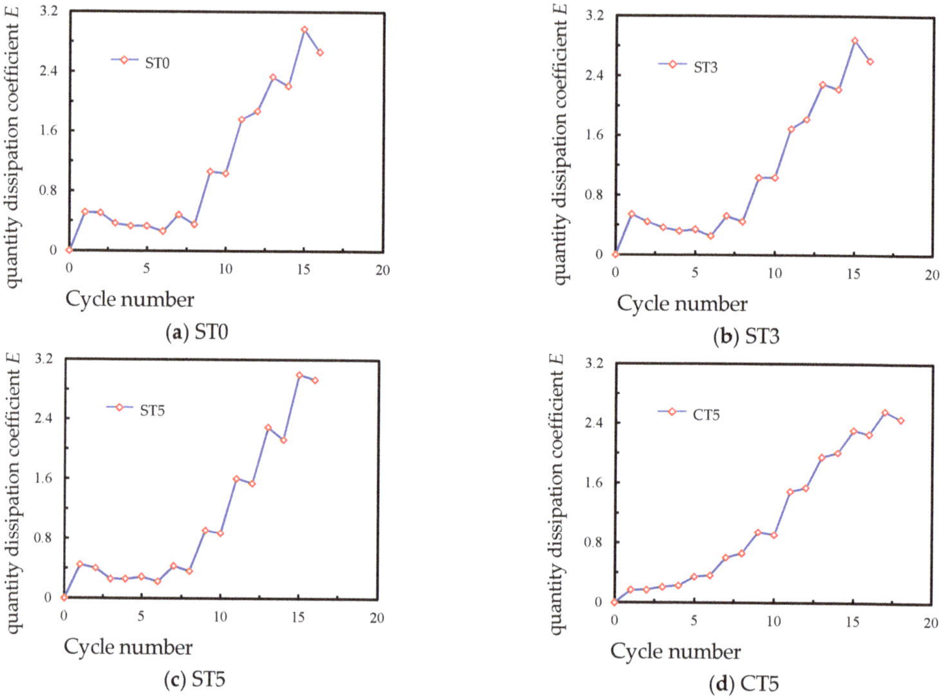

Figure 14. The quantity dissipation coefficient of specimens.

3.5. Ductility

In general, the ductility coefficient is used to reflect the strength of the structure's ductility. The displacement ductility coefficient is defined in the following equation [25].

$$\mu = \frac{\Delta_{max}}{\Delta_y} \tag{3}$$

Table 8 lists the displacement ductility coefficient of the joints during loading. It can be seen from Table 8 that the displacement ductility coefficient of the joints is increased, with the hollow ratio of the CFDST chord increasing from 0 to 0.5. The displacement ductility coefficient of the specimens could be improved by more than 50% by filling steel slag concrete with the 0.5 hollow ratio of the CFDST chord.

Table 8. The displacement ductility coefficient of specimens.

Number	Yield Displacement		Ultimate Displacement		Ductility Coefficient	
	Δ_y^+ (mm)	Δ_y^- (mm)	Δ_u^+ (mm)	Δ_u^- (mm)	μ^+	μ^-
ST0	2.57	−2.75	28.3	−26.8	11.01	9.75
ST3	2.74	−2.38	30.7	−28.0	11.20	11.76
ST5	3.27	−3.46	40.9	−43.7	12.51	12.63
CT5	3.73	−3.69	32.1	−22.8	8.61	6.18

4. Finite Element Analysis

4.1. Finite Element Model

In order to analyze the stress development and failure mechanism of the steel slag CFDST T-joint in detail, a finite element model (FEM) was established with Abaqus software, as shown in Figure 15. The steel tube and concrete were modelled using the C3D8R brick elements. The uniaxial stress–strain relationship of concrete adopts the relationship proposed by Han [26]. The dimension and boundary conditions of the model are the same as the experimental specimens.

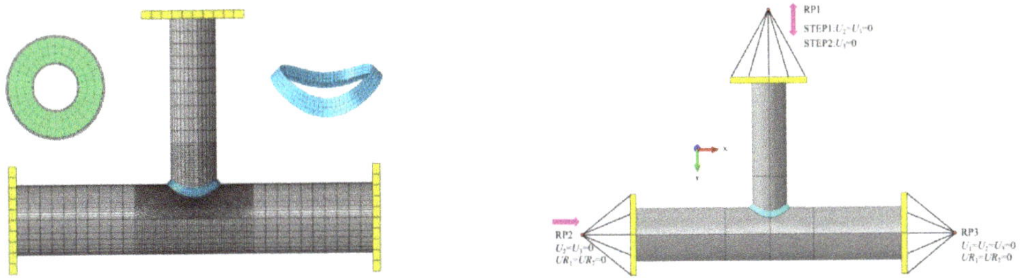

(**a**) The specimen meshing (**b**) The boundary condition of the FEM

Figure 15. The FEM of specimen.

4.2. Verification of the Finite Element Model

The comparison between the FEM results and the experimental results is shown in Figure 16. It can be seen that the stiffness and ultimate bearing capacity of the FEM and experimental specimens are generally consistent with each other.

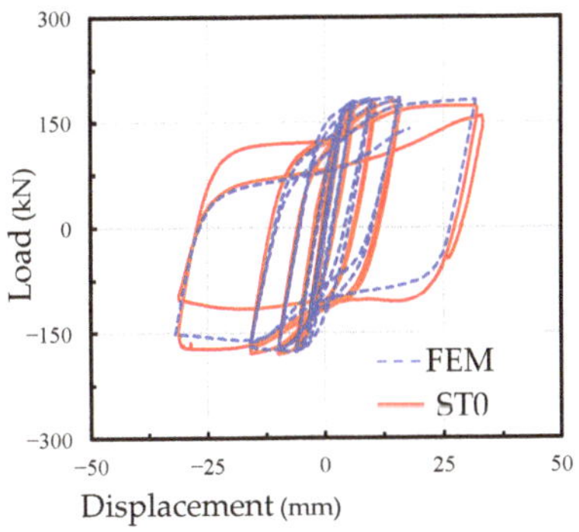

(**a**) ST0

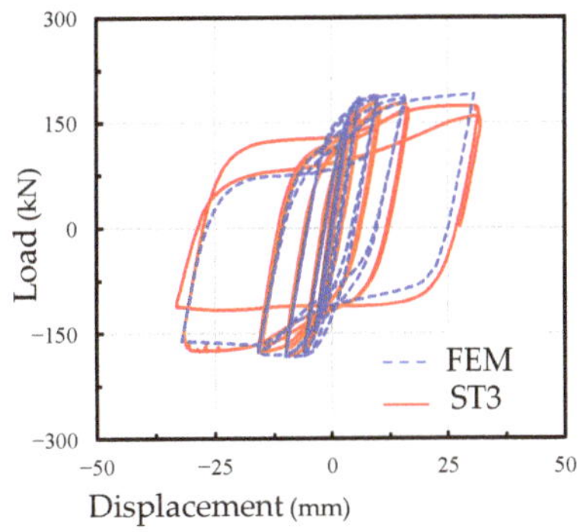

(**b**) ST3

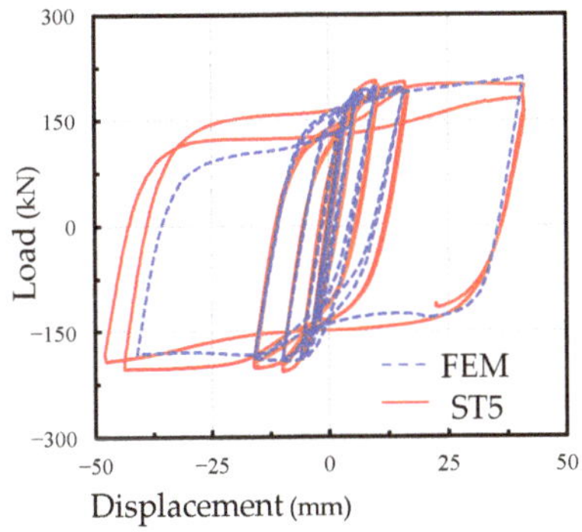

(**c**) ST5

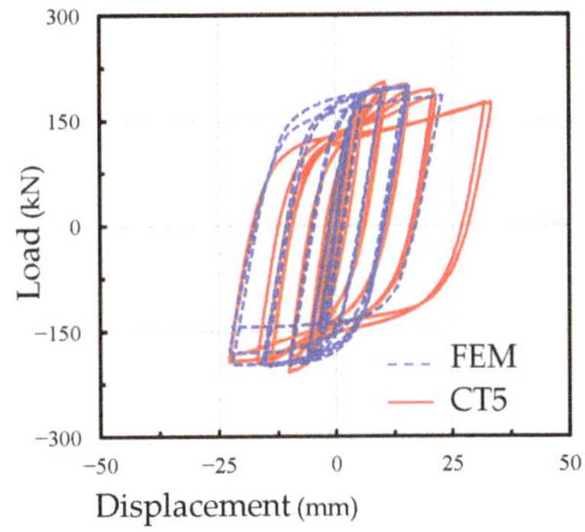

(**d**) CT5

Figure 16. Load–displacement curve comparison of FEM and experimental.

4.3. Failure Mechanism Analysis

From the experimental results above, there are two failure modes, with those being the main pipe fracture and joint area compression-bending failures. In order to clarify the failure mechanisms of the two typical failure modes, the ST3 specimen is considered to be the typical specimen for fracture failure and the ST5 specimen is considered to be the typical specimen for joint area compression-bending failure. Figure 17 shows the comparison of the failure modes between the two typical specimens from the FEM results and the experimental results. It can be seen that the failure modes between the FEM and experimental specimens are generally consistent with each other.

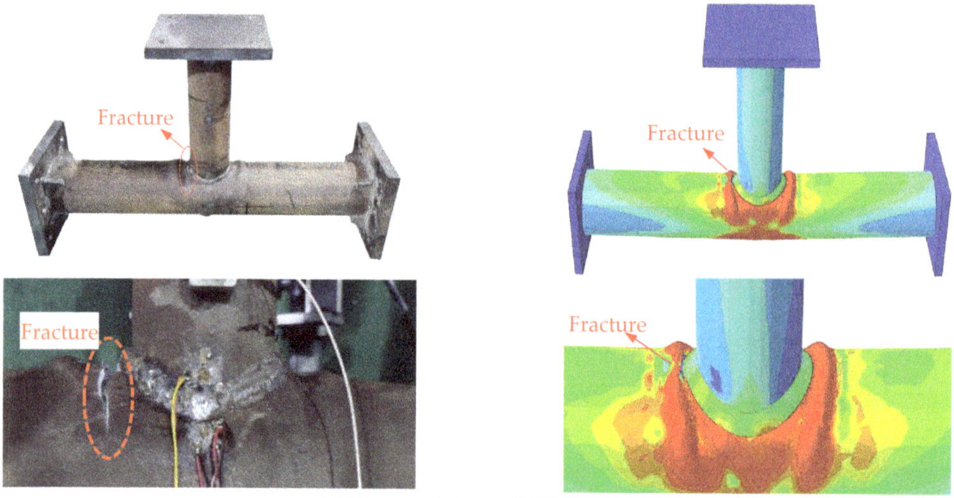

(**a**) Fracture failure of ST3 specimen

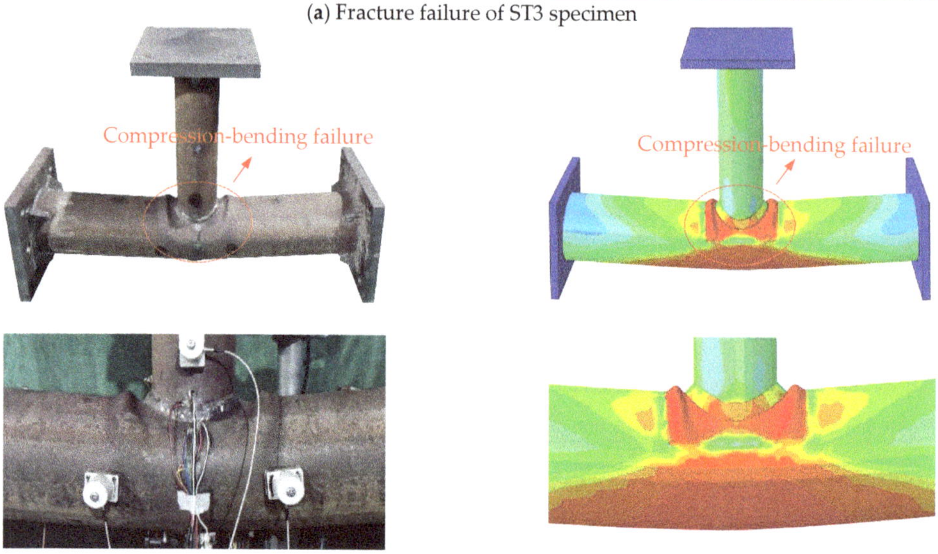

(**b**) Compression-bending failure of ST5 specimen

Figure 17. Failure simulation of specimens.

In order to analyze the contact stress development of the CFDST T-joint under the two modes of failure, the location of the span center section is studied, as shown in Figure 18. The contact stress development is shown in Figure 19. As shown in Figure 19, the contact stress with large hollow ratio distribution is more uniform, while for the specimens without hollow ratio, the stress is easier to concentrate. The reason for this is that the inter-reaction between the concrete and steel tube work better with a large hollow ratio. The location of the crown section is studied, as shown in Figure 20. The contact stress development is shown in Figure 21. As shown in Figure 21, the contact stress of the crown section is the stress concentration. The stress concentration at the crown point with the small hollow specimen is more serious. The failure mechanism of the steel slag CFDST T-joint concerns the stress concentration, leading to damage of the concrete and weakening effective support to the main pipe. Then, the main pipe will bear increasing stress until fracture. However, the stress concentration with the large hollow ratio specimen is not obvious. Therefore, the concrete will be damaged later, allowing the concrete to continue to provide good support for the main pipe in the later stages, with the final failure mode being compression-bending failure.

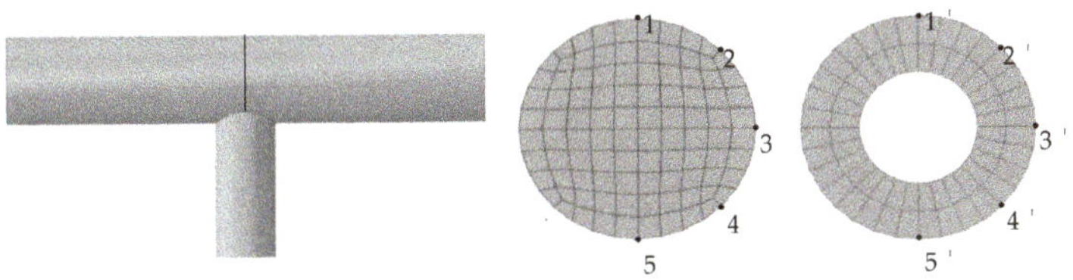

Figure 18. The diagram of the span center section.

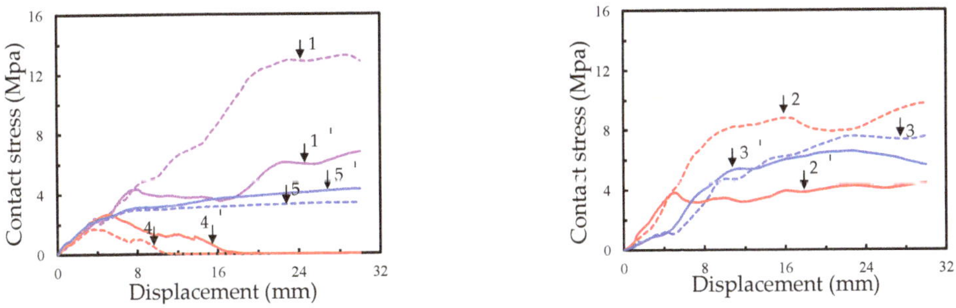

Figure 19. Contact stress of the span center section.

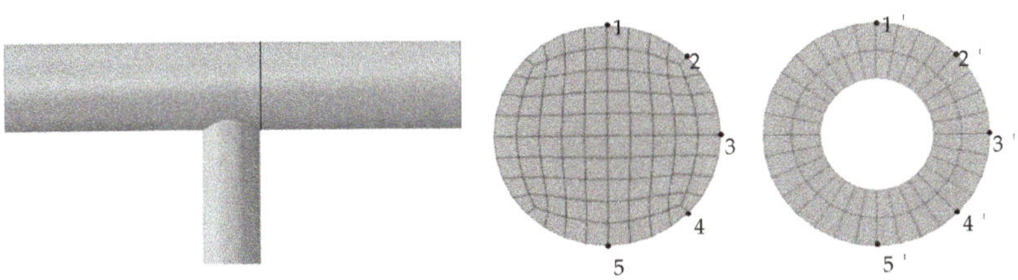

Figure 20. The diagram of the crown section.

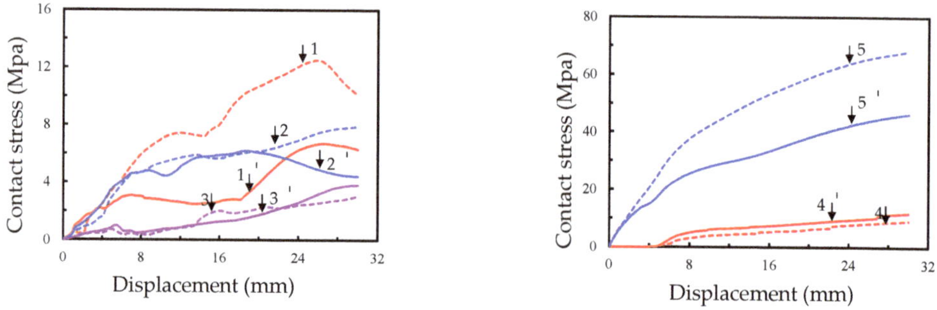

Figure 21. Contact stress of crown section.

4.4. Parameter Analysis

The finite element models were established to analyze the effect of the hollow ratio, axial compression ratio and brace-to-main pipe diameter ratio. The bearing capacity N_u of the CFDST T-joint specimens with different parameters calculated by the finite element models are shown in Table 9.

Table 9. Bearing capacity N_u of CFDST T-joint specimens with different parameters.

Parameter	$D \times T \times L$ (mm × mm × mm)	$D_0 \times T_0 \times L_0$ (mm × mm × mm)	n	β	χ (%)	N_u (KN)
Benchmark	300 × 6 × 2000	150 × 4 × 400	0.1	0.5	0	500.8
	300 × 6 ×2000	150 × 4 × 400	0.1	0.5	30	509.3
	300 × 6 ×2000	150 × 4 × 400	0.1	0.5	50	521.1
	300 × 6 ×2000	150 × 4 × 400	0.1	0.5	70	526.0
	300 × 6 ×2000	150 × 4 × 400	0.1	0.5	80	494.1
β	300 × 6 × 2000	120 × 4 × 400	0.1	0.4	0	495.7
	300 ×6 × 2000	120 × 4 × 400	0.1	0.4	30	499.4
	300 × 6 ×2000	120 × 4 × 400	0.1	0.4	50	508.1
	300 × 6 × 2000	120 × 4 × 400	0.1	0.4	70	508.4
	300 × 6 × 2000	120 × 4 × 400	0.1	0.4	80	438
	300 × 6 × 2000	180 × 4 × 400	0.1	0.6	0	503.1
	300 × 6 × 2000	180 × 4 × 400	0.1	0.6	30	508.8
	300 × 6 × 2000	180 × 4 × 400	0.1	0.6	50	523.4
	300 × 6 × 2000	180 × 4 × 400	0.1	0.6	70	535.5
	300 × 6 × 2000	180 × 4 × 400	0.1	0.6	80	516.0
n	300 × 6 × 2000	150 × 4 × 400	0.3	0.5	0	500.7
	300 × 6 × 2000	150 × 4 × 400	0.3	0.5	30	501.3
	300 × 6 × 2000	150 × 4 × 400	0.3	0.5	50	516.3
	300 ×6 × 2000	150 × 4 × 400	0.3	0.5	70	517.1
	300 × 6 × 2000	150 × 4 × 400	0.3	0.5	80	495.8
	300 × 6 × 2000	150 × 4 × 400	0.5	0.5	0	499.1
	300 × 6 × 2000	150 × 4 × 400	0.5	0.5	30	499.9
	300 × 6 × 2000	150 × 4 × 400	0.5	0.5	50	507.5
	300 × 6 × 2000	150 × 4 × 400	0.5	0.5	70	509.0
	300 × 6 × 2000	150 × 4 × 400	0.5	0.5	80	493.8

Figure 22 shows the standardization N_u of the parameters with the benchmark result. It can be seen that the bearing capacity of the specimens are increased with increases in the hollow ratio while the hollow ratio is below 70%. The bearing capacity of the specimens are decreased with increases in the hollow ratio when the hollow ratio is above 70%. According to the parameter analysis in this paper, the optimal hollow ratio is 70%. The reason for this is that when the hollow ratio is too large, the filled concrete will be damaged and unable to provide effective support before the steel tube reaches its ultimate strength.

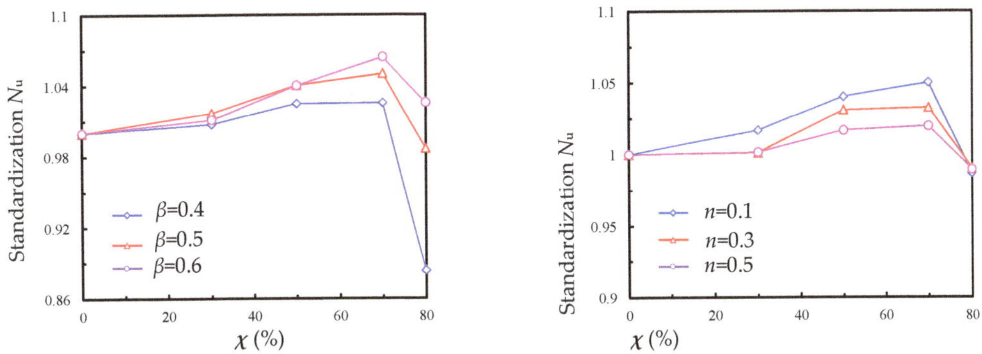

Figure 22. Parameters' effect on standardization N_u.

5. Conclusions

The seismic performance of the steel slag CFDST T-joint subjected to cyclic axial loading was investigated. Three parameters, including the effect of the hollow ratio, axial compression ratio and brace-to-main pipe diameter ratio, were considered. The following conclusions can be drawn from the analysis of the experimental results.

(1) The analysis shows that the ultimate capacity of the specimens could be improved with 54% displacement by filling steel slag concrete with a 0.5 hollow ratio of the CFDST main pipe.

(2) There are two failure modes for the CFDST T-joint specimens, with those being the main pipe fracture and joint area compression-bending failure. The hysteresis curves of all specimens have a full fusiform shape and good energy dissipation capacity.

(3) The steel slag CFDST T-joint provided in this paper that generated the displacement ductility coefficient results is strongly encouraged. The steel slag used in the structure can promote the reuse of solid waste and reduce project costs.

(4) The bearing capacity of CFDST T-joint specimens with different parameters calculated by the finite element models shows that the optimal hollow ratio is 70%. Further research should be conducted on the effect of different parameters, including the strength of materials and so on.

Author Contributions: Conceptualization, W.Z.; methodology, W.Z.; software, M.F.; validation, W.Z.; formal analysis, W.W.; investigation, Y.L.; resources, M.F.; data curation, M.F.; writing—original draft preparation, W.Z.; writing—review and editing, W.Z.; visualization, H.Z.; supervision, W.W.; project administration, W.Z.; funding acquisition, W.Z. All authors have read and agreed to the published version of the manuscript.

Funding: This research was funded by the Department of Science and Technology, Fujian Province, China (grant number: No. 2020J01389; 2022J011185).

Institutional Review Board Statement: Not applicable.

Informed Consent Statement: Not applicable.

Data Availability Statement: Data are contained within the article.

Conflicts of Interest: The authors declare no conflict of interest.

References

1. Rashad, A. A synopsis manual about recycling steel slag as a cementitious material. *J. Mater. Res. Technol.* **2019**, *8*, 4940–4955. [CrossRef]
2. Song, Q.; Guo, M.Z.; Wang, L.; Ling, T.C. Use of steel slag as sustainable construction materials: A review of accelerated carbonation treatment. *Resour. Conserv. Recycl.* **2021**, *173*, 105740. [CrossRef]

3. Tian, E.; Liu, Y.; Cheng, X.; Zeng, W. Characteristics of Pavement Cement Concrete Incorporating Steel Slag Powder. *Adv. Mater. Sci. Eng.* **2022**, *2022*, 6360301. [CrossRef]
4. Li, L.; Ling, T.C.; Pan, S.Y. Environmental benefit assessment of steel slag utilization and carbonation: A systematic review. *Sci. Total Environ.* **2022**, *806*, 150280. [CrossRef] [PubMed]
5. Wang, Q.; Kang, Q.; Cao, P. Flexural Strength for Negative Bending and Vertical Shear Strength of Composite Steel Slag-Concrete Beams. *Adv. Steel Struct.* **1999**, *1*, 385–389.
6. Qasrawi, H.; Shalabi, F.; Asi, I. Use of low CaO unprocessed steel slag in concrete as fine aggregate. *Constr. Build. Mater.* **2009**, *23*, 1118–1125. [CrossRef]
7. Hai, Y.; Zhan, Q.; Xiao, Y. Optimization of Mineralization Curing System for Efficient and Safe Utilization of Steel Slag Wastes. *J. Wuhan Univ. Technol. Mater* **2022**, *37*, 595–602.
8. Calmon, J. Effects of BOF steel slag and other cementitious materials on the rheological properties of self-compacting cement pastes. *Constr. Build. Mater.* **2013**, *40*, 1046–1053. [CrossRef]
9. Han, L.; Li, W.; Bjorhovde, R. Developments and advanced applications of concrete-filled steel tubular (CFST) structures: Members. *J. Constr. Steel Res.* **2014**, *100*, 211–228. [CrossRef]
10. Han, L.; Lam, D.; Nethercot, D. *Design Guide for Concrete-Filled Double Skin Steel Tubular Structures*; CRC Press: Boca Raton, FL, USA, 2018; pp. 1–7.
11. Yang, Y.; Shi, C.; Hou, C. Experimental and numerical study on static behaviour of uniplanar cfdst-chs t-joints. *J. Constr. Steel Res.* **2020**, *173*, 106250. [CrossRef]
12. Li, W.; Cheng, Y.; Wang, D.; Han, L.; Zhao, X. Behaviour of high-strength cfdst chord to chs brace t-joint: Experiment. *Eng. Struct.* **2020**, *219*, 110780. [CrossRef]
13. Wang, D. *Fatigue Performance of High-Strength Steel CFDST Chord to CHS Brace T-Joints*; Tsinghua University: Beijing, China, 2017.
14. Gao, D. *Experimental Study on Hysteretic Behavior of Concrete-Filled Double Skin Tube Overlapped K-Joints*; Ningxia University: Ningxia, China, 2018.
15. Ma, Y. *Static Behavior of Square Concrete-Filled Double-Skin Steel Tube T-Joints*; Dalian University of Technology: Dalian, China, 2018.
16. Shi, C. *Static Behavior of Concrete-Filled Double-Skin Steel Tube T-Joints*; Dalian University of Technology: Dalian, China, 2019.
17. Yang, Y.; Bie, X.; Fu, F. Static performance of square CFDST chord to steel SHS brace T-joints. *J. Constr. Steel Res.* **2021**, *183*, 1–20. [CrossRef]
18. Fang, Y.; Yu, F.; Zhang, Y. Mechanical behavior and bearing capacity calculation of self-stressing steel slag aggregate reinforced concrete filled circular steel tube columns. *Acta Mater. Compos. Sin.* **2020**, *37*, 1211–1220.
19. Shen, Q.; Gao, B.; Wang, J. Experimental study on performance of steel slag concrete filled elliptical steel tubular stub columns under axial load. *J. Build. Struct.* **2021**, *42*, 197–203.
20. Yu, F.; Cao, Y.; Fang, Y.; Zhang, Y.; Niu, K. Mechanical behavior of self-stressing steel slag aggregate concrete filled steel tubular short columns with different loading modes. *Structures* **2020**, *26*, 947–957. [CrossRef]
21. Yu, F.; Qin, Y.; Yao, C.; Bu, S.; Fang, Y. Experimental investigation on the seismic behavior of self-stressing steel slag CFST column. *Struct. Concr.* **2022**, *23*, 1492–1507. [CrossRef]
22. *GB/T228. 1-2021*; Metallic Materials-Tensile Testing-Part 1: Method of Test at Room Temperature. China Standard Press: Beijing, China, 2021.
23. *JGJ55-2011*; Specification for Mix Proportion Design of Ordinary Concrete. Architecture and Building Press: Beijing, China, 2011.
24. *GB/T 50081-2019*; Standard Test Method for Physical and Mechanical Properties of Concrete. China Standard Press: Beijing, China, 2019.
25. *JGJ/T 101-201*; Specification for Seismic Test of Buildings. China Architecture & Building Press: Beijing, China, 2015.
26. Han, L. *Concrete-Filled Steel Tube Structures—Theory and Practice*, 3rd ed.; Science Press: Beijing, China, 2016.

sustainability

Article

Effects of the Transverse Deck-Roadbed Pounding on the Seismic Behaviors of the Prefabricated Frame Bridge

Yuwei Wang [1], Jinli Zhang [1], Yingao Zhang [2], Rui Zuo [2], Liang Chen [2] and Tianyue Sun [2,*]

1 Anhui Transportation Holding Group Co., Ltd., Hefei 230088, China
2 Department of Civil Engineering, Hefei University of Technology, Hefei 230009, China
* Correspondence: 2019170551@mail.hfut.edu.cn

Abstract: Pounding effects on prefabricated frame bridges are not clear, which may influence seismic behaviors a lot in some cases. Prefabricated frame bridges are emerging structures designed to solve the problem of difficult land acquisition in highway expansion and reconstruction, the deck of the prefabricated frame bridge is adjacent to the original roadbed in the transverse direction, so the pounding potential exists under the earthquake ground motions. In this study, the artificial ground motions of the different seismic intensities are selected to carry out the nonlinear time history analyses, and the pounding effects on the prefabricated frame bridge are evaluated based on the pounding forces and the components' seismic response. It is found that the pounding effects are not obvious in all cases; some energy can be dissipated in the pounding process, which is also limited to some extent. Finally, the influences of the gap distance and seismic intensity are investigated according to the parameter sensitivity analysis. The results indicate that the gap distance and the seismic intensity are the two important factors related to the pounding effects, the seismic response of the components will decrease when the pounding effects are obvious, and the transverse deformation of the deck cannot influence the stress state of the superstructure.

Keywords: pounding; prefabricated frame bridge; seismic behaviors

check for updates

Citation: Wang, Y.; Zhang, J.; Zhang, Y.; Zuo, R.; Chen, L.; Sun, T. Effects of the Transverse Deck-Roadbed Pounding on the Seismic Behaviors of the Prefabricated Frame Bridge. *Sustainability* **2023**, *15*, 1554. https://doi.org/10.3390/su15021554

Academic Editor: Chunxu Qu

Received: 6 December 2022
Revised: 29 December 2022
Accepted: 30 December 2022
Published: 13 January 2023

1. Introduction

Prefabricated frame bridges are emerging structures designed to solve the problem of difficult land acquisition in highway expansion and reconstruction, which is based on the strategy of sustainable development and environmentally friendly construction [1]. As shown in Figures 1 and 2, the prefabricated frame bridge is very long, with 15 spans of 6 m each. The superstructure of the prefabricated frame bridge is different from the ordinary bridge girder, which is designed to be very thin. Some mild rebars are installed in the deck, but no pre-stressed strands are installed. Piers and piles all consist of prestressed columns [2]. In terms of the connection modes between the girder and the piers, bearings are usually applied on traditional bridges to dissipate seismic energy and restrict the deformation of the deck [3,4]. Considering that bearing maintenance is a hard task and a large number of bearings are needed, which costs a lot [5], an elastoplastic column-deck joint (EC-DJ) is developed to replace bearings to connect the deck and the prestressed columns. The structural style of the prefabricated frame bridge is similar to the wharf in a sense [2,6].

Pounding force is a complicated load under seismic excitation, which occurs in a short time. Many factors influence the pounding potential between the adjacent segments, such as friction, gap distance, ground motion intensity, structure stiffness, and segment stiffness, so the process of energy transfer is complex [7,8]. Earthquake-induced pounding tends to be detrimental to the structures, which may cause material stiffness degradation, sliding, and even girder collapse [9,10]. Furthermore, the seismic behaviors of the adjacent structures can be affected to varying degrees by the pounding effect under different conditions [11,12].

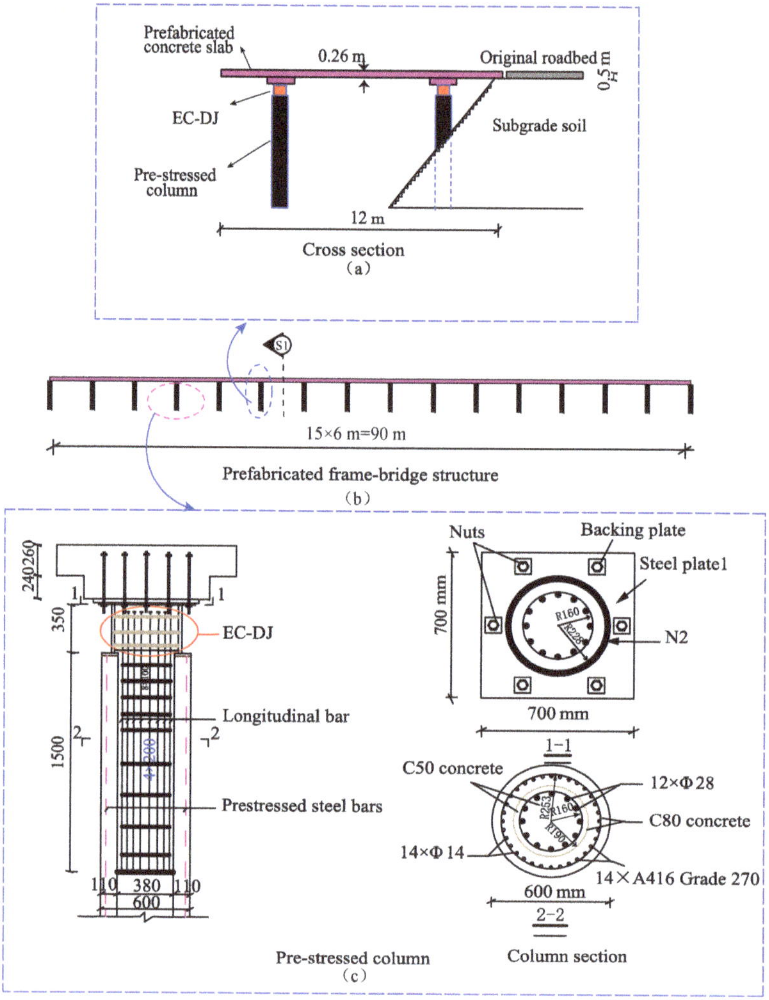

Figure 1. The configuration of the prefabricated frame bridge. (**a**) Cross section. (**b**) Pre-fabricated frame bridge structure. (**c**) Pre-stressed column.

Figure 2. The prefabricated frame bridge structure.

Many studies have been conducted to investigate the pounding effects on bridges under earthquake action. The pounding in the longitudinal direction was mostly between the girders and abutments or between adjacent girders, while poundings in the transverse direction were mostly between the retainers and the girders [8,13,14]. Some scholars also studied the bidirectional poundings of skew bridges and curved bridges [15,16]. Li B. et al. [17] found that the pier's response decreases with the consideration of the pounding force, and the pounding potential increases with the stiffness of the larger movable abutments. Chouw N. et al. [18] maintained that when the soil-structure interaction and ground motion spatial variation are considered, the pounding damage potentials between the adjacent decks may not be reduced under high seismic intensity. Some scholars proposed methods for mitigating the pounding damages. Guo A. et al. [13] present a magnetorheological damper on reducing the pounding of the adjacent superstructures. Abdel Raheem S.E. et al. [19] used a shock absorber to mitigate the pounding, so the pier forces could be smoothed due to the reduction of the pounding forces. The seismic performance can also be improved by retainers [14,20], which can restrict the transverse deformation of the girder due to pounding.

In seismic design, pounding potential is usually an important factor that affects the analysis results. Kun C. et al. [16,21] suggested that the seismic responses are significantly underestimated when the pounding effect is ignored for the skew bridges. Li B. and Chouw N. [22] investigated the impacts of spatial variation of ground motions on the pounding based on the shaking table and concluded that pounding damage may be overestimated for the elastic bridge studies. Huo Y. et al. [23] maintained a similar view for the skew bridges and provided valuable guidance for future bridge design.

However, in some cases, poundings can consume some seismic energy and limit the displacement of the structure, which is conducive to improving the seismic performance of the structure [13,24]. The prefabricated frame bridge is a new structure that is constructed based on the original roadbed to expand the road. The deck is likely to pound with the roadbed due to the transverse deformation under the seismic action, but the advantages and disadvantages of this to the overall seismic performance of the structure are unknown.

In this study, the artificial ground motions are generated by the SIMOKE code first, which are selected as the seismic excitation in the time history analysis. Then the seismic response law of the prefabricated frame bridge affected by the pounding effect is investigated. Finally, the influences of the interaction of the gap distance, the transverse deformation, and the ground motions intensity are considered to explore the pounding effect on the prefabricated frame bridge. The results indicate that the pounding between the deck and the roadbed can be ignored in cases where the intensity of the ground motions is between 0.2 g and 0.5 g. When the seismic intensity is moderate, seismic energy can be consumed, and the transverse deformation can be restricted obviously due to the pounding effect. Meanwhile, the response of the structure will also be enlarged with the decreasing gap distance and the increasing intensity of the pounding. According to the nonlinear time-history analysis and the parameter sensitivity analysis, the possible impact of the pounding on the prefabricated frame bridge is clear. Based on the analysis, some proper suggestions and evaluations for the seismic design are provided, which provides a reference for future research.

2. Prefabricated Frame Bridge Structure and Finite Element Model

2.1. Layout of the Prefabricated Frame Bridge

Prefabricated frame bridges are emerging structures for the reconstruction and expansion of the expressway. As shown in Figures 1 and 2, the fifteen span prefabricated frame bridge has a total length of 90 m and a width of 12 m, with spans of 6 m. The superstructure of the prefabricated frame bridge is comprised of decks, and the substructure is comprised of pre-stressed columns. An elastoplastic pile-deck joint (EP-DJ) is applied to connect the decks and the pre-stressed columns, which is comprised of the steel tube, confined concrete, and stirrups, as shown in Figure 1c. The thickness of the deck is 0.26 m, and the hunch is

0.24 m. The pre-stressed column is 9 m in height; the diameter is 0.6 m in the outer and 0.38 m in the inner. The depth of the column body buried in the soil is set as 3 m.

2.2. Structure Analysis Model

The dynamic finite element model was built using the OpenSees analysis software [25], in which the material nonlinearity is considered, as shown in Figure 3. Rigid connections were used between the prefabricated decks and EC-DJs. The nonlinear behavior of the pre-stressed columns was simulated by the nonlinear beam-column element, and the interactions between columns and soil were considered, which was simulated by a zero-length element. The shell elements were used to simulate prefabricated decks, which are expected to remain elastic under the earthquake actions, and the pounding effects between the decks and roadbed are considered. The pounding model uses the linear spring model, and the energy dissipation is considered by the damping material. The roadbed and the foundations are regarded as fixed boundaries. The material models are shown in Figure 3c,d, and the detailed introduction of the materials is referred to the reference [2,26,27].

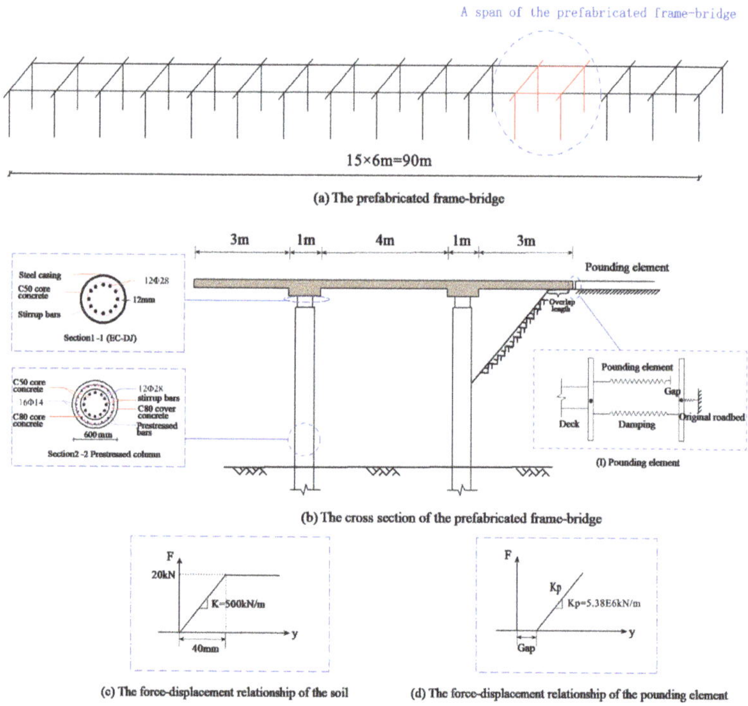

Figure 3. A span of the prefabricated frame bridge.

2.3. Consideration of the Pounding Element and the Soil

When the prefabricated frame bridge structure is under seismic excitation, passive earth pressure will be generated from roadbed soil, and the transverse deformation of the deck will be restricted. In this process, pounding force is generated, which influences the response of the structure to some extent. In this study, a bilinear model is applied to simulate the constitution of the roadbed soil, as shown in Figure 3. The initial stiffness is set as 500 kN/m, and the yield force is set as 20 kN [28,29].

Jankowski [30] maintained that the Kevin model can reflect the pounding effect accurately if the proper parameters are selected. In this study, the pounding effect between the deck and the roadbed is simulated by the Kevin model [31–33], which is composed of linear springs and dampers in parallel. The pounding force is simulated by the linear spring element, and the consumed energy is also considered in the model, which is simulated by the damping material.

Moreover, the pounding force can be calculated as follows:

$$F(y) = \begin{cases} K_P(d_0 + y) & d_0 + y < 0 \\ 0 & d_0 + y \geq 0 \end{cases} \tag{1}$$

where d_0 is the gap size, y is the relative displacement between the deck and roadbed, and K_P is the pounding stiffness. According to past studies [34–36], a large K_P value is usually assumed. In this analysis, the transverse stiffness of the deck is referred to define the K_P value, which is set as 5.38×106 kN/m.

2.4. Artificial Seismic Excitation

Earthquake ground motions are stochastic loads, which are difficult to calculate accurately. The actual earthquake data is lacking in the structure address, so artificial ground motions are selected as the input ground motions to simulate random earthquake actions [37,38]. In this paper, artificial ground motions are generated from the famous SIMOKE code (Vanmarcke and Gasparini [13]) based on the design response spectrum, which is according to the specification for the seismic design of highway bridges of the Chinese design code [6]. The PGA of the design response spectrums ranges from 0.1 g to 0.6 g, and 10 artificial earthquake waves are generated for each design response spectrum to conduct the nonlinear time history analyses. The time step size of these artificial earthquake waves is 0.02 s, and the period is 16 s. Figure 4 shows an example of the comparison of the artificial response spectrum and the given design response spectrum for a damping ratio of 5%. The result shows that the seismic excitation well matches the given design response spectrum.

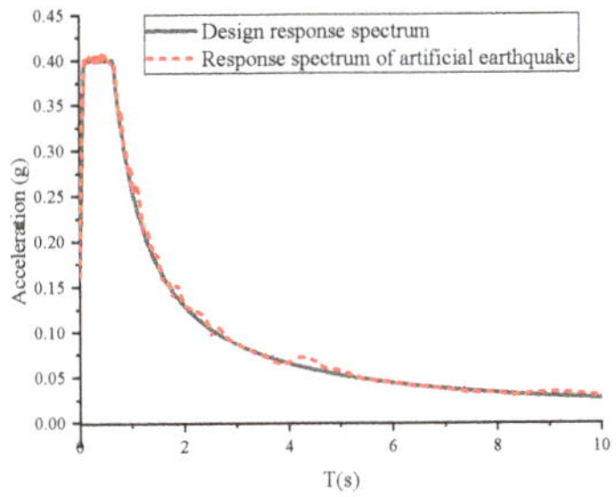

Figure 4. The response spectrum of the artificial ground motion.

3. Results and Observation

In order to study the dynamic behaviors of the prefabricated frame bridge under the ground motions with various intensities, seismic responses of the elastoplastic column-deck joints (EC-DJ) and the prestressed columns (PC) are investigated. Considering that the earthquake is a random process, so the maximum response of the structure is evaluated by random simulations, and the artificial seismic excitations are selected as ground motions input. In this section, the intensities of peak ground acceleration (PGA) are set as 0.1 g, 0.2 g, 0.3 g, 0.4 g, 0.5 g, and 0.6 g, respectively, to generate a total of six groups of design acceleration response spectra, and 10 artificial waves are generated for each design spectrum. Then the bending moment-curvature curves are extracted to analyze the energy consumption of the pounding effects. At last, the seismic intensity and the gap distance are investigated to explore the pounding effects on the seismic behaviors of the prefabricated frame bridge.

3.1. Seismic Response Analysis

Time history analysis can reflect the pounding effect on the prefabricated frame bridge clearly, so the responses of the EC-DJ and prestressed column (PC) are analyzed with considering pounding and without considering pounding. As is shown in Figures 5 and 6, the responses of EC-DJ and PC increased greatly at 6.6–7.8 s in the case of 0.3 g PGA, while the responses of the EC-DJ and PC increased a lot at 6.5–9.5 s in the case of 0.6 g PGA. As is shown in Figure 7, the pounding force surged at 5.8–7.0 s in the case of 0.3 g PGA. In the case of 0.6 g PGA, the pounding force surged from 5.8 s to 7.0 s. Moreover, pounding times in the case of 0.6 g PGA are more than that of 0.3 g PGA.

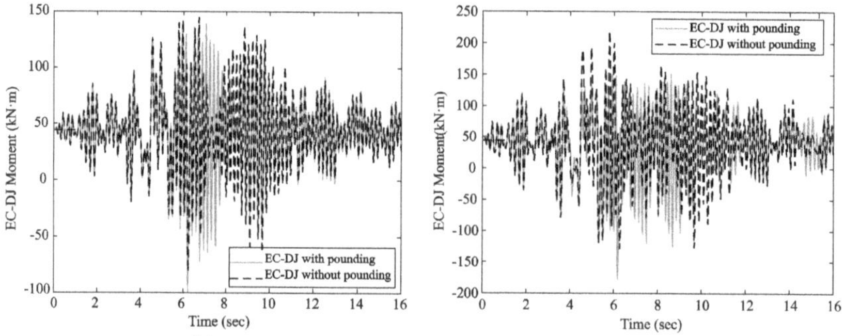

(a) (b)

Figure 5. EC-DJ force. (**a**) EC-DJs force for 0.3 g PGA. (**b**) EC-DJs force for 0.6 g PGA.

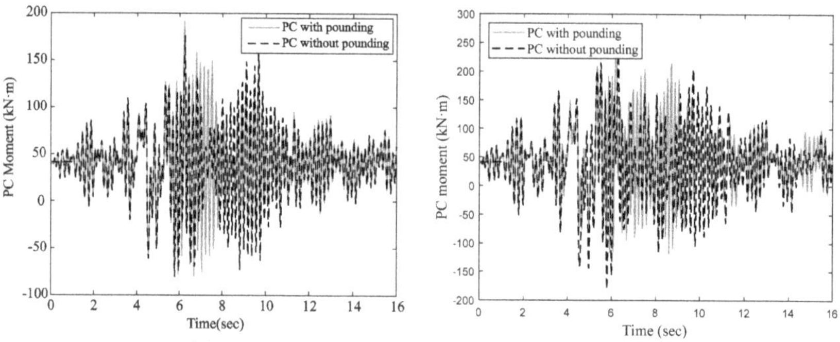

(a) (b)

Figure 6. Prestressed columns force. (**a**) Prestressed columns force for 0.3 g PGA. (**b**) Prestressed columns force for 0.6 g PGA.

Figures 8 and 9 show the comparison of the force-displacement relationship of EC-DJ and PC for 0.3 g PGA and 0.6 g PGA. The larger the enclosed area, the more seismic energy is dissipated. The areas enclosed by the dotted curve (without pounding) are smaller than the solid curve (with pounding), obviously, especially when the PGA is 0.3 g.

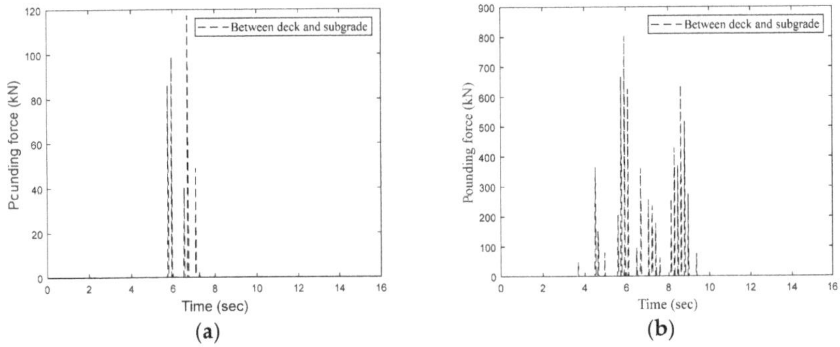

Figure 7. Pounding force. (**a**) Pounding force for 0.3 g PGA. (**b**) Pounding force for 0.6 g PGA.

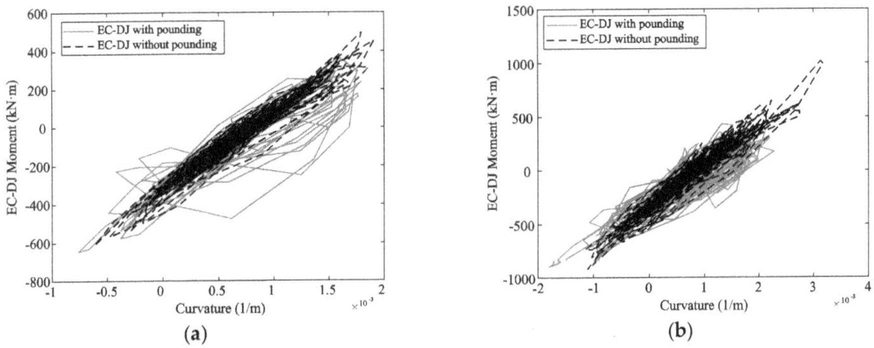

Figure 8. EC-DJ force-displacement relationship under different PGA. (**a**) EC-DJ force-displacement relationship for 0.3 g PGA. (**b**) EC-DJ force-displacement relationship for 0.6 g PGA.

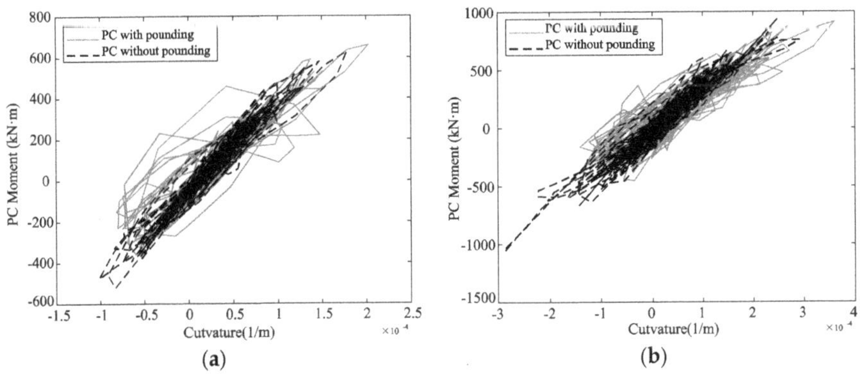

Figure 9. PC force-displacement relationship under different PGA. (**a**) PC force-displacement relationship for 0.3 g PGA. (**b**) PC force-displacement relationship for 0.6 g PGA.

Figures 10–13 show the bending moment-curvature relationship of the EC-DJ and the PC in different cases. The dotted line and the solid line nearly coincide when PGA is under 0.2 g, which indicates that pounding does not occur in those cases.

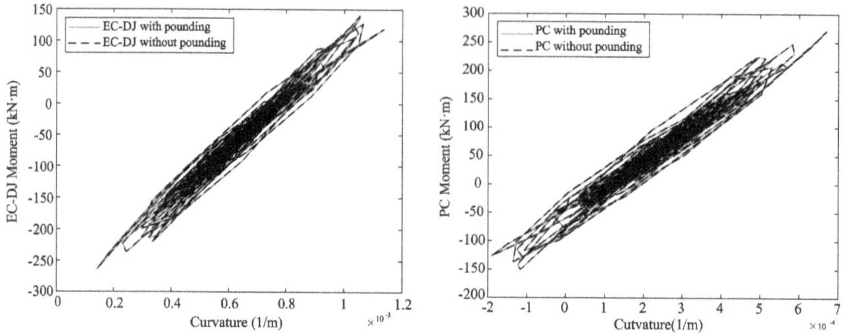

(a) (b)

Figure 10. Seismic responses under 0.1 g PGA. (**a**) EC-DJ force-displacement relationship for 0.1 g PGA. (**b**) PC force-displacement relationship for 0.1 g PGA.

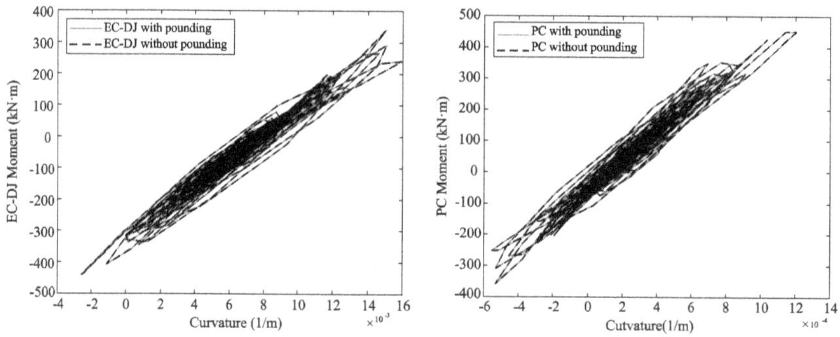

(a) (b)

Figure 11. Seismic responses under 0.2 g PGA. (**a**) EC-DJ response for 0.2 g PGA. (**b**) PC response for 0.2 g PGA.

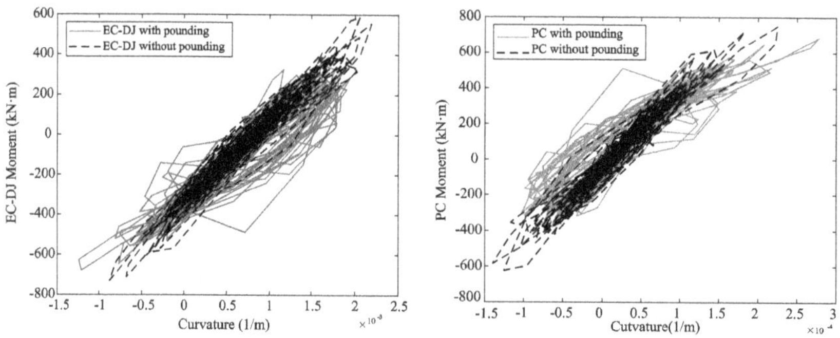

(a) (b)

Figure 12. Seismic responses under 0.4 g PGA. (**a**) EC-DJ response for 0.4 g PGA. (**b**) PC response for 0.4 g PGA.

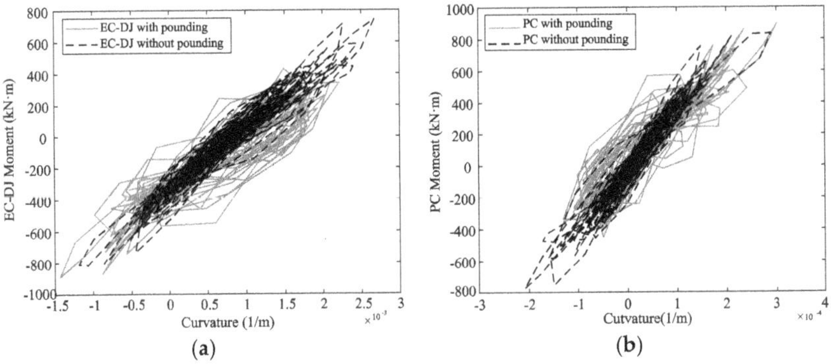

Figure 13. Seismic responses under 0.5 g PGA. (**a**) EC−DJ response for 0.5 g PGA. (**b**) PC response for 0.5 g PGA.

In the case of PGA is 0.3 g, 0.4 g, 0.5 g, and 0.6 g, compared with the enclosed areas of the dotted line and solid line, it can be found that the larger the PGA is, the less obvious the capacity of consuming seismic energy due to the pounding effect. Although much seismic energy can be consumed in the pounding process, it is limited to some extent.

When the PGA is not higher than 0.2 g, the EC-DJs and PCs are under the linear state, and no pounding occurs. When the PGA is higher than 0.2 g, the pounding times are more, and the pounding force is larger, which indicates that the pounding effects on the prefabricated frame bridge are more obvious. In this case, the EC-DJs and the PCs enter the nonlinear state but are recoverable, and some seismic energy is consumed in the pounding process.

3.2. Response Characteristics According to Various Gap Distances

Gap distance is a key factor that influences the pounding potential between the deck and the roadbed. In this section, seismic responses of the prefabricated frame bridge with different gap distances are recorded to analyze the pounding effect between the deck and the roadbed. To conduct the nonlinear time history analyses, six groups of artificial waves with different ground motion intensities are selected as ground motion input. The analysis, including the cases considering the pounding effect with the pounding gaps, is set as 0 mm, 10 mm, 20 mm, 30 mm, 40 mm, 50 mm, and 60 mm, and the pounding effect is ignored.

To describe the analysis results clearly, the direction of the deck close to the original roadbed is defined as the positive direction, and the reverse direction is defined as the negative direction. The maximum positive transverse deformation and the minimum negative transverse deformation for each case are shown in Figures 14 and 15.

The maximum positive deformation reflects the pounding between the deck and the original roadbed occurs or not. When the maximum positive deformation is larger than the gap distance, it can be concluded that the pounding exists. According to the comparative analysis of Figure 14, when the gap distance ranges from 0 to 20 mm, pounding exists under each seismic intensity; When the gap distance ranges from 30 to 40 mm, pounding exists in case the seismic intensity is greater than 0.3 g PGA; When the gap distance ranges from 50~60 mm, pounding exists in case the seismic intensity is not lower than 0.5 g PGA.

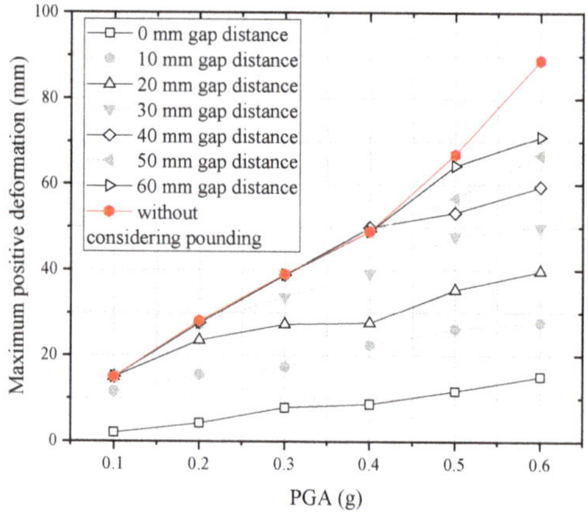

Figure 14. Maximum positive deformation of the deck.

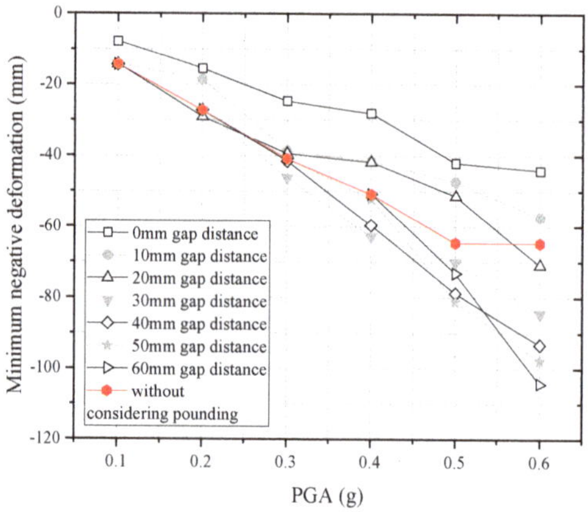

Figure 15. Minimum negative deformation of the deck.

The minimum negative deformation reflects the movement trend of the deck away from the original roadbed soil. When the negative deformation is large enough, the bridge deck may be separated from the original roadbed, resulting in the cantilever stress state of the bridge deck. The minimum negative deformation increases with the increasing seismic intensity, as can be seen in Figure 15. When the gap distance is not smaller than 40 mm, the minimum negative deformation of the deck is approximate for each seismic intensity; when the gap distance is smaller than 40 mm, it is clear that the minimum negative deformation increases with the increasing gap distance. When the pounding effect is considered, the minimum negative deformation value will increase by up to 200% at most.

In this study, it is found that the value of the minimum negative deformation reaches 104 mm. However, in the process of design and construction, the length of the deck overlapping on the original roadbed is designed as 1 m, which is large enough to prevent the stress state of the deck from changing, and the safety of the superstructure is guaranteed to some extent.

The mean responses of the EC-DJs and PCs are calculated from the bending moment during 10 nonlinear time history analyses for different PGAs, which are plotted in Figure 16. Because the EC-DJs and the prestressed columns are connected nearly rigidly, the responses of the two components are approximate.

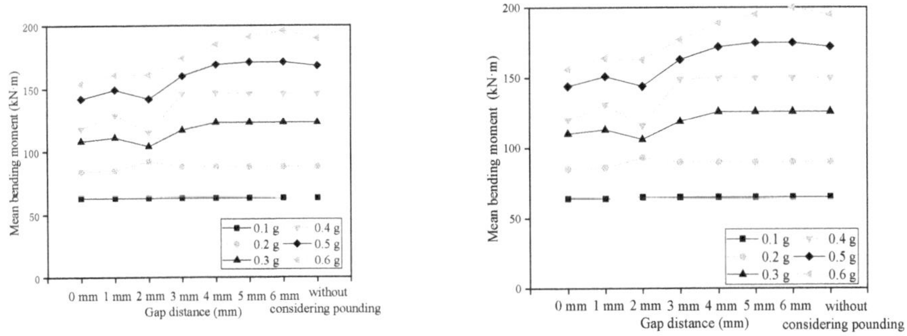

(a) (b)

Figure 16. Mean response of the components. (**a**) Mean response of the EC−DJ. (**b**) Mean response of the PC.

The deck is connected with piers firmly by the elastoplastic column-deck joint (EC-DJ), so the seismic responses of the components will be decreased. The seismic responses decrease obviously when the pounding is severe. Original roadbed soil plays a role in restricting the deck deformation, but this function depends on the soil stiffness. When the PGA is low and the gap distance is large, the pounding potential is low. When the PGA is too high and the gap distance is small, the soil will be destroyed due to pounding, and the original roadbed cannot limit the deck deformation or consume more seismic energy. For these two cases, the pounding effect can be ignored.

According to the comparative analysis, it can be seen from Figure 16 that when the seismic intensity is low, such as 0.1 g PGA and 0.2 g PGA, whether considering the pounding effect or not has little impact on the seismic response of EC-DJs and prestressed columns. Moreover, when the PGA is not lower than 0.2 g, the seismic response of the EC-DJs and prestressed columns is decreased with the increasing seismic intensity.

Compared to the system without considering the pounding effect, the seismic response of the system considering the pounding effect is reduced by 5% at most, with PGA is 0.1 g and 0.2 g; the seismic response is reduced by 15.6% at most with PGA is 0.3 g; the seismic response is reduced by 21.6% at most with PGA is 0.4 g; the seismic response is reduced by 15.5% at most with PGA is 0.5 g; the seismic response is reduced by 19.2% at most with PGA is 0.6 g.

4. Conclusions

Prefabricated frame bridges are emerging structures designed to solve the problem of difficult land acquisition in highway expansion and reconstruction. As shown in Figure 2, this new structure has been applied in the Wu-He expressway since 2019, which is proven to be practical, environmentally friendly, and economical.

Considering that the deck of the prefabricated frame bridge is adjacent to the original roadbed, the pounding between the deck and roadbed probably occurs under the earthquake ground motions, which will influence the seismic response to some extent. In this paper, the pounding effect is simulated by the Kelvin pounding model, and the pounding effect on the seismic response is investigated, accounting for two key parameters, including the gap distance and the seismic intensity. Then the pounding effect is assessed by comparing the seismic response of the two models, with and without consideration of pounding. The following trends can be concluded from the results:

(1) Pounding effects do not always exist under seismic excitation. When the pounding occurs, some seismic energy will be consumed, and the seismic response of the components will decrease;

(2) The smaller the gap distance and the higher intensity of the seismic excitation, the higher the pounding potential is. When the PGA is not higher than 0.2 g, the pounding effect can be ignored;
(3) The minimum negative deformation value will reach 104 mm at most when the pounding effect is considered, which is still much lower than the overlapping length between the bridge deck and the original roadbed, so the safety of the superstructure is ensured due to the cantilever state of the bridge deck is not occur.

The seismic response law and damage mechanism of prefabricated frame bridges are relatively complex, so theoretical analysis is not enough. However, experiments are lacking in this paper. For future studies, the corresponding shaking table test and pseudo-static test should be added to further verify and enrich the theoretical research results.

Author Contributions: Conceptualization, L.C.; Methodology, L.C.; Software, Y.W.; Validation, Y.W. and R.Z.; Formal analysis, T.S.; Investigation, J.Z. and T.S.; Resources, Y.Z.; Data curation, Y.W., J.Z., Y.Z. and T.S.; Writing—original draft, Y.W.; Writing—review & editing, L.C. and T.S.; Visualization, J.Z. and R.Z.; Supervision, L.C.; Project administration, J.Z. All authors have read and agreed to the published version of the manuscript.

Funding: This research was funded by the Anhui Province Natural Science Foundation of China grant number 2208085ME151.

Institutional Review Board Statement: Not applicable.

Informed Consent Statement: Informed consent was obtained from all subjects involved in the study.

Data Availability Statement: Data is unavailable due to privacy.

Conflicts of Interest: The authors declare no conflict of interest.

References

1. Zhong, J.; Shi, L.; Yang, T.; Liu, X.; Wang, Y. Probabilistic seismic demand model of UBPRC columns conditioned on Pulse-Structure parameters. *Eng. Struct.* **2022**, *270*, 114829. [CrossRef]
2. Chen, L.; Sun, T.; Hu, K.; Zhong, J. Probabilistic seismic assessment of a new elastoplastic column-deck joint on the prefabricated frame-bridge. *Structures* **2021**, *34*, 3099–3112. [CrossRef]
3. Hu, Z.; Wei, B.; Jiang, L.; Li, S.; Min, H. Track structural damage index for high-speed railway girder bridges considering residual deformations due to earthquake. *Bull. Earthq. Eng.* **2022**, *20*, 6587–6609. [CrossRef]
4. Zhong, J.; Wan, H.-P.; Yuan, W.; He, M.; Ren, W.-X. Risk-informed sensitivity analysis and optimization of seismic mitigation strategy using Gaussian process surrogate model. *Soil Dyn. Earthq. Eng.* **2020**, *138*, 106284. [CrossRef]
5. Orcesi, A.D.; Frangopol, D.M. Optimization of bridge maintenance strategies based on structural health monitoring information. *Struct. Saf.* **2011**, *33*, 26–41. [CrossRef]
6. *JTGT 2231-01-2020*; Specification for Seismic Design of Highway Bridges. China Communications Press: Beijing, China, 2020.
7. Jankowski, R. Earthquake-induced pounding between equal height buildings with substantially different dynamic properties. *Eng. Struct.* **2008**, *30*, 2818–2829. [CrossRef]
8. Crozet, V.; Politopoulos, I.; Yang, M.; Martinez, J.-M.; Erlicher, S. Sensitivity analysis of pounding between adjacent structures. *Earthq. Eng. Struct. Dyn.* **2018**, *47*, 219–235. [CrossRef]
9. Amjadian, M.; Agrawal, A.K. Rigid-Body Motion of Horizontally Curved Bridges Subjected to Earthquake-Induced Pounding. *J. Bridge Eng.* **2016**, *21*, 04016090. [CrossRef]
10. Li, N.; Xu, W.; Chen, Y.; Yan, W. Experimental research on adjacent pounding effect of midspan curved bridge with longitudinal slope. *Eng. Struct.* **2019**, *196*, 109320. [CrossRef]
11. Zhao, L.; Hao, H.; Bi, K.; Li, X. Numerical Study of the Seismic Responses of Precast Segmental Column Bridge under Spatially Varying Ground Motions. *J. Bridge Eng.* **2018**, *23*, 04018096. [CrossRef]
12. Won, J.-H.; Mha, H.-S.; Kim, S.-H. Effects of the earthquake-induced pounding upon pier motions in the multi-span simply supported steel girder bridge. *Eng. Struct.* **2015**, *93*, 1–12. [CrossRef]
13. Guo, A.; Li, Z.; Li, H.; Ou, J. Experimental and analytical study on pounding reduction of base-isolated highway bridges using MR dampers. *Earthq. Eng. Struct. Dyn.* **2009**, *38*, 1307–1333. [CrossRef]
14. Shrestha, B.; Hao, H.; Bi, K. Effectiveness of using rubber bumper and restrainer on mitigating pounding and unseating damage of bridge structures subjected to spatially varying ground motions. *Eng. Struct.* **2014**, *79*, 195–210. [CrossRef]
15. Dimitrakopoulos, E.G. Seismic response analysis of skew bridges with pounding deck–abutment joints. *Eng. Struct.* **2011**, *33*, 813–826. [CrossRef]

16. Kun, C.; Jiang, L.; Chouw, N. Influence of pounding and skew angle on seismic response of bridges. *Eng. Struct.* **2017**, *148*, 890–906. [CrossRef]
17. Li, B.; Bi, K.; Chouw, N.; Butterworth, J.W.; Hao, H. Effect of abutment excitation on bridge pounding. *Eng. Struct.* **2013**, *54*, 57–68. [CrossRef]
18. Chouw, N.; Hao, H. Significance of SSI and nonuniform near-fault ground motions in bridge response I: Effect on response with conventional expansion joint. *Eng. Struct.* **2008**, *30*, 141–153. [CrossRef]
19. Raheem, S.E.A. Pounding mitigation and unseating prevention at expansion joints of isolated multi-span bridges. *Eng. Struct.* **2009**, *31*, 2345–2356. [CrossRef]
20. Won, J.-H.; Mha, H.-S.; Cho, K.-I.; Kim, S.-H. Effects of the restrainer upon bridge motions under seismic excitations. *Eng. Struct.* **2008**, *30*, 3532–3544. [CrossRef]
21. Kun, C.; Yang, Z.; Chouw, N. Seismic performance of skewed bridges with simultaneous effects of pounding and supporting soil. *Eng. Struct.* **2018**, *174*, 26–38. [CrossRef]
22. Li, B.; Chouw, N. Experimental investigation of inelastic bridge response under spatially varying excitations with pounding. *Eng. Struct.* **2014**, *79*, 106–116. [CrossRef]
23. Huo, Y.; Zhang, J. Effects of Pounding and Skewness on Seismic Responses of Typical Multispan Highway Bridges Using the Fragility Function Method. *J. Bridge Eng.* **2013**, *18*, 499–515. [CrossRef]
24. Desroches, R.; Muthukumar, S. Effect of Pounding and Restrainers on Seismic Response of Multiple-Frame Bridges. *J. Struct. Eng.* **2002**, *128*, 860–869. [CrossRef]
25. Opensees Manual, O. *Open System for Earthquake Engineering Simulation User Command-Language Manual*; Pacifc Earthquake Engineering Research Centre University of California: Berkeley, CA, USA, 2009.
26. Linhai, H. *Concrete-Filled Steel Tubular Structures*; Science Press: Beijing, China, 2000.
27. *GB50017-2017*; Standard for Design of Steel Structures. China Architecture & Building Press: Beijing, China, 2017.
28. Yang, T.; Yuan, X.; Zhong, J.; Yuan, W. Near-fault Pulse Seismic Ductility Spectra for Bridge Columns Based on Machine Learning. *Soil Dyn. Earthq. Eng.* **2023**, *164*, 107582. [CrossRef]
29. Shamsabadi, A.; Rollins, K.M.; Kapuskar, M. Nonlinear Soil–Abutment–Bridge Structure Interaction for Seismic Performance-Based Design. *J. Geotech. Geoenviron. Eng.* **2007**, *133*, 707–720. [CrossRef]
30. Jankowski, R. Analytical expression between the impact damping ratio and the coefficient of restitution in the non-linear viscoelastic model of structural pounding. *Earthq. Eng. Struct. Dyn.* **2006**, *35*, 517–524. [CrossRef]
31. Bi, K.; Hao, H. Numerical simulation of pounding damage to bridge structures under spatially varying ground motions. *Eng. Struct.* **2013**, *46*, 62–76. [CrossRef]
32. Shi, Z.; Dimitrakopoulos, E.G. Comparative evaluation of two simulation approaches of deck-abutment pounding in bridges. *Eng. Struct.* **2017**, *148*, 541–551. [CrossRef]
33. Khatiwada, S.; Chouw, N.; Butterworth, J.W. A generic structural pounding model using numerically exact displacement proportional damping. *Eng. Struct.* **2014**, *62–63*, 33–41. [CrossRef]
34. Jankowski, R.; Wilde, K.; Fujino, Y. Pounding of superstructure segments in isolated elevated bridge during earthquakes. *Earthq. Eng. Struct. Dyn.* **2015**, *27*, 487–502. [CrossRef]
35. Jankowski, R.; Wilde, K.; Fujino, Y. Reduction of pounding effects in elevated bridges during earthquakes. *Earthq. Eng. Struct. Dyn.* **2015**, *29*, 195–212. [CrossRef]
36. Zhong, J.; Zheng, X.; Wu, Q.; Jiang, L.; He, M.; Dang, X. Seismic fragility and resilience assessment of bridge columns with dual-replaceable composite link beam under near-fault GMs. *Structures* **2023**, *47*, 412–424. [CrossRef]
37. Hu, Z.; Wei, B.; Jiang, L.; Li, S.; Yu, Y.; Xiao, C. Assessment of optimal ground motion intensity measure for high-speed railway girder bridge (HRGB) based on spectral acceleration. *Eng. Struct.* **2022**, *252*, 113728. [CrossRef]
38. Hu, Z.L.; Wei, B.; He, X.H.; Jiang, L.Z.; Li, S.S. Effects of spatial variation of ground motion (SVGM) on seismic vulnerability of ultra-high tower and multi-tower cable-stayed bridges. *J. Earthq. Eng.* **2022**, *26*, 8495–8524. [CrossRef]

sustainability

Article

Collision Analysis of Transverse Stops Considering the Vertical Separation of the Main Beam and Bent Cap

Wenjun An *, Lin Zhou, Ting Fang, Yiren Wu and Qi Li

School of Civil Engineering, Jiangxi University of Engineering, Xinyu 330046, China
* Correspondence: wjan@email.ncu.edu.cn

Abstract: In this study, a two-span continuous beam bridge is used as the model to calculate the theoretical solution of the lateral collision between the main beam and the retaining block under the vertical separation of the main beam and the bent cap, and the most unfavorable separation condition for the analysis is selected. The impact of the vertical separation of the bridge on the impact of the block is verified by calculating the lateral collision force of the block when the main beam and bent cap are separated. We calculate the changes in the vertical positions of the main beam and pier under multiple separation collision processes and select two characteristic parameters, the collision force and the collision position. According to the specifications, the impact of the structural collision caused by the separation conditions on the failure mode of the block is verified.

Keywords: small and medium bridge; near-fault earthquake; vertical separation; seismic responses; stopper damage

check for updates

Citation: An, W.; Zhou, L.; Fang, T.; Wu, Y.; Li, Q. Collision Analysis of Transverse Stops Considering the Vertical Separation of the Main Beam and Bent Cap. *Sustainability* **2023**, *15*, 2809. https://doi.org/10.3390/su15032809

Academic Editors: Kai Wei, Mingjin Zhang, Jian Zhong and Yutao Pang

Received: 30 November 2022
Revised: 30 January 2023
Accepted: 1 February 2023
Published: 3 February 2023

1. Introduction

Every year, earthquakes cause bridge damage, casualties, and property losses. Due to the significant difference in flexibility between the upper and lower structures of the bridge, seismic excitation can lead to large relative displacement of the structure, which will lead to beam falling failure of the bridge [1–4]. The impact between the main beam and the pier caused by the beam falling failure will cause secondary damage to the bridge. In the 2008 Wenchuan earthquake, many bridges fell, which hindered the rescue work [5]. To limit the excessive lateral relative displacement, restraining devices are usually set to prevent the beams from falling. Yang et al. [6–9] studied and developed the near-fault pulse seismic ductility spectrum based on machine learning, and suggested the seismic design of structures in near-fault areas.

Given the seismic damage forms for concrete blocks in previous earthquakes, to explore the failure mechanism of concrete blocks, Kazuhiko et al. [10,11] abroad, as well as Zheng et al. [12–16] in China, carried out a series of concrete blocks tests. It is proposed that the failure modes of concrete retaining blocks include inclined shear failure, horizontal shear failure, bending failure, and sliding failure, which are consistent with the failure modes that occur from actual earthquake damage. There are three main types of analytical models for retaining blocks that are commonly used in lateral seismic analyses of small- and medium-span beam bridges. The elastic block model mainly applies to the lateral collision analysis of bridges. The elastic–plastic block model applies to the seismic analysis of bridges with steel blocks, and [17–19] confirmed this mechanical model. The degenerated block model is mainly applicable to the concrete block structure, which is a highly recognized mechanical analysis model of concrete blocks. Based on a series of concrete block tests, Silva et al. [20] proposed for the first time that the contributions of concrete and reinforcements can be separated when calculating the block model parameters, thereby greatly simplifying the calculation of model parameters. Later, Goel and Chopra [21] simplified and improved the model and put forward an ideal mechanical analysis model of a trifold line stop. In recent years, a lot of progress has been made in the research on the structural forms of retaining blocks, such as the double-layer retaining block structure proposed by Wang et al. [22] and the X-shaped elastoplastic retaining block proposed by Li et al. [23]. The above studies used theory or experiments to analyze the failure mechanism of the blocks, and designed some new blocks based on the component form.

However, for the impact of lateral bridge blocks, the research is based on the structure always being in contact. However, some studies on near-fault vertical earthquakes indicate that vertical earthquakes may cause a separation collision coupling effect between the main beam and pier [24]. In the separation stage, the main beam and support are no longer connected, and the failure of the constraint results in a change in the dynamic response of the structure. Some studies have been carried out on pier damage caused by vertical separation [25,26]. However, the above research only considers the vertical and longitudinal two-way excitation, and the limiting effect of the lateral stop ignores the change in dynamic response caused by separation, which leads to the failure of the stop in a collision. The previous study pointed out that the separation may improve the piers.

Taking a double-span continuous beam bridge as an analysis model, this paper analyzes the lateral collision of the stop under the condition of vertical separation of the main beam and the bent cap. The bridge displacement is derived by the transient wave function method, and the extreme value of the first collision velocity under the separation condition is calculated. Used the indirect wave function expansion method and energy method to calculate the impact force of the bridge, the influence of the horizontal restraint failure of the bearing caused by the vertical separation of the main beam and the bent cap on the failure of the retaining block is verified. By calculating the displacement change and collision forces of piers with different heights, the influence of the pier height on the lateral collision of the retaining block is proposed when the vertical separation of the bridge is considered.

2. Bridge Calculation Model

To analyze the impact of vertical bridge separation on the block collision damage from near-field earthquakes, a two-span constant beam bridge is used in the study. The continuous beam bridge comprises four parts: the main beam, pier, bent cap, and bearing. The bent cap is set as a concentrated mass. The vertical displacement excitation of the bridge is D (t) and the lateral displacement excitation is B(t). The model diagram is shown in Figure 1. In the two-way dynamic calculation, the following assumptions are adopted for the analysis:

(1) Uniform excitation is adopted at the end of the main beam and pier, ignoring the possible traveling wave effect;
(2) In the process of collision, the structure is always in the elastic stage, ignoring the possible material damage to the structure;
(3) In the calculation of the bridge's dynamic response, both the vertical and horizontal ends of the main beam are assumed to be hinged and the bottom of the pier is assumed to be rigid;
(4) The dynamic response of the bent cap is ignored in the calculation of the vertical dynamic response. In the lateral analysis, the bent cap is assumed to be a concentrated mass;
(5) Two springs with different stiffness levels are used to simulate the compression elastic modulus and shear elastic modulus of the bearing, ignoring the damping of the bearing and the structure.

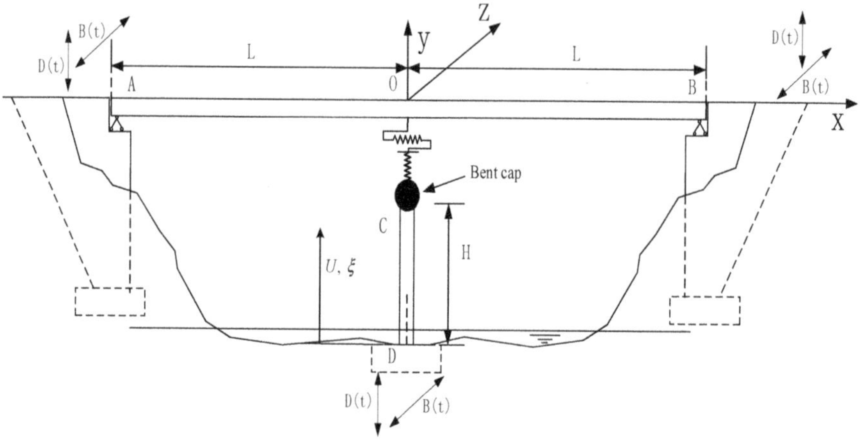

Figure 1. Bridge calculation model.

3. Bridge Lateral Displacement Response

3.1. Contact Stage

The wave functions of the bridge's lateral displacement response are as follows:

$$\begin{array}{l} OA: \frac{\partial^2(E_b I_b \partial^2 Z_1(x,t)/\partial x^2)}{\partial x^2} + \rho A_b \frac{\partial^2 Z_1(x,t)}{\partial t^2} = 0 \\ OB: \frac{\partial^2(E_b I_b \partial^2 Z_2(x,t)/\partial x^2)}{\partial x^2} + \rho A_b \frac{\partial^2 Z_2(x,t)}{\partial t^2} = 0 \\ CD: \frac{\partial^2(E_r I_r \partial^2 W(\xi,t)/\partial \xi^2)}{\partial \xi^2} + \rho A_r \frac{\partial^2 W(\xi,t)}{\partial t^2} = 0 \end{array} \tag{1}$$

wherein Z_1,Z_2 are the lateral displacements of the left and right spans of the main beam, respectively; W is the lateral displacement of the pier; ρ is the structural density; A_b is the main beam area; A_r is the pier area.

According to the assumptions of the previous model, it can be concluded that the boundary conditions of the bridge displacement are:

$$\begin{array}{c} Z_1(-L,t) = Z_2(L,t) = W(0,t) = B(t) \\ \frac{\partial^2 Z_1(-L,t)}{\partial x^2} = \frac{\partial Z_2(L,t)}{\partial x^2} = \frac{\partial W(0,t)}{\partial \xi} = 0 \end{array} \tag{2}$$

where *B(t)* is the lateral displacement excitation.

The continuity conditions in the middle of the main beam are:

$$Z_1(0,t) = Z_2(0,t) \quad \frac{\partial Z_1(0,t)}{\partial x} = \frac{\partial Z_2(0,t)}{\partial x} \quad \frac{\partial^2 Z_1(0,t)}{\partial x^2} = \frac{\partial^2 Z_2(0,t)}{\partial x^2} \tag{3}$$

The conditions of deformation continuity in the middle of the main beam, support, and cap beam are as follows:

$$\begin{array}{c} U(H,t) - Z_1(0,t) = \delta_1 = \frac{F_V}{K_V} = \frac{E_r I_r}{K_V} \frac{\partial^3 W(H,t)}{\partial \xi^3} \\ E_b I_b \left(\frac{\partial^3 Z_1(0,t)}{\partial x^3} - \frac{\partial^3 Z_2(0,t)}{\partial x^3} \right) = E_r I_r \frac{\partial^3 W(H,t)}{\partial \xi^3} = -\omega_n^2 M_a \frac{\partial^2 W(H,t)}{\partial t^2} \end{array} \tag{4}$$

where K_V is the shear stiffness of the bearing, $E_b I_b$ is the bending stiffness of the main beam, and $E_r I_r$ is the bending stiffness of the pier.

The lateral displacement response of the bridge structure can be expanded into the sum of the static displacement, rigid body displacement, and dynamic displacement:

$$\begin{aligned} Z(x,t) &= Z_s(x) + Z_g(x,t) + Z_d(x,t) \\ W(\xi,t) &= W_s(\xi) + W_g(\xi,t) + W_d(\xi,t) \end{aligned} \tag{5}$$

The subscripts s, g, and d represent the static, rigid, and dynamic displacements, respectively. The static displacement satisfies the continuity conditions and boundary conditions; the dynamic displacement satisfies the wave function, continuity condition, equilibrium differential equation, and force boundary condition.

The static displacements of the bridge are:

$$Z_{1s}(x,0) = Z_{2s}(x,0) = W_s(\xi,0) = 0 \tag{6}$$

The rigid body displacements of the bridge are:

$$Z_{1g}(x,t) = Z_{2g}(x,t) = W_g(\xi,t) = B(t) \tag{7}$$

The wave function of the dynamic bridge displacement can be obtained by substituting Equations (2)–(7) into Equation (1):

$$\begin{aligned} OA: &\quad \frac{\partial^2\left(E_b I_b \frac{\partial^2 Z_{1d}(x,t)}{\partial x^2}\right)}{\partial x^2} + \rho_b A_b \frac{\partial^2 Z_{1d}(x,t)}{\partial t^2} = -\rho A_b \frac{\partial^2 Z_{1g}(x,t)}{\partial t^2} \\ OB: &\quad \frac{\partial^2\left(E_b I_b \frac{\partial^2 Z_{2d}(x,t)}{\partial x^2}\right)}{\partial x^2} + \rho_b A_b \frac{\partial^2 Z_{2d}(x,t)}{\partial t^2} = -\rho A_b \frac{\partial^2 Z_{2g}(x,t)}{\partial t^2} \\ CD: &\quad \frac{\partial^2\left(E_r I_r \frac{\partial^2 W_d(\xi,t)}{\partial \xi^2}\right)}{\partial \xi^2} + \rho_r A_r \frac{\partial^2 W_d(\tau,t)}{\partial t^2} = -\rho A_r \frac{\partial^2 W_g(\xi,t)}{\partial t^2} \end{aligned} \tag{8}$$

The dynamic displacements of the bridge can be expanded as the superposition of the product of various modal functions and time functions:

$$\begin{aligned} Z_{1d}(x,t) = \sum_{n=1}^{\infty} \varphi_{nb1}(x) q_n(t),\ Z_{2d}(x,t) = \sum_{n=1}^{\infty} \varphi_{nb2}(x) q_n(t) \\ W_d(\xi,t) = \sum_{n=1}^{\infty} \varphi_{nr}(\xi) q_n(t) \end{aligned} \tag{9}$$

where $\varphi_{ib}(i = 1, 2)$ is the main beam wave mode function, φ_{nr} is the pier wave mode function, and q_{nt} is the time function.

The wave mode function of the lateral dynamic displacement of the main beam and pier can be expressed as:

$$\begin{aligned} \varphi_{nb1}(x) &= A_{n1} \sin k_{bn}x + B_{n1} \cos k_{bn}x + C_{n1}\sinh k_{bn}x + D_{n1} \cosh k_{bn}x \\ \varphi_{nb2}(x) &= A_{n2} \sin k_{bn}x + B_{n2} \cos k_{bn}x + C_{n2}\sinh k_{bn}x + D_{n2} \cosh k_{bn}x \\ \varphi_{nr}(\xi) &= A_{n3} \sin k_{rn}\xi + B_{n3} \cos k_{rn}\xi + C_{n3}\sinh k_{rn}\xi + D_{n3} \cosh k_{rn}\xi \end{aligned} \tag{10}$$

wherein k_{bn} and k_{rn} are the bending wave numbers of the bridge's lateral displacement, respectively, and A_{ni}, B_{ni}, C_{ni} and D_{ni} (i = 1, 2, 3) are the correlation coefficients of the modal wave function.

The boundary conditions of the wave mode function meet the following requirements:

$$\varphi_{nb1}(-L) = 0,\ \varphi_{nb2}(L) = 0,\ \varphi''_{nb1}(-L) = 0,\ \varphi''_{nb2}(L) = 0,\ \varphi_{nr}(0) = 0,\ \varphi'_{nr}(0) = 0 \tag{11}$$

The wave function continuity conditions of the bridge are:

$$\begin{aligned} \varphi_{nb1}(0) = \varphi_{nb2}(0),\ \varphi'_{nb1}(0) = \varphi'_{nb2}(0),\ \varphi''_{nb1}(0) = \varphi''_{nb2}(0) \\ \varphi_{nr}(H) = \varphi_{nb1}(0) + \frac{E_b I_b(\varphi'''_{nb1}(0) - \varphi'''_{nb2}(0))}{K_v} \\ E_b I_b(\varphi'''_{nb1}(0) - \varphi'''_{nb2}(0)) + E_r I_r \varphi'''_{nr}(H) = -\omega_n{}^2 M_a \varphi_{nr}(H) \end{aligned} \tag{12}$$

wherein M_a is the concentrated mass of the bent cap and ω_n is the transverse natural frequency of the bridge.

By introducing Equations (11) and (12) into Equation (10), it can be concluded that the wave mode functions are:

$$\begin{aligned} \varphi_{nb1}(x) &= A_{n1}(\sin k_{bn}x + \tan k_{bn}L\cos k_{bn}x - \mathrm{sinh}k_{bn}x - \mathrm{tanh}k_{bn}L\cosh k_{bn}x) \\ \varphi_{nb2}(x) &= A_{n1}(-\sin k_{bn}x + \tan k_{bn}L\cos k_{bn}x + \mathrm{sinh}k_{bn}x - \mathrm{tanh}k_{bn}L\cosh k_{bn}x) \\ \varphi_{nr}(\xi) &= M_{n1}A_{n1}(\sin k_{rn}\xi - \mathrm{sinh}k_{rn}\xi) + M_{n2}A_{n1}(\cos k_{rn}\xi - \cosh k_{rn}\xi) \end{aligned} \tag{13}$$

where M_{n1} and M_{n2} are wave-function-related parameters.

In the calculation of the structural dynamic response, it is required to meet the orthogonal consistency, which is:

$$\int_{-L}^{0} \rho A\varphi_{mb1}\varphi_{nb1}dx + \int_{0}^{L} \rho A\varphi_{mb2}\varphi_{nb2}dx + \int_{0}^{H} \rho A_r\varphi_{mr}\varphi_{nr}d\xi = \delta_{mn} \tag{14}$$

We can substitute Equation (13) into Equation (14) to obtain parameter A_{n1}.

The lateral natural frequency ω_n of the structure can be obtained using Equation (15):

$$\begin{aligned} &(1+\tfrac{M_a\omega_n{}^2}{K_V})[M_{n1}(\sin k_{rn}H - \mathrm{sinh}k_{rn}H) + M_{n2}(\cos k_{rn}H - \cosh k_{rn}H)] - \tan k_{bn}L + \mathrm{tanh}k_{bn}L \\ &= \tfrac{E_rI_rk_{rn}{}^3[M_{n1}(\cos k_{rn}H + \cosh k_{rn}H) + M_{n2}(-\sin k_{rn}H + \mathrm{sinh}k_{rn}H)]}{K_V} \end{aligned} \tag{15}$$

Through the Laplace transformation, we can find that the expansion of the time function $q_n(t)$ is:

$$\begin{aligned} q_n(t) = {} & e^{-\zeta\omega_n t}(q_n(0)\cos\omega_d t + \tfrac{\dot{q}_n(0)+\zeta\omega_n q_n(0)}{\omega_d}\sin\omega_d t) \\ & + \tfrac{1}{\omega_d}\int_0^t e^{-\zeta\omega_n\tau}\ddot{Q}_n(\tau)\sin(\omega_d(t-\tau))d\tau \end{aligned} \tag{16}$$

where ω_d is the circular frequency of the bridge structure under dampened vibration and ζ is the material damping of the bridge, where ζ = 5%.

3.2. Dynamic Response of the Bridge in the Separation Stage

When the vertical displacements of the main beam and pier are $Y(0,t_1) - U(H,t_1) = 0$ and $\dot{Y}(0,t_1) - \dot{U}(H,t_1) > 0$, respectively, the main beam presents a throw upstate at this time, and the main beam and pier have their own frequency of dynamic response.

Ignoring the bearing restraint, the main beam moves in the separation stage. The wave functions of girder AB and pier CD are:

$$\begin{aligned} AB: \tfrac{\partial^2(E_bI_b\partial^2\overline{Z}(x,t)/\partial x^2)}{\partial x^2} + \rho_bA_b\tfrac{\partial^2\overline{Z}(x,t)}{\partial t^2} = 0 \\ CD: \tfrac{\partial^2(E_rI_r\partial^2\overline{W}(\xi,t)/\partial\xi^2)}{\partial\xi^2} + \rho_rA_r\tfrac{\partial^2\overline{W}(\xi,t)}{\partial t^2} = 0 \end{aligned} \tag{17}$$

where $\overline{Z}(x,t)$ and $\overline{W}(\xi,t)$ are the lateral deformations of the main beam and pier, respectively, in the separated state.

In the separation stage, the lateral displacements of the bridge are:

$$\begin{aligned} \overline{Z}(x,t) = \overline{Z}_s(x) + \overline{Z}_g(x,t) + \overline{Z}_d(x,t) \\ \overline{W}(\xi,t) = \overline{W}_s(\xi) + \overline{W}_g(\xi,t) + \overline{W}_d(\xi,t) \end{aligned} \tag{18}$$

The calculation of the static displacement and rigid body displacement is consistent with the contact stage.

The dynamic displacement is:

$$\overline{Z}_d(x,t) = \sum_{n=1}^{\infty}\overline{\varphi}_{nb}(x)q_{nb}(t),\ \overline{W}_d(\xi,t) = \sum_{n=1}^{\infty}\overline{\varphi}_{nr}(\xi)q_{nr}(t) \tag{19}$$

Here, $\overline{\varphi}_{nb}$ and $\overline{\varphi}_{nr}$ are the wave mode functions of the main beam and pier in the separation stage, respectively.

The bending wave mode function of the main beam and pier is:

$$\begin{aligned} \overline{\varphi}_{nb}(x) &= A_{n4}\sin\overline{k}_{bn}x + B_{n4}\cos\overline{k}_{bn}x + C_{n4}\text{sinh}\overline{k}_{bn}x + E_{n4}\cosh\overline{k}_{bn}x \\ \overline{\varphi}_{nr}(\xi) &= A_{n5}\sin\overline{k}_{rn}\xi + B_{n5}\cos\overline{k}_{rn}\xi + C_{n5}\text{sinh}\overline{k}_{rn}\xi + E_{n5}\cosh\overline{k}_{rn}\xi \end{aligned} \tag{20}$$

where $A_{nj}, B_{nj}, C_{nj}, D_{nj}$ (j = 4,5) is the flexural wave modal function coefficient of the bridge, and $\overline{k}_{bn}$ and $\overline{k}_{rn}$ are the wave numbers of the main beam and pier at the separation stage, respectively.

The boundary conditions of the wave mode functions are consistent with Equation (11). It can be concluded that the wave mode functions of the main beam and pier are:

$$\begin{aligned} \overline{\varphi}_{nb}(x) &= \overline{A}_{nb}\sin\overline{k}_{bn}(x+L) \\ \overline{\varphi}_{nr}(\xi) &= \overline{A}_{nr}((\cosh\overline{k}_{rn}\xi - \cos\overline{k}_{rn}\xi) + M_{n3}(\text{sinh}\overline{k}_{rn}\xi - \sin\overline{k}_{rn}\xi)) \end{aligned} \tag{21}$$

$$\overline{k}_{bn} = n\pi/2L, \quad \overline{k}_{rn} = \sqrt{\omega_{nr}/c} \tag{22}$$

wherein M_{n3} is the wave-function-related parameter and c is the pier bending wave velocity; ω_{nr} is the natural automatic frequency of the pier, which can be solved by Equation (23):

$$E_r I_r \overline{k}_{nr}{}^3(1+\cos\overline{k}_{nr}H\cosh\overline{k}_{nr}H) = \omega_{nr}{}^2 M_a(\sin\overline{k}_{nr}H\cosh\overline{k}_{nr}H - \cos\overline{k}_{nr}H\text{sinh}\overline{k}_{nr}H) \tag{23}$$

The respective time functions q_{nb}(t) and q_{nr}(t) of the main beam and pier after separation are:

$$\begin{aligned} q_{nb}(t) = \; & e^{-\zeta\omega_{nb}t}(q_{nb}(t_1)\cos\omega_{bd}(t-t_1) + \tfrac{1}{\omega_{bd}}\dot{q}_{nb}(t_2)\sin\omega_{bd}(t-t_1)) \\ & + \tfrac{1}{\omega_{bd}}\int_{t_1}^{t} e^{-\zeta\omega_{nb}t}\ddot{Q}_{nb}(\tau)\sin\omega_{bd}(t-\tau)d\tau \end{aligned} \tag{24}$$

$$\begin{aligned} q_{nr}(t) = \; & e^{-\zeta\omega_{nr}t}(q_{nr}(t_1)\cos\omega_{rd}(t-t_1) + \tfrac{1}{\omega_{rd}}\dot{q}_{nb}(t_2)\sin\omega_{rd}(t-t_1)) \\ & + \tfrac{1}{\omega_{rd}}\int_{t_1}^{t} e^{-\zeta\omega_{nr}t}\ddot{Q}_{nr}(\tau)\sin\omega_{rd}(t-\tau)d\tau \end{aligned} \tag{25}$$

Here, ω_{bd} and ω_{rd} is the circular frequencies of the girder and pier under damped vibration, respectively.

3.3. Structure Collision

In the process of lateral bridge collision, compared with the main beam, the bending stiffness of the stop is more significant. During the lateral collision of the bridge, the bent cap is set as a rigid body, and the rubber cushion is designated as a spring (see Figure 2 for the crash model). The following assumptions are adopted in the calculation using the model in Figure 2:

(1) The structure is always in the elastic stage during the collision, ignoring the possible material damage;
(2) The deflection deformation caused by the collision of the stop and the rotation of the bent cap is ignored;
(3) The bent cap and pier are rigidly connected.

Through the Duhamel integral, the collision deformation of the bridge structure can be calculated as:

$$\begin{aligned} X_F(x,t) &= -\sum_{n=1}^{\infty}\overline{\varphi}_{nb}(x)\int_{t_{2k}}^{t^*}F_{nb}h_{nb}d\tau \\ W_F(\xi,t) &= \sum_{n=1}^{\infty}\overline{\varphi}_{nr}(\xi)\int_{t_{2k}}^{t^*}F_{nr}h_{nr}d\tau \end{aligned} \tag{26}$$

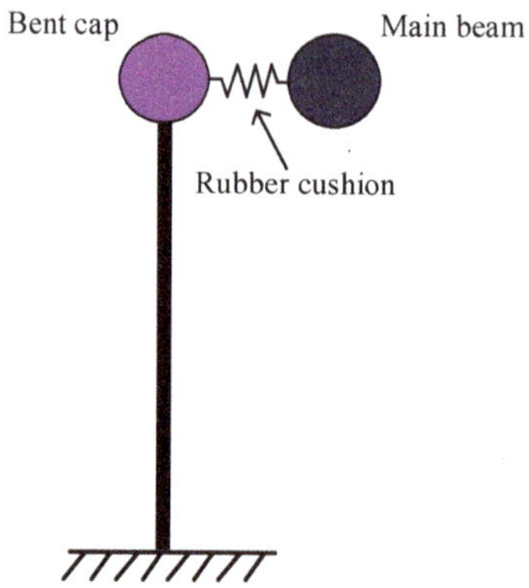

Figure 2. Lateral impact model.

The collision excitation response equals the convolution of its unit collision impulse response function and the generalized force. Here, $F_{nb} = F_k(t)\varphi_{nb}(x_0)$, $F_{nr} = F_k(t)\varphi_{nr}(\xi_0)$ are the generalized forces and $F_k(t)$ is the collision force; x_0 and ξ_0 coordinate the collision points of the main beam and pier, respectively. The positive and negative signs in Equation (26) represent the relationships between the force and the direction of displacement, respectively; h_{nb} and h_{nr} are the collision pulse response functions.

$$h_{nb} = \frac{1}{M_{nb}\omega_{nb}} \sin\omega_{nb}(t-\tau),\ \ h_{nr} = \frac{1}{M_{nr}\omega_{nr}} \sin\omega_{nr}(t-\tau)$$

where $M_{nb} = \int_{-L}^{L} \rho A_b \overline{\varphi}_{nb}^2(x)dx,\ \ M_{nr} = \int_0^H \rho A_r \overline{\varphi}_{nr}^2(\xi)d\xi + M_a$ are the modal masses of the bent cap and pier, respectively.

At the beginning of the collision, the relative vertical displacement of the main beam and pier gradually decreases, and the bearing is compressed and deformed. Therefore, $X(O,t) - U(H,t) = F_P(t)/K_c$ occurs in the collision process.

It can be calculated that:

$$\begin{aligned}
&\sum_{n=1}^{\infty} \overline{\varphi}_{nb}(0)\Big\{q_{nb}(t_2)\cos\omega_{nb}(t-t_2) + \tfrac{\dot{q}_{nb}(t_2)}{\omega_{nb}}\sin\omega_{nb}(t-t_2) \\
&\quad -\tfrac{1}{\omega_{nb}}\int_{t_2}^{t} F_p(t)\overline{\varphi}_{nb}(0)h_{nb}d\tau + \tfrac{1}{\omega_{nb}}\int_{t_2}^{t} \ddot{Q}_{nb}(\tau)\sin\omega_{nb}(t-\tau)d\tau\Big\} \\
&\quad -\sum_{n=1}^{\infty} \overline{\varphi}_{nr}(H)\sum_{n=1}^{\infty} \overline{\varphi}_{nb}(0)\Big\{q_{nr}(t_2)\cos\omega_{nr}(t-t_2) + \tfrac{\dot{q}_{nr}(t_2)}{\omega_{nr}}\sin\omega_{nr}(t-t_2) \\
&\quad -\tfrac{1}{\omega_{nr}}\int_{t_2}^{t} F_p(t)\overline{\varphi}_{nr}(H)h_{nr}d\tau + \tfrac{1}{\omega_{nr}}\int_{t_2}^{t} \ddot{Q}_{nr}(\tau)\sin\omega_{nr}(t-\tau)d\tau\Big\} \\
&\quad = F_p(t)/K_c
\end{aligned} \tag{27}$$

4. Impact Response

4.1. Bridge Model

The calculation model is a double-span continuous beam bridge. The main beam is a prestressed reinforced box girder, and the pier is the circular reinforced concrete column. The bent cap and main beam are connected by plate rubber (see Figure 3 for the sectional dimensions; all dimensions in the drawing are mm). The bridge block is 0.65 m high, 1.5 m thick, and 0.8 m wide, and C40 concrete is used.

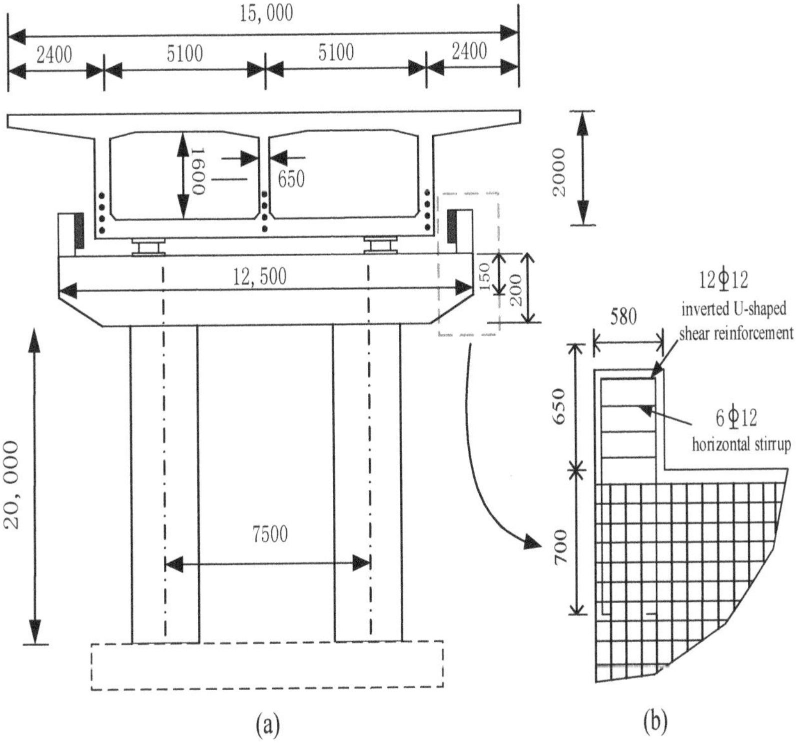

Figure 3. The bridge's structural dimensions: (**a**) cross-section; (**b**) stopper (unit: mm).

To verify the reliability of the natural frequency of the structure calculated by the theoretical solution in Formula (15), the calculation results of the theoretical solution and the finite element solution are compared and verified. The finite element model is as follows. The two ends of the main beam are articulated points with three-dimensional constraints, and the bottom of the pier adopts rigid joints. The BEAM188 unit is used for the main beam and pier, and the MASS21 unit is used for the capping beam. For the bridge bearings, different types of units are used in the vertical and longitudinal directions. The vertical direction is set as the LINK10 unit, and the height of the bearing is measured when it is not stressed. In the longitudinal direction, the COMBIN14 element is adopted, and the spring stiffness is the shear stiffness of the support. The bearing is bonded to the pier and connected to the main beam by an overlap. To simplify the model, the bridge structure is a single pier. From the comparison and analysis, the first-order vertical (transverse) natural vibration frequency of the finite element model is 23.6 (11.9) rad/s, which has a small error with the theoretical solution of 23.3 (10.9) rad/s. The natural frequency of the structure calculated by Formula (15) has a certain level of reliability.

Considering the influence of material damping on high-order vibration, the first five natural frequencies of the bridge are selected for the analysis in this paper. The vertical and horizontal natural frequencies of the bridge can be obtained through calculations, as shown in Table 1.

Table 1. Natural frequency of the bridge.

Natural Frequency	1	2	3	4	5
Vertical (rad/s)	23.3	40.8	58.2	75.4	92.9
Lateral (rad/s)	10.9	19	27.3	35.6	43.6

4.2. Influence of Separation on Bridge Displacement Response

For the response under the excitation of an actual seismic wave, the FFT analysis can be carried out using MATLAB to obtain the frequency, amplitude, and phase angle of each simple harmonic wave, and then the first n simple harmonics with the largest amplitude can be selected via sequencing to calculate the dynamic response of the bridge under the superimposed seismic excitation. Figure 4 shows some actual Kobe seismic waves.

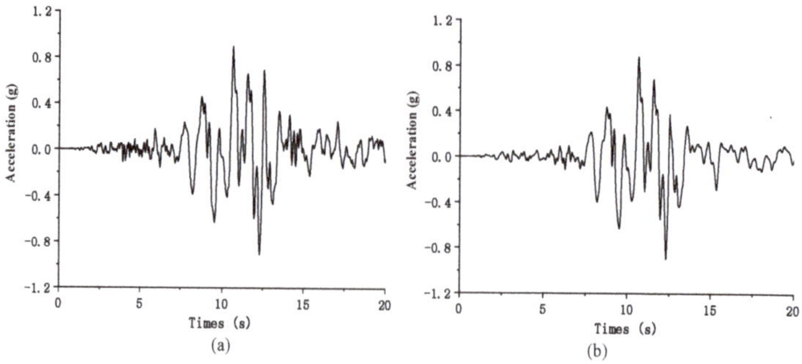

Figure 4. Partial recordings of Kobe seismic waves: (**a**) time history record; (**b**) artificial synthesis.

However, the use of an FFT usually results in high variance and bias, resulting in inaccurate spectrum estimation results. Some scholars have adopted multi-taper spectral estimations for stationary processes [27,28]. This study mainly studies the bridge responses under different frequencies and amplitude excitation; therefore, this paper uses a sine wave instead of a seismic wave for the analysis.

To study the impact of separation on the structural impact, the selected excitation period is close to the first vertical natural period of the bridge. Figure 5 shows the lateral dynamic response of the bridge when T = 0.25 s, where T is the period of seismic excitation. The seismic fortification is set as 8 degrees, and the excitation acceleration is 0.5 g. At this time, the vertical seismic excitation is ignored, and the structure is always in contact. In the analysis, the maximum lateral relative displacement of the middle of the main beam and the top of the pier is 29.56 mm, which is less than the 40 mm gap between the stop blocks, meaning the structures do not collide.

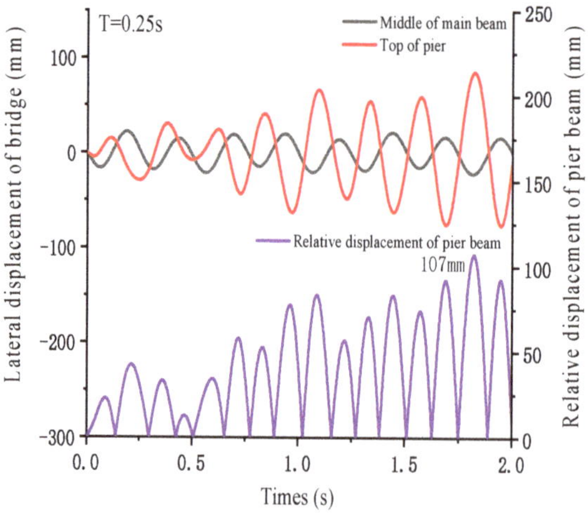

Figure 5. Lateral displacement of the bridge in the contact state.

When considering the separation condition, the selected excitation conditions are T = 0.25 and V/H = 0.67. Figure 6 shows the vertical displacement response of the bridge. When t = 0.71 s, the main beam and pier are vertically separated. It should be noted that the actual seismic wave can be expanded into multiple simple harmonic waves through a fast Fourier transform. Therefore, the initial vertical time and initial separation velocity have a certain randomness. The different initial vertical rate of the separation affects the separation–recontact process. In addition, for the actual seismic clock, the bridge may have multiple separation collision processes.

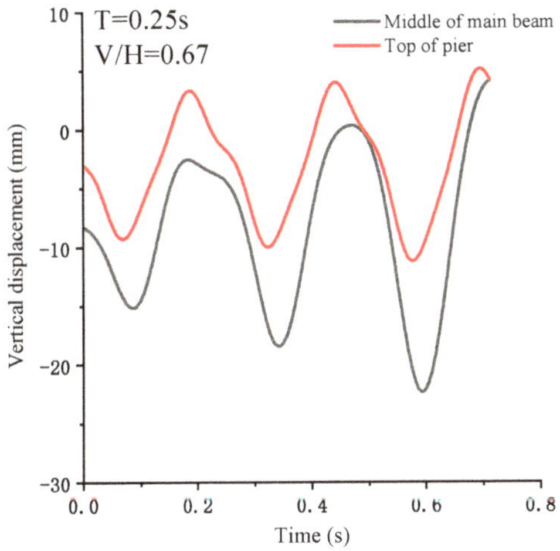

Figure 6. The bridge's vertical displacement.

To simplify the calculation, the time when the maximum relative displacement will occur in the lateral direction is selected as the initial separation time, and the separation of the main beam and the pier occurs in the collision phase. Equation (28) below provides the specific calculations:

$$\begin{aligned}\overline{W}_d(x,t) = &\left[\sum_{n=1}^{5}\overline{\varphi}_{nr}(\xi)q_{nr}(t_1)+\sum_{n=1}^{5}\frac{A_{nr}B_{nr}\omega_0^2B_0}{(\omega_0^2-\omega_{rd}^2)}\sin\omega_0t_1\right]e^{-\zeta\omega_{nr}t}\cos\omega_{nr}(t-t_1)\\ &+\frac{\left[\sum_{n=1}^{5}\overline{\varphi}_{nr}(\xi)\dot{q}_{nr}(t_1)+\sum_{n=1}^{5}\frac{A_{nr}B_{nr}\omega_0^2B_0}{\omega_{rd}(\omega_0^2-\omega_{rd}^2)}\omega_0\cos\omega_0t_1\right]}{\omega_{nr}}e^{-\zeta\omega_{nr}t}\sin\omega_{rd}(t-t_1)\\ &-\sum_{n=1}^{5}\frac{A_{nr}B_{nr}\omega_0^2B_0}{\omega_{rd}(\omega_0^2-\omega_{rd}^2)}\omega_{rd}\sin\omega_0t\end{aligned} \tag{28}$$

where A_{nr} and B_{nr} are the correlation coefficients of the wave mode function.

Figure 7 shows the lateral displacement of the bridge. It can be seen that the relative lateral displacement of the main beam and pier increases significantly after separation. When t = 1.77 s, the relative lateral displacement of the pier beam increases significantly from 107 mm to 204 mm.

4.3. Calculation of Collision Force

The first collision is selected for the analysis. When the pier beams are not separated, the lateral relative velocity of the main beam and the cover beam is 0.62 m/s, and the lateral impact force is 1.24 MN. When pier beam separation is considered, the lateral relative velocity and the lateral impact force are 1.42 m/s and 2.81 MN, respectively. The time–history response of the first impact force is shown in Figure 8.

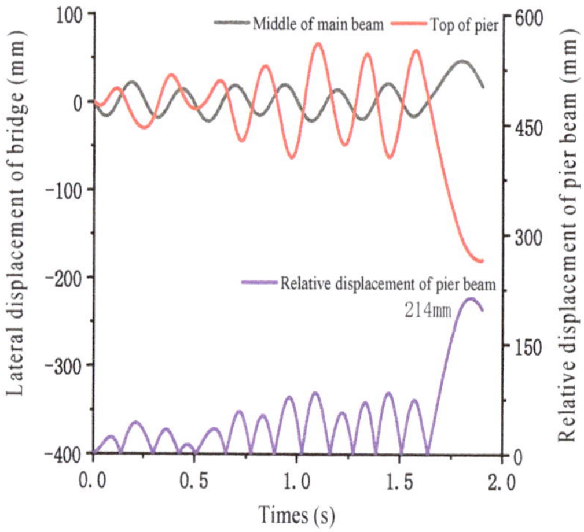

Figure 7. The bridge's lateral displacement.

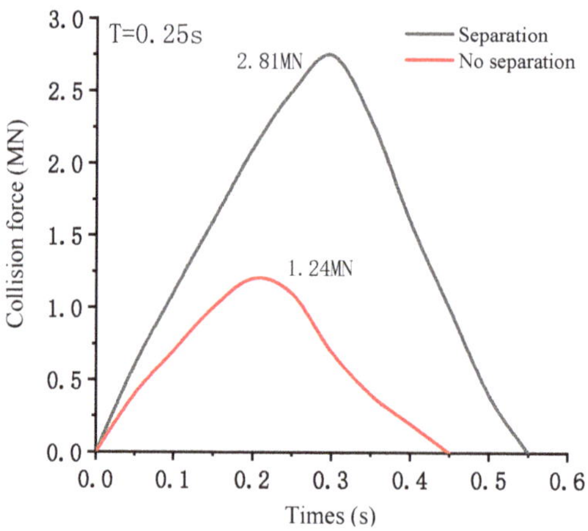

Figure 8. First collision force between the main beam and bent cap.

4.4. Stopper Damage

When the main beam collides with the stop, the stop may be damaged by tensile and compressive stresses and rotational friction. Equations (29) and (30) provide the calculations of the shear bearing capacity, while [29] provides the specific parameter meanings. Here, a is the height between the collision force position and the bottom of the stop when in contact. In the subsequent calculation, the position of the central axis of the collision contact surface between the main beam and the stop is taken for the analysis:

$$V_n = 0.2\sqrt{f_c{}'}A_{cv} + (A_{sh}f_{yh}g + \frac{n_h h^2 A_{ss} f_{ys}}{2s})\frac{1}{h+a} \tag{29}$$

$$V_n = 0.2\sqrt{f_c{}'}A_{cv} + A_s f_y(\mu_f \cos\theta_s + \sin\theta_s) \tag{30}$$

In the calculation, the angle of the inclined crack is taken as θ = 45°. To avoid the influence of the reinforcement ratio on the failure mode, we take $\frac{A_{sh}f_{yh}}{A_s f_y} = 1$, $\rho_{sv} = 0.91$. We follow [29] for the specific classification basis.

Comparing the different failure modes, it can be found that for tensile and compressive stress failures, the change in the collision position caused by the separation of the main beam and bent cap will affect the bearing capacity of the stop. In the time history of the seismic responses, the main beam and bent cap may have multiple separation coupling collisions. The 2 s displacement response is calculated here (see Figure 9 for the vertical displacement response). During the calculation, the main beam and bent cap are separated 6 times, and the maximum separation distance is 7.8 mm.

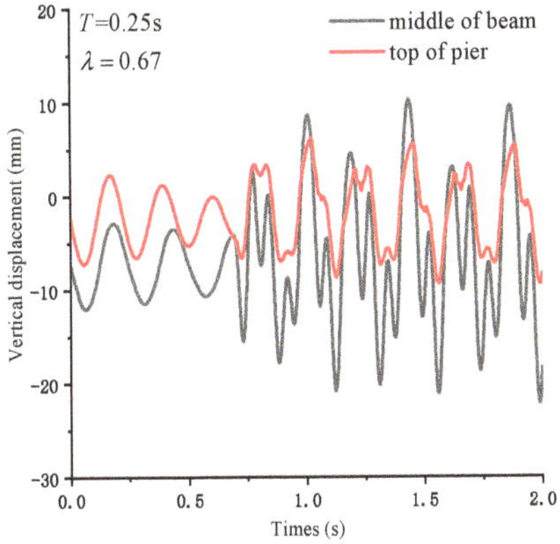

Figure 9. Vertical displacement response.

Under the two failure modes, the anti-bearing capacity levels of the retaining block show little difference at 2.37 MN and 2.4 MN, respectively. The evolution of the collision position between the main beam and the stop caused by separation has little influence on the bearing capacity of the stop.

Figure 10a shows the vertical displacement responses under different excitation acceleration rates. The value range of seismic excitation horizontal acceleration rates is $0.2\times g$~$1.0\times g$. The ratio of vertical acceleration to horizontal acceleration is 0.67. With the increase in vertical acceleration, the middle of the main beam and the top of the bent cap increase monotonically. Figure 10b shows the collision between the main beam and the stop. Comparing the two failure modes, it is concluded that the impact position change caused by the vertical seismic action has little impact on the bearing capacity of the retaining block. The change in lateral displacement caused by separation will significantly increase the risk of block failure. In the case of no separation, the stop will be damaged when the horizontal acceleration exceeds $1.0\times g$, while in the case of separation, the stop will be damaged when the acceleration is $0.4\times g$.

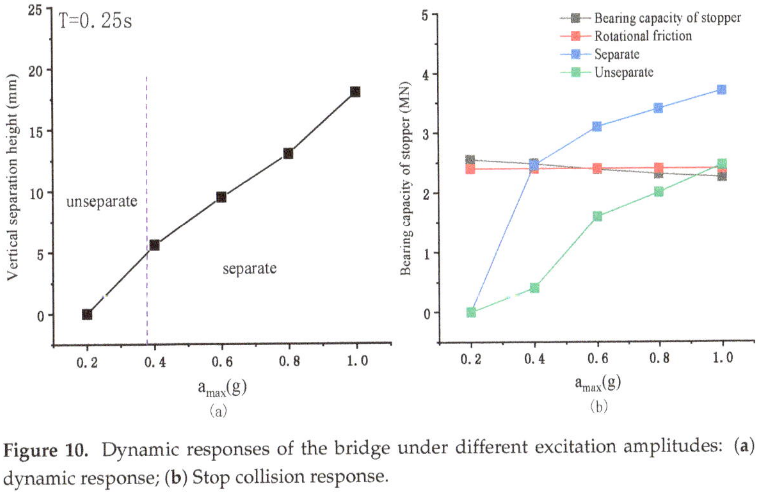

Figure 10. Dynamic responses of the bridge under different excitation amplitudes: (**a**) Vertical dynamic response; (**b**) Stop collision response.

5. Conclusions

In this paper, a double-span continuous beam bridge is selected as the analysis model and the lateral displacement response of the bridge is derived theoretically. By introducing the separation contact working condition, the extreme value of the longitudinal deformation of the pier under the separation condition is derived theoretically. Through the theoretical calculation of the changes in longitudinal deformation and the collision force of the piers under the conditions of separation or no separation, as well as the fluctuations of the relative vertical displacements of the main beam and capping beam caused by multiple separations, the following conclusions are drawn by comparing the specifications:

1. The excitation period is close to the vertical natural period of the bridge, which will cause the separation of the main beam and the bent cap. The separation condition may increase the initial collision speed and collision force of the main beam and the stop;
2. The change in the collision position of the main beam and the block of the vertical separation guide has little impact on the bearing capacity of the block, and the block's resistance to lateral collision is mainly based on the structural size and reinforcement;
3. At the same excitation frequency, with the increase in excitation duplication, the separation or lack of separation will increase the bridge's lateral displacement response and collision force. However, by considering the separation condition, the excitation acceleration required for failure is less than when the separation is ignored.

Author Contributions: Methodology, W.A.; Software, L.Z.; Investigation, T.F.; Writing—review & editing, Y.W.; Supervision, Q.L. All authors have read and agreed to the published version of the manuscript.

Funding: This research was funded by Jiangxi Provincial Department of Education Science and Technology [CJJ202915] and The APC was funded by [CJJ202908].

Institutional Review Board Statement: Not applicable.

Informed Consent Statement: Not applicable.

Data Availability Statement: The datasets used and analysed during the current study are available from the corresponding author on reasonable request.

Conflicts of Interest: The authors declare no conflict of interest.

References

1. Xiang, N.; Goto, Y.; Obata, M.; Alam, M.S. Passive seismic unseating prevention strategies implemented in highway bridges: A state-of-the-art review. *Eng. Struct.* **2019**, *194*, 77–93. [CrossRef]
2. Chouw, N.; Hao, H. Pounding damage to building and bridge in the 22 February 2011 Christchurch earthquake. *Int. J. Prot. Struct.* **2012**, *3*, 123–140. [CrossRef]
3. Yu, M.; Liu, B.; Wang, W. Damage characteristics of nearfault bridges. *J. Inst. Disaster Prev.* **2016**, *18*, 41–53.
4. Du, X.-L.; Han, Q.; Li, Z.-X.; Li, L.-Y. The seismic damage of bridge in the 2008 Wenchuan earthquake and lessons from its damage. *J. Beijing Univ. Technol.* **2008**, *34*, 1270–1279.
5. Lin, Q.; Lin, J.; Liu, J. A study on damage matrix of highway bridges based on Wenchuan earthquake investigation. *Earthq. Eng. Eng. Dyn.* **2018**, *38*, 118–126.
6. Yang, T.; Yuan, X.; Zhong, J.; Yuan, W. Near-fault Pulse Seismic Ductility Spectra for Bridge Columns Based on Machine Learning. *Soil Dyn. Earthq. Eng.* **2023**, *164*, 107582. [CrossRef]
7. Zhong, J.; Shi, L.; Yang, T.; Liu, X.; Wang, Y. Probabilistic seismic demand model of UBPRC columns conditioned on Pulse-Structure parameters. *Eng. Struct.* **2022**, *270*, 114829. [CrossRef]
8. Zhong, J.; Zheng, X.; Wu, Q.; Jiang, L.; He, M.; Dang, X. Seismic fragility and resilience assessment of bridge columns with dual-replaceable composite link beam under near-fault GMs. *Structures* **2023**, *47*, 412–424. [CrossRef]
9. Zhong, J.; Mao, Y.; Yuan, X. Lifetime seismic risk assessment of bridges with construction and aging consideration. *Structures* **2023**, *47*, 2259–2272. [CrossRef]
10. Kawashima, K.; Unjoh, S.; Hoshikuma, J.I.; Kosa, K. Damage of bridges due to the 2010 maule, chile, earthquake. *J. Earthq. Eng.* **2011**, *15*, 1036–1068. [CrossRef]
11. Kottari, A. Horizontal Load Resisting Mechanisms of External Shear Keys in Bridge Abutments. Ph.D. Thesis, University of California San Diego, San Diego, CA, USA, 2016.
12. Zheng, W.; Tang, G. Quasi-static test research on seismic blocks of bridges. *Technol. Highw. Transp.* **2013**, *4*, 54–58.

13. Xu, L.; Li, J. Seismic strength prediction of reinforced concrete retainers based on rigid body rotation model. *Eng. Mech.* **2014**, *31*, 143–150.
14. Han, Q.; Hu, M.H.; Wen, J.N.; Du, X.L. Seismic capacity evaluation of interior shear keys for highway bridges. *J. Earthq. Eng.* **2018**, *24*, 972–987. [CrossRef]
15. Han, Q.; Zhou, Y.; Ou, Y.; Du, X. Seismic behavior of reinforced concrete sacrificial exterior shear keys of highway bridges. *Eng. Struct.* **2017**, *139*, 59–70. [CrossRef]
16. Han, Q.; Zhou, Y.; Zhong, Z.; Du, X. Seismic capacity evaluation of exterior shear keys of highway bridges. *J. Bridge Eng.* **2016**, *22*, 04016119. [CrossRef]
17. Vasseghi, A. Energy dissipating shear key for precast concrete girder bridges. *Sci. Iran.* **2011**, *18*, 296–303. [CrossRef]
18. Li, J.; Xiang, N.; Tang, H.; Guan, Z. Shake-table tests and numerical simulation of an innovative isolation system for highway bridges. *Soil Dyn. Earthq. Eng.* **2016**, *86*, 55–70. [CrossRef]
19. Tang, Z. Experimental Study and Calculation Method on Bridge Seismic Stoppers. Master's thesis, Chongqing Jiaotong University, Chongqing, China, 2015.
20. Silva, P.F.; Megally, S.; Seible, F. Seismic performance of sacrificial exterior shear keys in bridge abutments. *Earthq. Spectra* **2009**, *25*, 643–664. [CrossRef]
21. Goel, R.K.; Chopra, A.K. Role of shear keys in seismic behavior of bridges crossing fault-rupture zones. *J. Bridge Eng.* **2008**, *13*, 398–408. [CrossRef]
22. Wang, K.H.; Hui, Y.X.; Wu, G. Study on seismic performance of seismic double-layer stopper. *Earthq. Eng. Eng. Dyn.* **2014**, *34*, 505–510.
23. Li, J.; Tang, H.; Guan, Z. A new isolation system for small and medium span bridges on laminated rubber bearings. *China J. Highw. Transp.* **2015**, *28*, 35–43.
24. Yang, H.B.; Yin, X.C.; Hao, H. Theoretical Investigation of Bridge Seismic Responses with Pounding under Near-Fault Vertical Ground Motions. *Adv. Struct. Eng.* **2015**, *18*, 452–468. [CrossRef]
25. An, W.; Song, G.; Chen, S. Near-Fault Seismic Response Analysis of Bridges Considering Girder Impact and Pier Size. *Mathematics* **2021**, *9*, 704. [CrossRef]
26. An, W.; Song, G. Influence of Bearing on Pier Failure Considering the Separation Condition under Near-Fault Earthquake. *Symmetry* **2021**, *13*, 692. [CrossRef]
27. Huang, Z.; Xu, Y.-L. A Multi-Taper S-Transform Method for Spectral Estimation of Stationary Processes. *IEEE Trans. Signal Process.* **2021**, *69*, 1452–1467. [CrossRef]
28. Zhang, Y.-M.; Huang, Z.; Xia, Y. An adaptive multi-taper spectral estimation for stationary processes. *Mech. Syst. Signal Process.* **2023**, *183*, 109629. [CrossRef]
29. *CJJ 166-2011*; Code for Seismic Design of Urban Bridges. Ministry of Housing and Urban-Rural Development of the People's Republic of China: Beijing, China, 2011.

sustainability

MDPI

Article

Failure Mode Analysis of Bridge Pier Due to Eccentric Impact Based on Separation of Pier and Beam

Wenjun An *, Lin Zhou, Meilan Kang, Hailin Yang and Lanyan Mo

School of Civil Engineering, Jiangxi University of Engineering, Xinyu 330046, China
* Correspondence: wjan@email.ncu.edu.cn

Abstract: By considering the near-field vertical seismic spectrum and calculating the change in vertical contact force between the main beam and the pier, the possible vertical separation contact condition of a bridge is deduced. By calculating the extreme value of the pier–beam vertical contact force and the longitudinal deformation of the pier under the structural separation, the influence of the separation on the failure of the pier is determined. Separation increases the risk of pier failure under compression, bending, and shear, and different separation times lead to different longitudinal responses from the pier, and the first failure mode is different. Therefore, it is of great significance to reasonably design bridges near faults.

Keywords: bridge; near-fault earthquake; vertical separation; eccentric impact; failure mode

Citation: An, W.; Zhou, L.; Kang, M.; Yang, H.; Mo, L. Failure Mode Analysis of Bridge Pier Due to Eccentric Impact Based on Separation of Pier and Beam. *Sustainability* **2023**, *15*, 1435. https://doi.org/10.3390/su15021435

Academic Editors: Kai Wei, Mingjin Zhang, Jian Zhong and Yutao Pang

Received: 7 December 2022
Revised: 5 January 2023
Accepted: 11 January 2023
Published: 12 January 2023

1. Introduction

As the main bearing component of a bridge, damage to the pier not only destroys traffic, but also affects rescue after a disaster [1,2]. Near-fault earthquakes not only have great velocity pulse effects; in addition, the proportion of vertical earthquake and horizontal earthquake exceeds the standard 2/3. These seismic features can affect the failures of piers [3,4].

Many scholars have carried out a series of studies on bridge structure damage caused by near-fault earthquakes. Jiang and Yang [5,6] used the vulnerable surface model to study the sensitive regions under different damage states and discuss the mechanism of structural brittleness changes caused by earthquake damage. Furthermore, many scholars have studied the influence of vertical seismic action on bridge structures. Sung [7] took the Santa Monica bridge as the research object and compared the experiment with the Zeus-NL hybrid simulation method to study the influence of axial force changes on the bridge structure. Rodrigues [8] applied different two-way horizontal loads and variable axial forces to reinforced concrete columns through experiments, and the test results showed that the horizontal deformation of concrete columns under irregular axial forces was less than under constant pressures. Quantitative studies have been carried out on the effects of tomography on complex structural anomalies [9–11]. Zuo and Yang [12,13] studied and developed a near-fault pulse seismic ductility spectrum based on machine learning and put forward suggestions for the seismic design of structures in near-fault areas. Zhong [14–16] analyzed the vulnerability of bridges through algorithms considering a near-field earthquake, especially pulse excitation. The above studies considered the situation of bridge impairment under bidirectional excitation for different models and obtained some results. However, for vertical seismic excitation, the numerical value specified in the code is more used for analysis, ignoring the large vertical excitation in the near-fault area.

In China, most girder bridges adopt rubber bearings, which lack tensile capacity. A near-fault vertical earthquake can cause the separation of the girder and the support. For the bridge collision, the current stage mainly considers the transverse adjacent beam collision [17], while the research on the vertical collision is less. The vertical collision is close to the beam rod collision. Ruan [18] analyzed the dynamic behavior of a moving free beam impacting the tip of a cantilever beam using the modal approximation technique; however, this collision condition belongs to a single collision, and the actual situation is that the pier and beam have multiple separation contacts. Yang [19] used theoretical methods to solve the vertical impact force of the pier beam under multiple separation conditions. However, Yang's study only considered the vertical excitation and ignored the horizontal displacement of the bridge caused by the separation, and the analysis of collision process takes the structure as a whole. When the compressive stiffness of the bearing is low, the calculated force is unreasonable. The authors of [20,21] analyzed the influence of separation on the failure of a bridge pier under vertical and longitudinal excitation, but did not provide a qualitative discussion on separation conditions and only considered the flexural failure of the bridge pier.

In this study, the influences of pier–beam separation on bridge force and the displacement response are calculated by establishing a reliable theoretical method for dynamic bridge response. By using the mode superposition method [22], we calculate the limit solution of the vertical impact force of the pier beam and the longitudinal deformation of the pier top after the first pier–beam separation. Through calculation, we analyze the bridge separation under different excitation frequencies and amplitudes, and study the influence of dynamic response changes caused by separation on the failure mode of the bridge pier.

2. Structural Model

This paper mainly studies the separation of a pier and a beam and the impact of the separation on pier failure, which lays a foundation for the study of the combined influence of multispan structures in the future. Therefore, a double-span prestressed box girder bridge was selected in this paper, as shown in Figure 1. Both ends of the main beam are hinged with the ground, and the pier bottom is fixed. The displacements of horizontal excitation and vertical excitation are *D*(t) and *B*(t), respectively. The axial stiffness of the bearing is K_C and the shear stiffness is K_V. Three assumptions have been made, as follows:

(1) It is assumed that the structure is permanently elastic, and the plastic effect is ignored;
(2) It is assumed that the bridge is rigidly connected to the ground, and the coupling effect of soil and foundation is ignored;
(3) Vertical and longitudinal dynamic responses are calculated separately, ignoring the coupling effect between them.

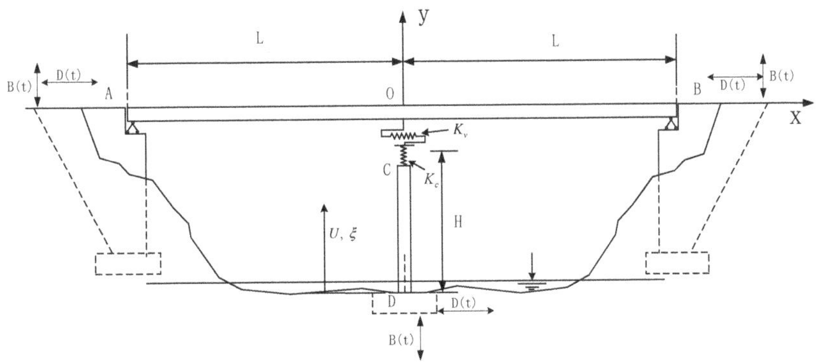

Figure 1. Theoretical solution model of the bridge structure.

3. Bridge Dynamic Response Calculation

3.1. Vertical Dynamic Response

The vertical dynamic response of the bridge refers to [20,21]. Near-fault earthquakes may contain pulse-type earthquakes, which transfer most of the energy of the ground motion to structures in a short time. The impulse excitation is usually expressed as the superposition of two or more simple excitations, including rectangular excitation, sinusoidal excitation, and triangular excitation, as shown in Figure 2. To study the influence of different forms of pulse excitation on the vertical dynamic response of a bridge under vertical pulse excitation, the impulse of three kinds of pulse excitation is the same.

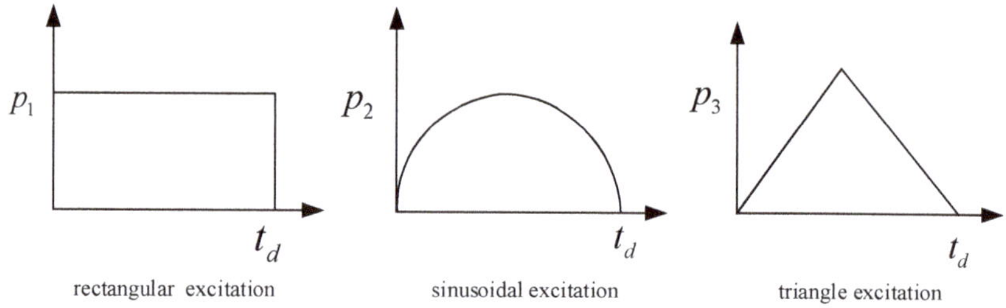

Figure 2. Acceleration curves of different pulse excitations.

When the impulse generated by the three kinds of pulse excitation is the same, the ratio of the peak force is $p_1 : p_1 : p_1 = 2/\pi : 1 : 4/\pi$. For different levels of pulse excitation, $\ddot{B}(\tau)$ is

$$\ddot{B}(\tau) = \begin{cases} -\frac{2}{\pi}\alpha_{\max} & \text{rectangular excitation} \\ \alpha_{\max} \times \sin\omega_0\tau & \text{sin usoidal excitation} \\ -\frac{8\alpha_{\max}\tau}{\pi t_d} & \text{triangle excitation}, 0 < \tau \le 0.5t_d \\ -\frac{8\alpha_{\max}}{\pi t_d}(t_d - \tau) & \text{triangle excitation}, 0.5t_d < \tau \le t_d \end{cases} \quad (1)$$

Figure 3 shows the relationship between pier–beam vertical contact force and excitation conditions. The range of excitation duration is $0.25 \sim 2T_1$, and the range of excitation acceleration amplitude is 0.25~1.0 g. The value T_1 is the vertical first-order natural period of the bridge. Under the rectangular pulse excitation, the contact force changes the least. When the pulse excitation time exceeds T_1, the contact force remains unchanged, and the minimum contact force is 8.6 MN. Under triangular pulse excitation, the vertical dynamic response of the pier beam increases monotonically with the increase in excitation duration and amplitude of excitation acceleration. When the excitation duration is $2T_1$ and the excitation acceleration amplitude is 1.0 g, the minimum contact force is 4.3 MN.

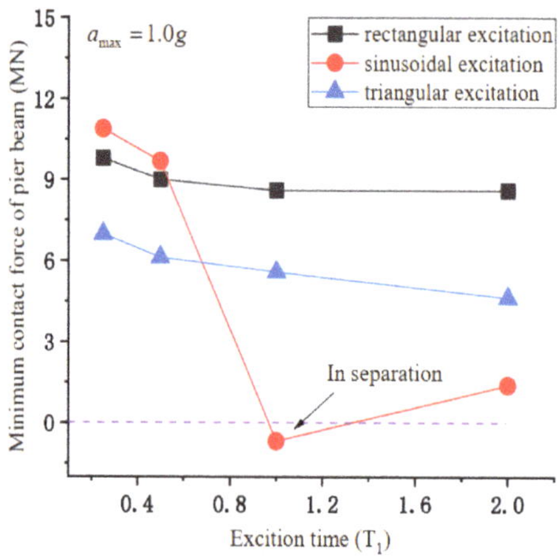

Figure 3. The minimum force of pier beam under different vertical pulse excitations.

Under sinusoidal pulse excitation, the length of the pulse excitation and the amplitude of the excitation acceleration have more influence on the vertical dynamic response than rectangular and triangular pulse excitation. When the pulse excitation duration is T_1, the vertical contact force is the lowest, which is because the vertical pulse excitation frequency is equal to the natural vertical frequency at this time. The resonance causes the vertical dynamic response of the bridge to surge. When the pulse excitation duration is T_1, the vertical contact force is −0.58 MN, which means that the vertical separation of the pier beam occurs at this time.

According to the vertical dynamic responses of different pulse excitation forms, it can be concluded that the vertical separation is most easily caused by sinusoidal pulse excitation. However, a single sinusoidal pulse causes a significant excitation acceleration requirement for the pier–beam separation, which exceeds the specification. At the same time, the pulse excitation lasts for a short time, while the periodic excitation can last for a long time. When the excitation frequency is close to the natural vertical frequency, there is a large dynamic response. Next, the dynamic response of the bridge is analyzed theoretically by simple harmonic excitation.

For the near-fault seismic response, the maximum value of the acceleration response spectrum is concentrated in the range T = 0.1~0.5 s. The high-order frequency has little influence on the excitation response of the structure. According to the conventional engineering calculation, the first five orders of the modal truncation of the dynamic response are taken.

Figure 4 shows the separation of the piers and beams under different excitation frequencies and accelerations. When the ω_1 (vertical first-order natural frequency) is close to the ω_0 (excitation frequency), a slight vertical acceleration results in pier–beam separation. When the ω_0 deviates from the ω_1, the vertical excitation acceleration required for the separation increases significantly. When ω_0 = 40 rad/s, the vertical excitation acceleration needs to reach 2.2 g to cause pier–beam separation. When the ω_0 is close to the ω_2 (vertical second-order natural frequency), the smaller excitation acceleration also causes pier–beam separation. Compared with the separation interval near the ω_1, the separation interval of the second-order vertical natural frequency is narrower. The excitation acceleration required for the separation is greater, essentially exceeding the standard value. The separation of the bridge piers and beams mainly occurs when $\omega_0 \approx \omega_1$.

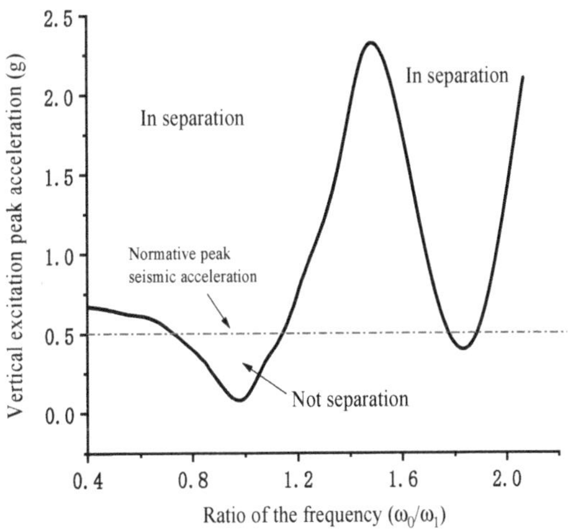

Figure 4. Vertical separation conditions of pier beam.

When pier and beam separation occurs, the natural frequency of the structure changes, resulting in a different dynamic response from when there is no separation. When the pier and beam collide in the vertical direction, there may be a large deviation in the longitudinal direction of the pier top. The longitudinal deformation of the pier top directly affects the degree of eccentric vertical collision. Therefore, it is necessary to analyze the extreme value of longitudinal pier deformation under the condition of pier–beam separation.

3.2. Longitudinal Dynamic Response

For the calculation of the longitudinal dynamic response of the bridge, refer to [20,21], where the time of the longitudinal separation is taken as the time of the vertical separation. In actual earthquakes, there is a time difference between horizontal and vertical earthquakes due to their different transmission wave velocities. In addition, many separation–collision phenomena occur during the seismic excitation time. By referring to the longitudinal displacement solution in [20], the theoretical solution of the longitudinal displacement response of a bridge pier after separation can be obtained through calculation, as shown in Equation (2). Because the natural frequency of the pier decreases after separation, the longitudinal deformation of the structure increases. Considering the phase difference, the extreme value of the longitudinal deformation of the bridge pier after separation is the sum of the extreme value of the initial state (the first two terms) and the excitation state (the third term).

$$
\begin{aligned}
\overline{W}_d(x,t) = {} & \left[\sum_{n=1}^{5} \overline{\varphi}_{nr}(\xi) q_{nr}(t_1) + \sum_{n=1}^{5} \frac{A_{nr} B_{nr} \omega_0^2 B_0}{(\omega_0^2 - \omega_{nr}^2)} \sin \omega_0 t_1 \right] \cos \omega_{nr}(t - t_1) \\
& + \frac{\left[\sum_{n=1}^{5} \overline{\varphi}_{nr}(\xi) \dot{q}_{nr}(t_1) + \sum_{n=1}^{5} \frac{A_{nr} B_{nr} \omega_0^2 B_0}{\omega_{nr}(\omega_0^2 - \omega_{nr}^2)} \omega_0 \cos \omega_0 t_1 \right]}{\omega_{nr}} \sin \omega_{nr}(t - t_1) \\
& - \sum_{n=1}^{5} \frac{A_{nr} B_{nr} \omega_0^2 B_0}{\omega_{nr}(\omega_0^2 - \omega_{nr}^2)} \omega_{nr} \sin \omega_0 t
\end{aligned} \tag{2}
$$

Here, $\overline{\varphi}_{nr}$ is the pier wave mode function in the separation phase, q_{nt} is the pier time function in the separation phase, ω_0 is the excitation frequency, ω_{nr} is the pier natural frequency in the separation phase, and A_{nr} and B_{nr} are the correlation coefficients of the wave mode function.

The first part vibrates at the natural frequency of the separated pier, and the amplitude is not affected by the frequency. By calculation, the maximum absolute deformation value is 16.7 mm. The second part vibrates with the natural frequency of the separated pier, and the amplitude is affected by the frequency. By calculation, the maximum absolute deformation value is 32.6 mm. The maximum deformation obtained by the first two calculations is 34.5 mm. The third part vibrates at the excitation frequency. The amplitude is affected by the frequency, and the maximum value is 8.2 mm. Considering the different vibration frequencies of the first, second, and third terms, when considering the separation, the absolute value of the maximum deformation at the top of the pier increases to 42.7 mm, an increase of 120%. This situation is caused by the numerical difference between the ω_{nr} and the ω_n after separation.

Figure 5 shows the longitudinal deformation of the pier top under different excitation frequencies. When the peak excitation acceleration is 0.2 g, the pier and beam are always in contact, and the maximum deformation of the pier top occurs in the interval $\omega_1 = \omega_0$. When the peak excitation acceleration of the excitation is 0.5 g and 0.8 g, vertical excitation causes pier–beam separation. When $\omega_0 \approx \mu_1$, the peak value of the pier roof deformation appears; μ_1 is the first longitudinal natural frequency of the structure.

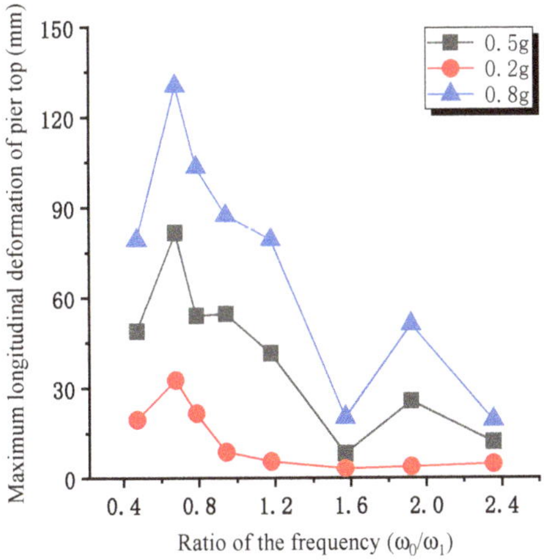

Figure 5. Longitudinal relative displacements of pier beams during different excitation periods.

3.3. Dynamic Response during a Collision

In the previous study, the vertical impact force of the structure was calculated using the model in Figure 6. In the actual collision process, the pier is in a longitudinal bending state. The eccentric distance between the vertical collision position of the pier beam and the pier axis is Δx. To simplify the calculation, the structural collision model is simplified as Figure 6b. In the process of a vertical collision, the structure is affected by three forces: the vertical collision force, the bearing shear force, and the inertial force of the forced vibration of the piers. The indirect mode method [22] is adopted for the calculation, and the change in the bending strain energy of the bridge pier caused by the initial offset of the bridge pier during collision is ignored.

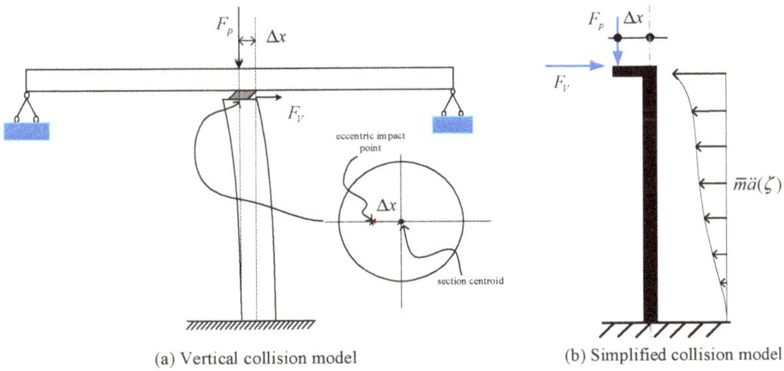

Figure 6. Vertical collision calculation model.

When the longitudinal deformation of the pier reaches its extreme value, the inertial force of the pier is very small; therefore, it is not urgent to ignore it here. During eccentric collision, the kinetic energy and strain energy of the main beam are converted into the axial compression strain energy of the bearing, and to the bending strain energy of the main beam and the pier. The kinetic energy and bending strain energy of the main beam at the time of collision are

$$\begin{aligned} T_b &= \frac{1}{2}\int_{-L}^{L} \rho A_b \left[\sum_{n=1}^{5} \overline{\varphi}_{nb}(x)\dot{q}_n(t_1) \right]^2 dx \\ U_b &= \frac{1}{2}\int_{-L}^{L} E_b I_b \left[\sum_{n=1}^{5} \frac{\partial^2 \overline{\varphi}_{nb}(x)}{\partial x^2} q_n(t_1) \right]^2 dx \end{aligned} \tag{3}$$

where t_1 is the initial time of the collision between the main beam and the support.

The vertical collision diagram of the bridge is shown in Figure 7. During the collision, the strain changes of the bridge include the bending strain energy of the main beam and the bending strain energy of the pier.

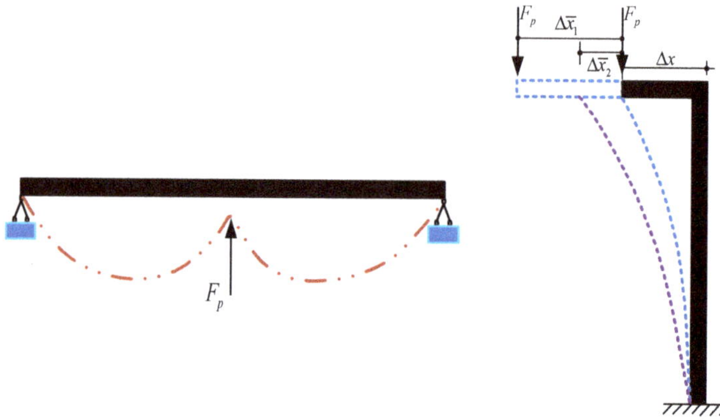

(a) Diagram of main beam deformation (b)Diagram of pier deformat

Figure 7. Deformation of vertical impact bridge.

The strain energy generated by the self-weight of the main beam during the collision process is

$$W_b = \frac{L^3}{E_b I_b}\left(\frac{2q^2L^2}{15} - \frac{F_p{}^2}{24}\right) \tag{4}$$

The strain energy generated by the self-weight of the main beam during the collision process is

$$W_r = \frac{H}{6E_r I_r}\left(3F_p{}^2\eta^2\Delta x^2 - F_v{}^2H^2\right) \tag{5}$$

where η is the eccentric impact amplification coefficient; for the coefficient value, we refer to [21]. The value K_v is the bearing shear force.

According to the energy conservation in the collision process, it can be concluded that

$$T_b + U_b + W_b = \frac{1}{2}K\Delta y^2 + W_r \tag{6}$$

where $K = \frac{K_1K_2}{K_1+K_2}$. K_1 is the bearing axial compression stiffness and K_2 is the axial stiffness of the pier.

Under specific excitation conditions, piers and beams are separated and vertical collisions occur. The first collision is selected here for study. Figure 8 is the calculation flow chart of the vertical displacement response of the bridge under vertical excitation. When pier–beam separation does not occur, the calculation ends when the calculation time is 2 s. When the vertical contact force of the pier beam is zero, the analysis ends.

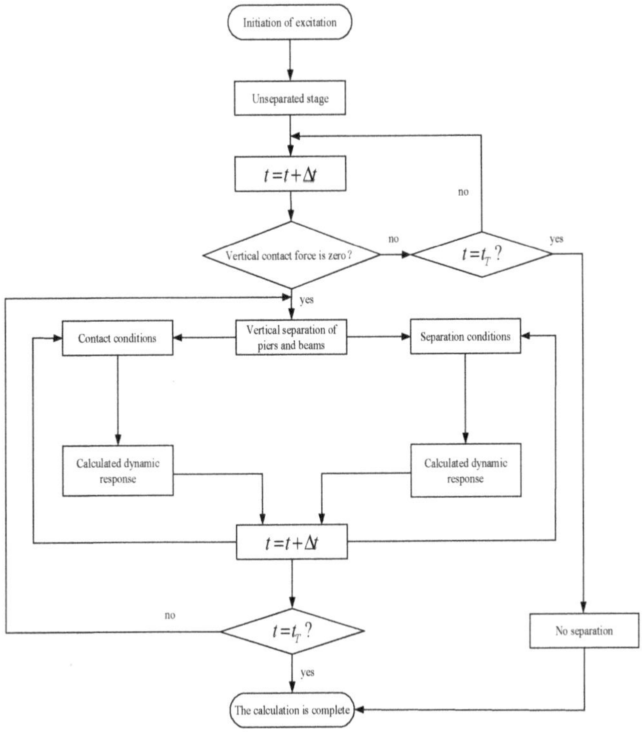

Figure 8. Flow chart of the bridge vertical displacement calculation program.

To ensure the accuracy of the calculation, an appropriate time-step increment must be selected. For the choice of time-step increment, since it is necessary to clearly express the transmission characteristics of the bending wave and the longitudinal wave in the main beam and pier, the time-step increment should not be too large. The longitudinal wave velocity of the pier is $c_r = \sqrt{E_r/\rho_r}$ = 3492 m/s, and the bending wave velocity of the pier is $a_r = \sqrt{E_r I_r/\rho_r A_r}$ = 1060 m/s. The maximum time-step increment is less than the time taken by the bending wave and the longitudinal wave to pass through the whole pier. A time-step increment of 0.001 s was selected for the calculation.

Figure 9 shows the change in the vertical collision force in the process of the first impact of the pier calculated by the indirect mode superposition method [2]. The contact force fluctuates with the first peak of 28.92 MN and the approximate maximum impact force of 30.84 MN. The maximum value does not appear in the first-wave peak because the collision process is also affected by the vertical excitation force vibration.

Figure 10 shows the vertical impact force under different levels of vertical peak excitation acceleration and excitation frequency. From the analysis, we concluded the following: (1) the maximum impact force of the pier beam increases monotonically with the increase in vertical excitation acceleration, and the slight fluctuation in the middle is due to the fact that the upthrow velocity in the middle of the main beam has a certain contingency at the vertical separation moment; (2) the collision mainly occurs when $\omega_0 \approx \omega_1$, and the closer the two frequencies, the greater the collision force; (3) when the excitation frequency/natural vertical frequency is <0.8 or >1.2, the change in excitation frequency has little influence on the maximum impact force of the pier beam.

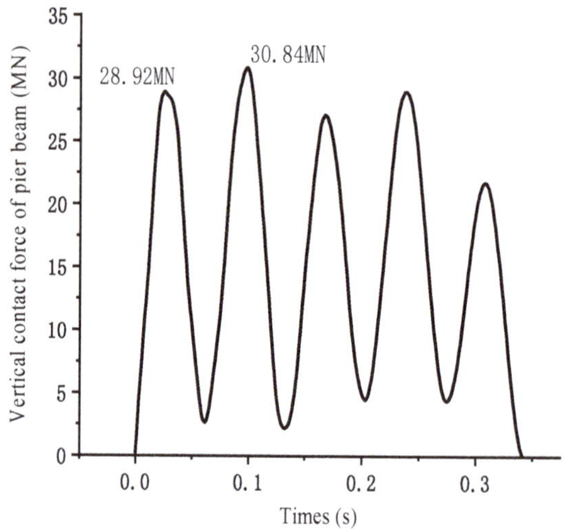

Figure 9. The vertical contact force of pier beam under the first impact.

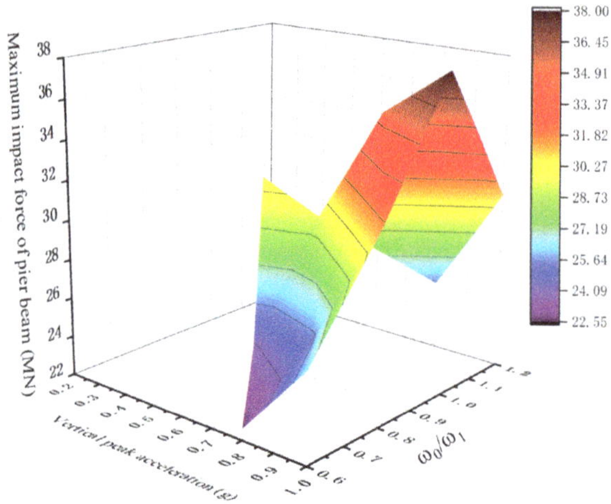

Figure 10. Vertical contact force of pier beam.

4. Pier Failure

The previous research only considered the influence of the dynamic response change of the pier top and the increase in the vertical contact force on the bending failure of the pier under separation—collision coupling conditions. However, pier failure includes compression failure, bending failure, and shear failure. Nevertheless, the influence of the change in the dynamic response of bridges caused by separation on the failure mode of bridge piers still requires research.

4.1. Compression Failure

Under longitudinal excitation, a pier is always under eccentric compression. The previous studies indicate that when the excitation frequency is close to the natural vertical frequency of the bridge, the vertical impact force of the pier beam increases monotonically with the increase in the vertical excitation peak value. Moreover, the separation may increase the longitudinal deformation at the top of the pier and affect the allowable pressure of the pier. Three typical seismic waves from Imperial Valley, Northridge, and Kobe were used to analyze the time history of bridge pier damage. The excitation acceleration in the vertical direction is the excitation acceleration in the horizontal direction multiplied by the ratio parameter α(V/H). To simplify the calculation, α = 1 is taken here.

The maximum and allowable pressure at the bottom of the pier under different excitation amplitudes and the maximum pressure at the collision stage adopt the limit solution. In the case of eccentric compression, the bearing capacity of the bridge pier should conform to the requirements of Equation (7), as specified in the code [23].

$$\gamma_0 N_d \leq n_u A f_{cd} \tag{7}$$

Figure 11 shows the changes in pier bottom pressure under different excitation amplitudes. When the peak excitation acceleration is less than 0.55 g, the vertical contact force increases monotonically. Nevertheless, the longitudinal deformation of the pier top changes little because there is no pier–beam separation. When the excitation peak exceeds 0.55 g, vertical separation occurs, the extreme value of the longitudinal deformation at the top of the pier increases significantly, and the allowable pressure of the pier decreases. With the increase in excitation amplitude, the maximum values of the vertical contact force of the pier beam and the longitudinal deformation of the pier top increase monotonically. When the peak excitation acceleration is 0.65 g, the pier is destroyed under compression.

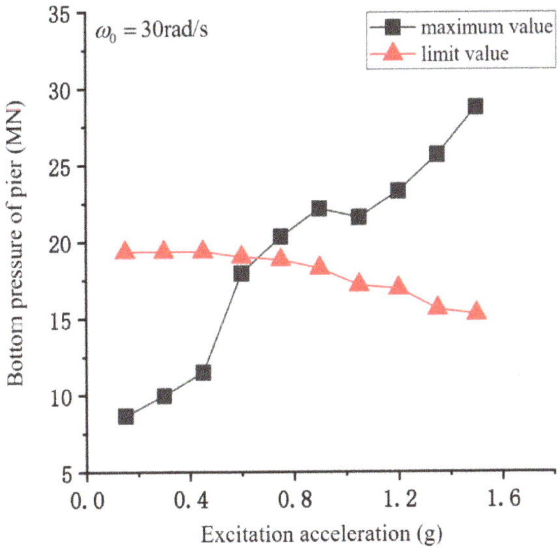

Figure 11. Bottom pressure of pier when ω_0 = 30 rad/s.

Figure 12 shows the change in the maximum and allowable pressure under different excitation peaks when ω_0 = 50 rad/s. When ω_0 is far away from ω_1, the vertical contact force increases monotonically with the decrease in the excitation peak. Since the structure does not separate, and ω_0 is far away from μ_1, the allowable value of the pier pressure remains constant. The maximum vertical contact force of the pier beam is always less than the allowable value, and the structure is in the safe range.

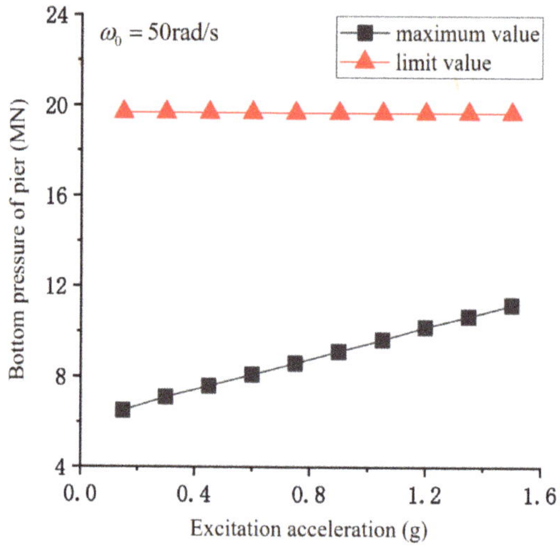

Figure 12. Bottom pressure of pier when ω_0 = 50 rad/s.

Figure 13 shows the changes in the maximum and allowable values of pier bottom pressure under different excitation amplitudes when ω_0 = 15 rad/s. At this point, the excitation frequency deviates from the natural vertical frequency of the bridge but approaches the longitudinal excitation frequency of the bridge. When the peak seismic acceleration is less than 1.2 g, the maximum pressure at the pier bottom increases monotonically with the increase in excitation. By contrast, the extreme deformation value at the pier top changes little. When the peak excitation acceleration exceeds 1.2 g, the structure separates and undergoes a small mutation, leading to a further decrease in the allowable pier pressure. When the peak excitation acceleration exceeds 1.5 g, the pier is destroyed under compression.

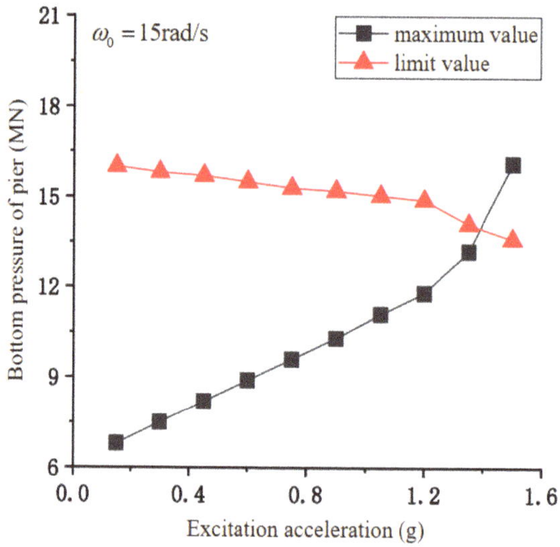

Figure 13. Bottom pressure of pier when ω_0 = 15 rad/s.

4.2. Bending Failure

To calculate the maximum bending moment at the bottom of the first impact pier, the calculation model is simplified as follows:

1. The maximum deformation time at the top of the pier is selected for the calculation. At this time, the bending moment generated by the horizontal vibration inertia force of the pier is zero;
2. To calculate the most unfavorable condition of the structure, it is assumed that the maximum eccentricity and the maximum impact force occur simultaneously;
3. Since the dead weight of the pier is minimal, the bending moment generated by its dead weight at the bottom is ignored;
4. The bending failure is only considered at the bottom of the pier, and the possible instability failure of the pier is ignored;
5. When the shear deformation of the rubber bearing exceeds the allowable deformation value, it is assumed that the rubber bearing fails, and the increase in deformation cannot improve the horizontal binding force;
6. The possible plastic changes at the bottom of the pier are ignored and it is assumed that the pier is always in an elastic stage.

The allowable bending moment calculation of an eccentric compression pier is shown in Equation (8), and the specific parameters are shown in the specification [23]:

$$\begin{aligned} N &= \alpha f_{cd} A_r \left(1 - \frac{\sin 2\pi\alpha}{2\pi\alpha}\right) + (\alpha - \alpha_t) f_{sd} A_s \\ M_{ud} &= \frac{2}{3} f_{cd} A_r \frac{\sin^3 \pi\alpha}{\pi} + f_{sd} A_s r_s \frac{\sin \pi\alpha + \sin \pi\alpha_t}{\pi} \end{aligned} \quad (8)$$

Figure 14 shows the changes in the total and acceptable bending moments under different excitation amplitudes. With the increase in the excitation amplitude, the allowable bending moment continues to decrease. By contrast, the total bending moment continues to increase and surges when the amplitude is 0.55 g. At this point, the sudden change in the maximum bending moment is due to the sharp increase in the longitudinal displacement caused by the separation.

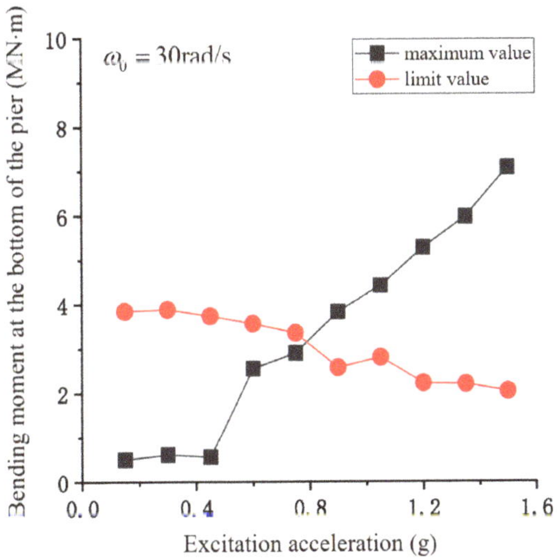

Figure 14. Bending moment at the bottom of the bridge pier under different levels of vertical excitation acceleration.

4.3. Shear Failure

The change in a pier's longitudinal displacement and vertical force may also cause pier shear failure. See Equation (9) for the check calculation of the shear strength, and see the specification [23] for the meaning of each parameter.

$$\begin{array}{c} V_c \leq \phi(V_c + V_s) \\ V_c = 0.1 v_c A_e \end{array} \quad (9)$$

Since the elastic model is adopted in this paper and the plastic hinge interval is ignored, H is the height from the top to the bottom of the pier. In order to study the shear condition at the bottom of the pier when the bending moment is at its maximum, P is selected as the maximum pressure at the bottom of the pier. According to Equation (9), the shear strength of the bridge pier increases monotonically with the increase in the pier bottom pressure. The increase in pressure can not only increase the shear strength of the pier, but also lead to the increase in the bending moment at the bottom of the pier. Figure 15 shows the shear changes of the bridge pier under different vertical excitation amplitudes. When there is no separation between the main beam and the pier, both the shear design value and the allowable shear value of the pier increase with the increase in the vertical excitation amplitude, and the pier is always in the safe range. When the structure is separated vertically, the bending moment caused by the horizontal excitation and the eccentric collision increases greatly due to the increase in the pier deformation caused by the separation. However, the increase in the pier's axial force can improve the pier's shear design value. However, this increase is less than the shear force generated by the seismic excitation, and the bottom of the pier is damaged by the shear.

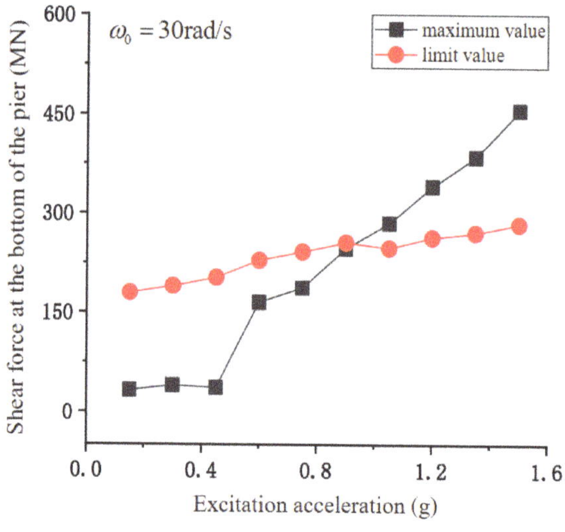

Figure 15. Bottom shear of pier under different excitation amplitudes.

4.4. Influence of Dynamic Response on the Failure Mode of Bridge Pier

Under the conditions of pier–beam separation, there may be multiple modes of failure under high-amplitude excitation. However, the limit solution is adopted in the calculation of pier failure in the front, while the dynamic response of the bridge under the actual excitation is within the limit solution. Therefore, it is necessary to analyze the first failure mode of the bridge pier under the condition of separation.

The dynamic response of the bridge after pier–beam separation is calculated and compared with the standard formula. Two parameters affecting the failure of the pier are proposed: the vertical impact force of the pier beam and the longitudinal deformation of the pier top. Figure 16 shows the failure of the pier under different parameters. When the pier top is slightly deformed, the increase in the vertical collision force of the pier beam leads first to the compression failure of the structure, followed by shear failure, and, finally, bending failure. The failure modes of the bridge pier change with the increase in the maximum deformation. The failure gradually changes from compression failure to shear failure, and the pier bottom pressure required for failure gradually decreases. When large longitudinal deformation occurs at the pier top, bending failure occurs first, and only a small axial force is needed to cause it.

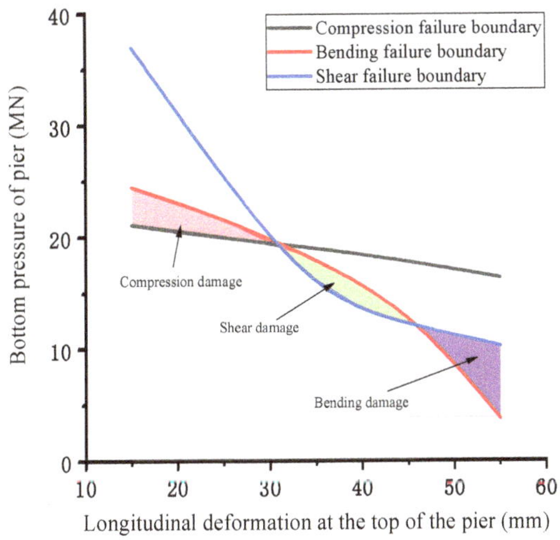

Figure 16. Failure of bridge pier under different axial forces.

Regarding the bridge model in Figure 2, when $\omega_0 \approx \mu_1$, the top of the pier is deformed significantly, but the excitation frequency deviates greatly from the natural vertical frequency, the pier–beam contact force is low, and the pier is mainly subjected to bending failure. When the excitation frequency deviates significantly from the longitudinal frequency of the bridge and is close to the natural vertical frequency of the bridge, the deformation at the top of the pier gradually decreases, the pressure gradually increases, and shear failure gradually occurs. When $\omega_0 \approx \omega_1$, the pier is subjected to a large vertical impact force. Through the analysis of Figure 16, it can be found that compression failure is the most likely failure to be caused at this time. The failure modes of the pier are different with different excitation frequencies.

5. Conclusions

In this paper, a double-span continuous beam bridge is selected to analyze the influence of near-field vertical earthquake on pier failure. Analysis of the longitudinal displacement and vertical contact force response of the bridge is undertaken. Different types of possible damage of the pier under seismic excitation are solved. It is calculated that:

(1) Under simple harmonic excitation, there is a risk of mound separation when the excitation frequency is close to the first-order vertical intrinsic frequency of the first-order of the bridge, and it is mainly focused on the excitation frequency being close to the first-order vertical intrinsic frequency interval.

(2) The vertical separation of the mounds increases the longitudinal deformation on the top of the piers, and the frequency difference is larger. At the same time, the collision that the separation would cause would greatly increase the mound vertical contact force, and the closer the excitation frequency is to the natural vertical frequency, the greater the vertical excitation amplitude, and the greater the impact force.

(3) With pier and beam separation, the increase in excitation amplitude causes a monotonic increase in dock longitudinal deformation. The dock will be damaged by compression, bending, and shearing. With the longitudinal deformation value of the pier top as the failure basis, when the vertical contact force of the pier beam is low, the flexural failure of the structure occurs first. With the increase in pier bottom pressure, the first failure mode gradually changes into shear failure. When the vertical impact force of the pier beam is large, only a small longitudinal deformation is required to cause compression failure of the pier beam.

Author Contributions: Formal analysis, W.A.; Data curation, M.K.; Writing—original draft, L.Z.; Writing—review & editing, L.M.; Supervision, H.Y. All authors have read and agreed to the published version of the manuscript.

Funding: This research was funded by Jiangxi Provincial Department of Education Science and Technology [CJJ202915] And The APC was funded by [CJJ202908].

Institutional Review Board Statement: Not applicable.

Informed Consent Statement: Not applicable.

Data Availability Statement: The datasets used and analysed during the current study are available from the corresponding author on reasonable request.

Conflicts of Interest: The authors declare no conflict of interest.

References

1. Li, H.N.; Xiao, S.Y.; Huo, L.S. Damage investigation and analysis of engineering structures in the Wenchuan earthquake. *J. Build. Struct.* **2008**, *29*, 10–19.
2. Zhuang, W.L.; Liu, Z.Y.; Jiang, J.S. Earthquake-induced damage analysis of highway bridges in Wenchuan earthquake and countermeasures. *Chin. J. Rock Mech. Eng.* **2009**, *28*, 1377–1387.
3. Bozorgnia, Y.; Niazi, M.; Campbell, K.W. Characteristics of free-field vertical ground motion during the Northridge earthquake. *Earthq. Spectra* **1995**, *11*, 515–526. [CrossRef]
4. Bozorgnia, Y.; Campbell, K.W. The Vertical -to-horizontal response spectral ratio and Entative procedures for developing simplified V/H and vertical design spectra. *J. Earthq. Eng.* **2004**, *8*, 175–207. [CrossRef]
5. Jiang, Z.; Tao, Y.; Yu, T.P.; Yuan, W. A Novel Structure-Pulse Coupled Model for Quantifying the Column Ductility Demand under Pulse-Like GMs. *J. Earthq. Eng.* **2022**, *26*, 8185–8203.
6. Shao, Y.; Wei, Y.; Yang, T.; Ni, M.; Zhong, J. Empirical models of bridge seismic fragility surface considering the vertical effect of near-fault ground motion. *Structures* **2021**, *34*, 2962–2973. [CrossRef]
7. Sung, J.K.; Curtis, J.H.; Amr, S.E. Analytical Assessment of the Effect of Vertical Earthquake Motion on RC Bridge Piers. *J. Struct. Eng.* **2011**, *137*, 252–260.
8. Rodrigues, H.; Furtado, A.; Arêde, A. Behavior of rectangular reinforced-concrete columns under biaxial cyclic loading and variable axial loads. *J. Struct. Eng.* **2016**, *142*, 1–8. [CrossRef]
9. Long, J.D.; Xiao, J.T.; Ju, M. Quantitative investigation of tomographic effects in abnormal regions of complex structures. *Engineering* **2021**, *7*, 1011–1022.
10. Zhang, Y.-B.; Yao, X.-L.; Liang, P.; Wang, K.-X.; Sun, L.; Tian, B.-Z.; Liu, X.-X.; Wang, S.-Y. Fracture evolution and localization effect of damage in rock based on wave velocity imaging technology. *J. Cent. South Univ.* **2021**, *28*, 2752–2769. [CrossRef]
11. Nguyen Luan, T.; Ryan, T. Ultrasonic wavefield inversion and migration in complex heterogeneous structures: 2D numerical imaging and nondestructive testing experiments. *Ultrasonics* **2018**, *82*, 357–370. [CrossRef] [PubMed]
12. Zuo, Y.; Sun, G.J.; Li, H.J. Comparison and Research of Unseating Prevention Measures in Seismic Codes of China and Foreign Countries. *J. Disaster Prev. Mitig. Eng.* **2016**, *36*, 617–623, 639.
13. Tao, Y.; Xinzhe, Y.; Jian, Z.; Yuan, W. Near-fault Pulse Seismic Ductility Spectra for Bridge Columns Based on Machine Learning. *Soil Dyn. Earthq. Eng.* **2023**, *164*, 107582.
14. Zhong, J.; Shi, L.; Yang, T.; Liu, X.; Wang, Y. Probabilistic seismic demand model of UBPRC columns conditioned on Pulse-Structure parameters. *Eng. Struct.* **2022**, *270*, 114829. [CrossRef]
15. Zhong, J.; Zheng, X.; Wu, Q.; Jiang, L.; He, M.; Dang, X. Seismic fragility and resilience assessment of bridge columns with dual-replaceable composite link beam under near-fault GMs. *Structures* **2023**, *47*, 412–424. [CrossRef]
16. Zhong, J.; Mao, Y.-H.; Yuan, X. Lifetime seismic risk assessment of bridges with construction and aging consideration. *Structures* **2023**, *47*, 2259–2272. [CrossRef]
17. Khatiwada, S.; Chouw, N.; Butterworth, J.W. A generic structural pounding model using numerically exact displacement proportional damping. *Eng. Struct.* **2014**, *62–63*, 33–41. [CrossRef]

18. Ruan, H.H.; Yu, T.X. Local deformation models in analyzing beam-on-beam collisions. *Int. J. Mech. Sci.* **2003**, *45*, 397–423. [CrossRef]
19. Yang, H.B.; Yin, X.C.; Hao, H. Theoretical Investigation of Bridge Seismic Responses with Pounding under Near-Fault Vertical Ground Motions. *Adv. Struct. Eng.* **2015**, *18*, 452–468. [CrossRef]
20. An, W.; Song, G.; Chen, S. Near-Fault Seismic Response Analysis of Bridges Considering Girder Impact and Pier Size. *Mathematics* **2021**, *9*, 704. [CrossRef]
21. An, W.; Song, G. Influence of Bearing on Pier Failure Considering the Separation Condition under Near-Fault Earthquake. *Symmetry* **2021**, *13*, 692. [CrossRef]
22. Xing, Y.F.; Wang, L.J. Analysis of wave propagation in the built-up structures of rod-beam and beam-beam. *J. Beijing Univ. Aeronaut. Astronaut.* **2013**, *30*, 520–523.
23. *CJJ 166-2011*; Code for Seismic Design of Urban Bridges. Ministry of Housing and Urban-Rural Development of the People's Republic of China: Beijing, China, 2011.

sustainability

Article

Research on Elevation Survey Method of Sea-Crossing Bridge under Adverse Conditions

Jun Xiao [1,2], Jianping Xian [1,2], Song Li [1] and Shuai Zou [2,3,*]

1 CCCC Second Highway Engineering Co., Ltd., Xi'an 710065, China
2 Shanxi Union Research Center of University and Enterprise for Bridge Intelligent Construction, Xi'an 710199, China
3 School of Civil Engineering, Chongqing Jiaotong University, Chongqing 400074, China
* Correspondence: zs2448602237@163.com; Tel.: +86-186-8169-2538

Abstract: Aiming to survey scenarios of offshore projects with difficult horizontal elevation transmission and long-distance, all-weather elevation monitoring operations, a long-distance, total station, trigonometric leveling based on dynamic compensation is proposed. The feasibility of this method was verified by an outdoor survey experiment, and the range of transverse coverage and accuracy reached by this method was quantitatively analyzed. The results indicate that this method shows a good correction effect on the survey results of test points under different environmental conditions, which proves that this method is feasible. The correction effect of this method is affected by the distance between the test point and the datum point; within the range of 60 m horizontally from the datum point, an assurance rate of about 90% can be achieved for the error range of 20 mm. Combining with the built-in ATR (Automatic Target Recognition) technology of the total station, this method can make the elevation survey result reach the accuracy of millimeter level under the range of about 1000 m, by obtaining multiple groups of data and then calculating the mean value. This paper provides a new method for the elevation transfer of sea-crossing bridges under long-distance conditions and harsh environmental conditions.

Keywords: sea-crossing bridges; triangle elevation; leveling method; dynamic compensation; total station; harsh environment; long distance

check for updates

Citation: Xiao, J.; Xian, J.; Li, S.; Zou, S. Research on Elevation Survey Method of Sea-Crossing Bridge under Adverse Conditions. *Sustainability* **2022**, *14*, 11641. https://doi.org/10.3390/su141811641

Academic Editors: Kai Wei, Mingjin Zhang, Jian Zhong and Yutao Pang

Received: 28 July 2022
Accepted: 15 August 2022
Published: 16 September 2022

1. Introduction

In recent years, there has been an upsurge in sea-crossing traffic engineering construction around the world, and the length of sea-crossing bridges has been refreshed again and again. The earliest sea-crossing bridge in China is the Xiamen Bridge, which was built in 1987. The total length of the bridge is 6.695 km. It was finally completed in 1991 and opened to traffic in May. Since then, the curtain of China's sea-crossing bridge construction has been lifted [1]. The following Table 1 lists the top 10 typical sea-crossing bridges that have been built or are under construction in China.

To meet the increasing traffic demand, the construction of the sea-crossing bridge is gradually developing towards the strong-wave, deep-water area. Its length ranges from several kilometers to tens of kilometers. Large-scale sea-crossing projects under harsh environmental conditions will inevitably cause many severe technical problems in construction and maintenance [2,3]. In the maintenance of large-scale sea-crossing bridges, to meet the high-standard testing requirements, it has become an important research direction in this field to innovate the testing methods by combining the traditional non-destructive testing (NDT) methods with advanced techniques [4,5]. Khedmatgozar Dolati et al. [4] compiled in one place all the NDT techniques, including the application of drones, sensors, or robots for rapid and efficient assessment of damage on small and large scales, which kept researchers up-to-date with existing methods and paved the way for further innovations in this regard. In the construction of sea-crossing bridges, long-distance elevation surveys in a complex marine environment is one of the important technical problems. In the project, some offshore survey platforms are often established through offshore test pile projects, and some bridge piers are under priority construction so that the survey distance can be shortened to less than 2 km, which provides practical conditions for using high-precision total stations to transfer the elevation data to offshore buildings or structures and to realize the direct connection of the sea-crossing elevation [6,7]. Affected by waves and fluctuating tides, the offshore survey platform and piers are always subject to slight shaking, and the vertical angle of the total station placed on the platform or pier can be changed for tens of seconds or even minutes [8,9]. To ensure the survey effect, the stiffness and size of the offshore survey platforms are often designed to be very large, and the distance between the survey platforms will be increased based on the consideration of construction cost.

Table 1. The top 10 typical sea-crossing projects that have been built or are under construction in China.

Order	Bridge Name	Span/km	Built/Building
1	Hong Kong–Zhuhai–Macao Bridge	55.0	Built
2	Jiaozhou Bay Bridge	36.5	Built
3	Hangzhou Bay Bridge	36.0	Built
4	East Sea Bridge	32.5	Built
5	Quanzhou Bay Bridge	26.7	Built
6	Shenzhen–Zhongshan Bridge	24.0	Building
7	Jintang Bridge	18.5	Built
8	Huangmao Hai Link—Cross-sea section	14.4	Building
9	Jiashao Bridge	10.1	Built
10	Nan'ao Bridge	9.3	Built

For high-precision total station trigonometric leveling, the main factors affecting the accuracy of the survey are the size of the zenith distance, the observation error, and the measurement error of the vertical atmospheric refraction coefficient, if the stability of the station is guaranteed [10,11]. The influence degree of each factor gradually increases with the increase in the distance between the survey station and the target prism. Therefore, it is of great significance to improve the accuracy of the survey by either improving the survey method or compensating and correcting the survey results to reduce the deviation caused by various influencing factors the survey results.

Qiu Yang et al. [12] used auxiliary equipment to measure the height of the total station and the target plate, appropriately increased the number of angular survey rounds to improve the accuracy of vertical angle, and carefully planned the survey steps to reduce the vertical atmospheric refraction error and other means to effectively eliminate and reduce the impact of the main error source. Yaming Xu et al. [13] proposed a linear structure survey method based on the traditional method, combined with relatively strict synchronous opposite observation, which reduced the complexity and observation amount of the original method. References [14–16] improved and optimized the survey method proposed in reference [13] and carried out long-distance, sea-crossing elevation transmission in the Chuandao area of Jiangmen Taishan City. Zhongping Wang et al. [17] improved the point layout structure based on the trigonometric elevation observation method proposed in reference [13] and eliminated and screened the data of each survey section in combination with the specification and reasonable restriction of the elevation difference between each group. The requirements of the national second-class leveling accuracy were met through project testing.

Heng Zhang et al. [18] have greatly weakened the influence of atmospheric refraction, vertical deviation, and Earth curvature by using two high-precision total stations of the same model to observe the opposite direction of the high–low biprism group at the same time. Based on the least square theory, Mingbo Liu et al. [19] gave a new method to calculate the refraction coefficient of the survey area. It was demonstrated in production practice that using this method can produce triangular elevations with accuracy close to the second-class level. Jianzhou Li et al. [20] deduced the error formula of the survey result through the error propagation law according to the rigorous calculation formula of precision trigonometric leveling and proved that controlling the vertical angle and survey distance of the instrument can improve the survey accuracy and achieve second-class leveling accuracy. Jianzhou Li et al. [21] developed TriLevel, a precision triangulation elevation survey system that realizes the automatic measurement and quality control of precision triangulation elevation. Refs. [22–24] weakened the deviation caused by atmospheric refraction on elevation survey results by avoiding measuring the total station height and realizing a high-precision elevation survey. Peibing Yang et al. [25] designed a special observation mark to solve the problem of long-distance surveys, aiming at the need for high-precision elevation transmission in wide sea areas, and also responded to the impact of the shaking of the survey platform on the survey accuracy by carrying out rigorous and rapid observation operations under favorable survey conditions.

At present, there are many pieces of research on the improvement of the survey method of total station trigonometric leveling and the methods of compensation and correction of the survey results, which can make the accuracy of the survey results reach a high level, but to ensure the accuracy requirements of the survey results, the improved survey method or the method of correction of the survey results are generally cumbersome. However, there is relatively little research on developing targeted, simple, and efficient survey methods or correction methods under some specific conditions and harsh environments.

Because of this, this paper proposes a long-distance total station elevation survey method based on dynamic compensation, which is aimed at survey scenarios of offshore projects with difficult horizontal elevation transmission under long-distance, all-weather elevation monitoring operations and harsh environmental conditions. The feasibility of the proposed elevation survey method was verified by an outdoor survey experiment. At the same time, through the processing and analysis of a large number of surveyed data at different test points, a quantitative study was carried out within the scope that this method can cover in the horizontal direction and the accuracy that the survey results can reach. This method can be used as an efficient and high-precision method for the elevation survey of sea-crossing bridges under adverse conditions.

2. Principles and Methods

2.1. Engineering Background

The elevation survey method proposed in this paper mainly serves the elevation survey operation in the construction of the Peljesac Bridge. The Peljesac Bridge is located at the southern end of Croatia, bordering Bosnia and Herzegovina. It is located in the Malostonski Bay protection area on the west side of the Adriatic Sea, with the Peljesac Peninsula and mainland on either side of the bridge. The main Peljesac Bridge is a low-tower, cable-stayed bridge with a total length of 2404 m, which has a central single cable plane and a steel box girder, and its span group is (84 + 108 + 108 + 189.5 + 5 × 285 + 189.5 + 108 + 108 + 84) m. The bridge has 14 piers and abutments, of which 4 piers and abutments are arranged on land on both banks, and the remaining 10 pier towers are in the water. The layout of the Peljesac Bridge is shown in Figure 1.

During the construction of the Peljesac Bridge, no sea survey platform was set up. By setting survey stations on both banks, the elevations of the bridge piers and tower columns in the sea were surveyed. The survey operations are shown in Figure 2.

2.2. Basic Principle of Total Station Trigonometric Leveling

Trigonometric leveling is a method to determine the height difference between two points by surveying the horizontal distance and vertical angle of the two control points. The basic formula for the calculation of one-way survey height difference is shown in Formula (1), which is generally derived from Figure 3:

$$h_{AB} = Scos\alpha + \frac{1-K}{2R}S_0^2 + i_A - v_B \tag{1}$$

where S is the oblique distance between two points A and B, m; α is the vertical angle between the observation point of the total station and the target prism, (°); R is the average curvature radius of the earth, $R \approx 6371$ km; K is the local atmospheric refraction coefficient (related to temperature, pressure, and atmospheric density), which can be taken as 0.10~0.15 for general mountainous areas and 0.5 for flat areas near the ground; S_0 is the horizontal distance between two points, m, $S_0 = Scos\alpha$; i_A is the height of the total station, m; v_B is the height of the prism, m.

Figure 1. The layout of the Peljesac Bridge.

Figure 2. The process of survey operations.

The calculation formula of the mean square error in the elevation difference of a one-way survey can be derived from the total differential of Formula (1) [26] and then from the error propagation rate, as shown in Formula (2):

$$M_{h_{AB}}^2 = S^2 cos^2\alpha \frac{m_\alpha^2}{\rho^2} + sin^2\alpha m_S^2 + \frac{1}{4R^2} S_0^4 m_{K_i}^2 + m_{i_A}^2 + m_{v_B}^2 \tag{2}$$

where m_α is the mean square error of vertical angle observation; m_S is the mean square error of the oblique distance survey; m_K is the mean square error of the atmospheric refraction coefficient measurement, generally ±0.03~±0.05 [27]; m_{i_A} is the mean square error of the total station height measurement; m_{v_B} is the mean square error of the target height measurement; ρ is the constant used for angle unit conversion, $\rho = 206265$.

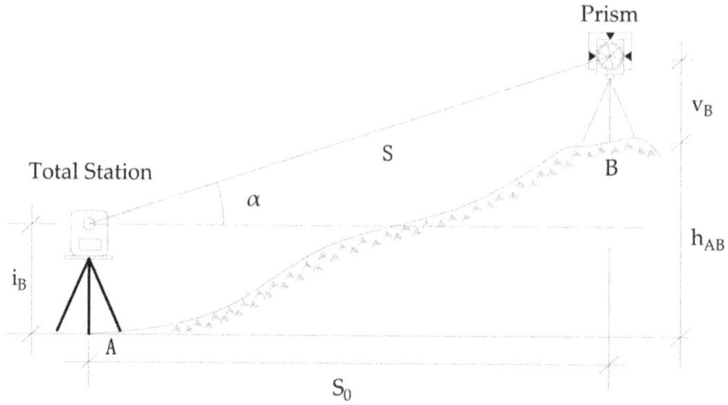

Figure 3. Principle of triangle elevation.

2.3. *Principle of Dynamic Compensation*

According to the calculation formula of the mean square error of the survey height difference shown in Formula (2), the high-precision total station needs to be compensated and corrected due to the influence of atmospheric refraction, topographic conditions, the earth's curvature, etc., during long-distance trigonometric leveling. As shown in Figure 4, the line of sight *e1* of the datum point in the figure and the line of sight *e2* of the test point to be surveyed should be the same at the same time because the distance between the datum point and the test point is short.

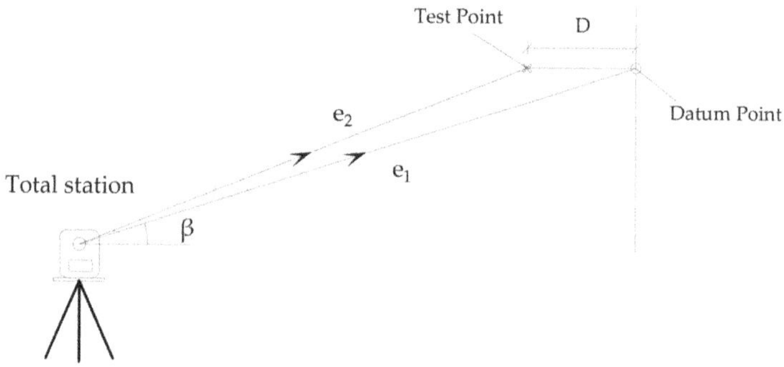

Figure 4. The principle of dynamic compensation.

If the accurate elevation data of the datum point is known, the elevation can be transmitted to the datum point by densifying the datum point on land or structure under the meteorological conditions required in the survey specification, such as night or morning with good weather conditions, and finally determined by taking the mean value through multiple surveys; when the survey operation conditions allow, the accurate elevation value of the datum point can also be obtained through leveling. When the elevation survey operation is carried out, the elevation of the datum point is surveyed first, and the difference between its accurate elevation value and the current elevation survey value is taken as the dynamic compensation correction value under the current condition. Then, the elevation of the test point is surveyed, and the obtained dynamic compensation correction value is used to correct the elevation survey result of the test point, to reduce or eliminate the influence of external factors. The correction formula is as follows:

Dynamic compensation value:

$$\mu = H_0 - H_0' \quad (3)$$

Correction value of current survey result:

$$H_i' = H_i + \mu \quad (4)$$

where H_0 is the precise elevation value of the datum point; H_0' is the elevation survey result of the datum point during the current survey; μ is the dynamic compensation value of the current elevation survey result; H_i is the elevation survey result of the point to be surveyed in the current survey; H_i' is the elevation correction value of the point to be surveyed in the current survey.

2.4. Introduction of Experiment

To verify and carry out quantitative research on the trigonometric leveling method based on dynamic compensation proposed in this paper, an outdoor field survey experiment was carried out. The specific contents of the experiment are as follows.

2.4.1. Scheme Design

Considering the ultimate purpose of this experiment, the selection of the experimental site should meet the following requirements:

(1) The distance between the survey station and the datum point should be far enough; there should be no obstructions within the line of sight between the survey station, the datum point, and the test point; and the intervisibility condition should be good.
(2) The survey site should allow the convenient use of the level gauge to transmit the elevation of the survey station to the datum point and the test point station by station;
(3) A certain range around the datum point should be open, which is convenient for the layout of the test points under different conditions.

To meet the requirements of the above experimental site, an open construction site was selected for this experiment, and the layout of the site is shown in Figure 5. To quantify the range that the method can cover horizontally over long distances, the distance between the datum point (JZ01) and the survey station (CZ01) was selected as 1000 m. According to the different distances between the datum point and the test point, a total of seven test points were proposed, as shown in Table 2. Due to the long-distance survey, the efficiency of the target prism aiming manually is low, and large observation errors may be caused by human factors. Therefore, the built-in ATR function of the total station was used for target collimation in this experiment. At the same time, to verify the applicability of this method under different environmental conditions, this experiment chose to continuously observe for a day, and the observation frequency was 1 group every 10 min.

Table 2. Test point details.

Test Point	Distance from Datum Point L/m	Collimation Mode	Measurement Frequency
DC-30	30	ATR	1 Group/10 Min
DC-50	50	ATR	
DC-70	70	ATR	
DC-60	60	ATR	
DC-80	80	ATR	
DC-90	90	ATR	
DC-100	100	ATR	

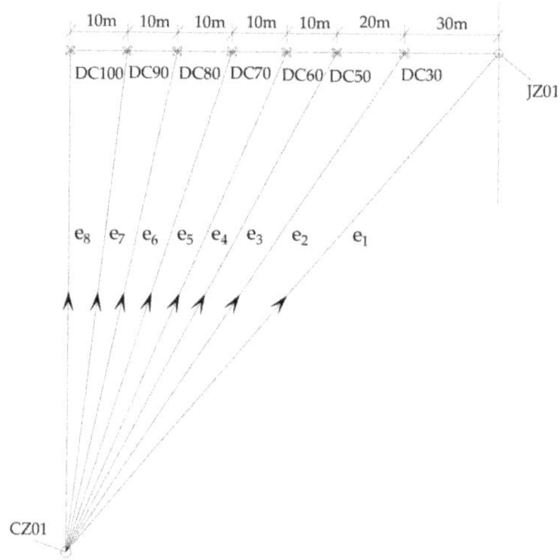

Figure 5. The layout of the site.

2.4.2. Experiment Content

In order to carry out the experiment orderly and ensure the reliability of the survey results, the experiment should be carried out according to the following operating steps:

(1) In the night or morning with good weather conditions, the elevation of the survey station should be transmitted to the datum point and the test points station by station by using the precision level, and the surveyed elevation values should be drawn up as the accurate elevation values of each point.
(2) During the actual survey, set up a total station at the station, input the current environmental parameters (such as temperature, air pressure, humidity, etc.) into the total station for preliminary correction, and make corresponding records at the same time;
(3) Survey the elevation value of the datum point first, and take the difference between this survey value and the accurate elevation value of the datum point surveyed in step (1) as the correction value of the current survey (errors caused by atmospheric refraction, earth curvature, etc.).
(4) Survey the elevation values of the test points under each working condition in turn, and correct the elevation survey results of each test point by using the correction value in step (2) according to Formula (4) in Section 2.3.
(5) Repeat steps (2), (3), and (4) within a fixed time interval to continuously survey the elevation of the test points.

The test procedure is shown in Figure 6.

The following points should be paid attention to during the experiment to ensure the accuracy of the experimental results.

a. In step (1), the accurate elevation values of the datum point and the test points are very important for the correction of the survey results and the evaluation of the experimental results. Therefore, the accuracy of the accurate elevation value survey of each point should be guaranteed.
b. As the environmental conditions in a day are changing at all times, the corresponding parameters (temperature, air pressure, humidity, etc.) should be input into the total station for preliminary correction before each data observation.
c. To ensure that the survey process is completed under basically the same environmental conditions, the survey results of each group should be completed within two minutes.

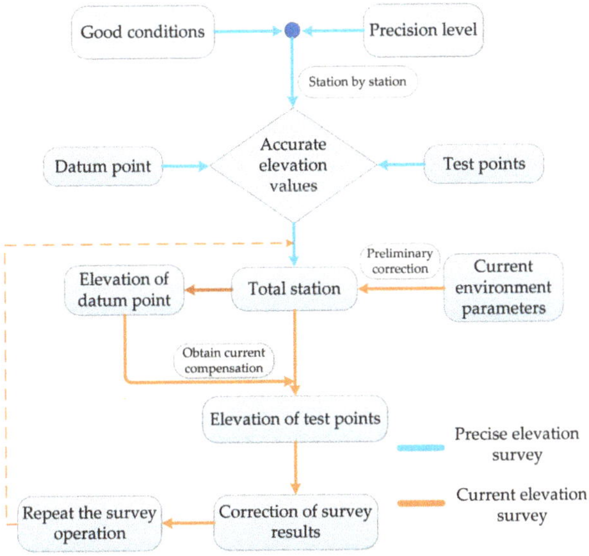

Figure 6. The procedure of the outdoor survey test.

2.4.3. Experimental Equipment

The orderly conduct of the experiment is closely related to the preliminary preparations. Therefore, the test equipment shall be prepared in advance and shall be within the qualification period. The equipment required for this experiment includes a high-precision level, a high-precision total station (Leica TS60), a thermometer, a barometer, a humidity meter, some measuring nails, a manual painting, a test record book, etc. A Leica TS60 total station with an angle survey accuracy of 0.5″ and a distance survey accuracy of 0.5″ was used for this experimental survey, with a survey range of 3500 m in ATR mode, to ensure the testing effect.

Before the experiment, the specific position of the datum point and each test point should be determined through the distance survey function of the observation instrument, and the datum point and the test points should be marked by measuring nails and manual painting.

3. Results

Because of the constant changes in external factors (temperature, atmospheric refraction, and so on), the results of each survey varied to varying degrees. According to the basic principle of the elevation survey method proposed in this paper, the difference between the elevation survey result of the datum point and the accurate elevation value of the datum point was taken as the current dynamic compensation value μ, which was used to correct the elevation survey results of each test point to be surveyed at that time.

Through the analysis of multiple groups of data at different test points, the assurance rate of different survey accuracies and the overall correction effect is used to evaluate the applicability of the method under different distances between the datum point and the test point.

For simplicity of expression, the following applicable scope refers to the scope covered by the distance between the datum point and the test point, which can be used to make a relatively ideal correction of the elevation survey results. This scope is the circular area with the datum point as the center and the distance between the datum and the test point as the radius.

3.1. Experimental Result

According to the basic principle of the elevation survey method proposed in this paper, the accuracy of the survey results was evaluated by the following Formula (5):

$$\Delta H = \left(H_i' - H_{i0}\right) \times 1000 \tag{5}$$

where ΔH is the deviation between the corrected survey result and the precise elevation value of the test point, mm; H_{i0} is the precise elevation survey value of the test point, m; and H_i' is the elevation correction value of the test point in the current survey, m.

Figure 7 shows the changes in accuracy curves before and after the correction of elevation survey results for each test point at different times. The figure shows:

(1) From the uncorrected data results of each test point in each figure, it can be seen that the elevation survey results directly using the total station are greatly affected by the environmental conditions at different times, showing an obvious survey window period. In the early morning with good weather conditions, the light, humidity, temperature, and intervisibility conditions are all within the scope specified in the survey, and the deviation between the survey results and the accurate value is relatively small. Under the ranging condition of about 1000 m, the maximum deviation between the survey result and the accurate value is about 24 mm. With the change in environmental conditions, the deviation between the surveyed results and the accurate value gradually increases.

(2) When the elevation is continuously surveyed in a day, the effect of the continuous change in external environmental conditions on the total station can cause the survey results to shift in a certain direction. Under the experimental conditions described in this paper, the maximum deviation between the elevation survey results and the accurate value can even reach about 25 cm for a test point, and at twilight, when the environmental conditions are relatively mild, the effect of the previous environmental conditions on the total station will continue to affect the elevation survey results. After the survey results are corrected, they can return to the accurate value.

(3) From the variation in the deviation between the survey results of each test point in the figure before and after correction and the accurate value, it can be seen that each test point at different distances from the datum point, after the elevation survey method proposed in this paper is used to correct the elevation survey results at any time, the corrected elevation results can fluctuate within a certain range near the accurate value. At the test point, which is 30 m away from the datum point, the deviation between the corrected survey result and the accurate value can be kept within 20 mm.

(4) Under the same environmental conditions, with the increase in the distance between the test point and the datum point, the fluctuation range of the corrected survey results generally shows a gradually increasing trend. At the test point at a distance of 30 m from the datum point, the maximum deviation between the corrected elevation survey result and the accurate value is 18.0 mm; at the test point at a distance of 100 m from the datum point, the maximum deviation between the corrected elevation survey result and the accurate value is 34.5 mm.

3.2. Data Process

3.2.1. The Accuracy Assurance Rate

The quantitative statistical analysis of the survey accuracy of multiple groups of survey data at the same test point can be completed by the assurance rate under the corresponding accuracy conditions, and the accuracy assurance rate of survey results η is calculated with the following formula:

$$\eta = \frac{N}{n} \times 100\% \tag{6}$$

$$N = number(|\Delta H| \leq \alpha) \tag{7}$$

or

$$N = number(|\Delta H| > \alpha) \quad (8)$$

where n is the number of s data groups; N is the number of groups for which the absolute value of the the survey result's correction value accuracy meets a certain condition; and α is the accuracy condition of the survey results, mm.

Figures 8 and 9 show the distribution of different accuracy assurance rates of the corrected survey results and the variation trend of the same accuracy ratio with the increase in the distance between the test point and the datum point, respectively.

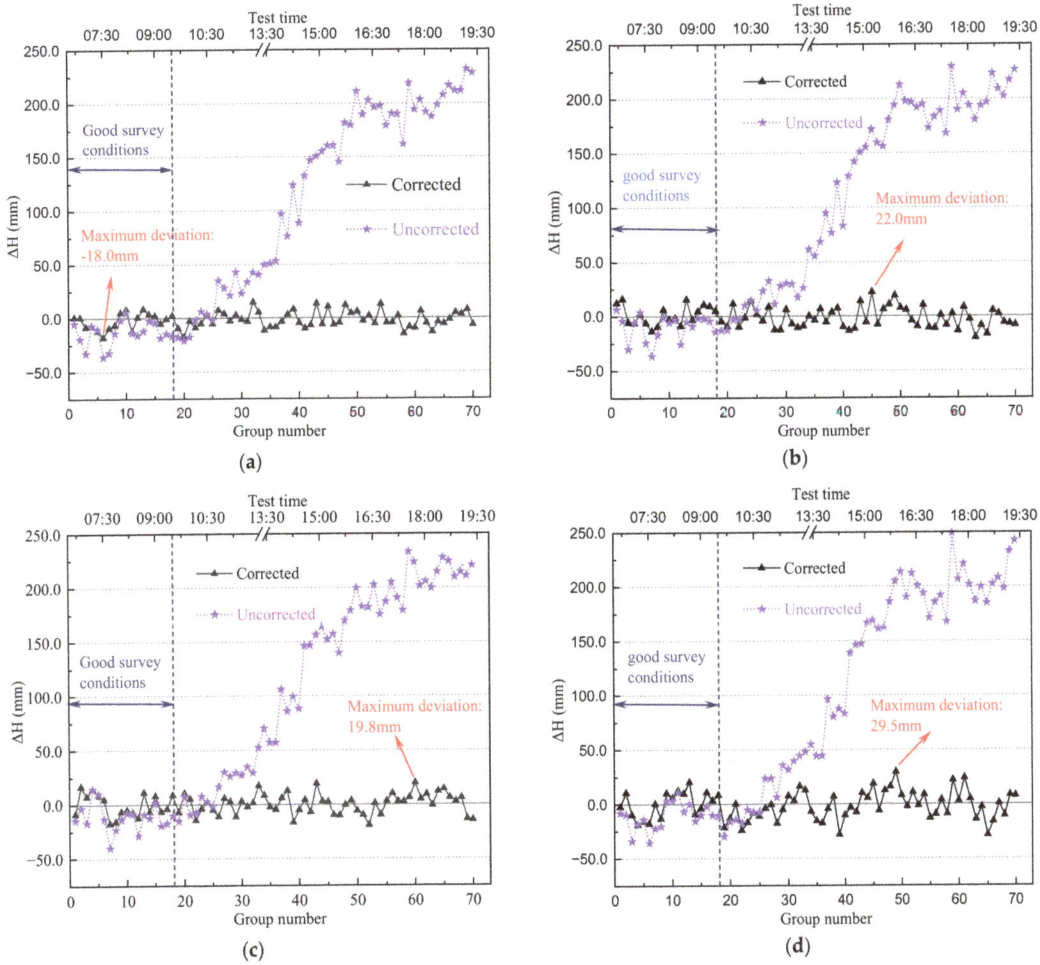

Figure 7. *Cont.*

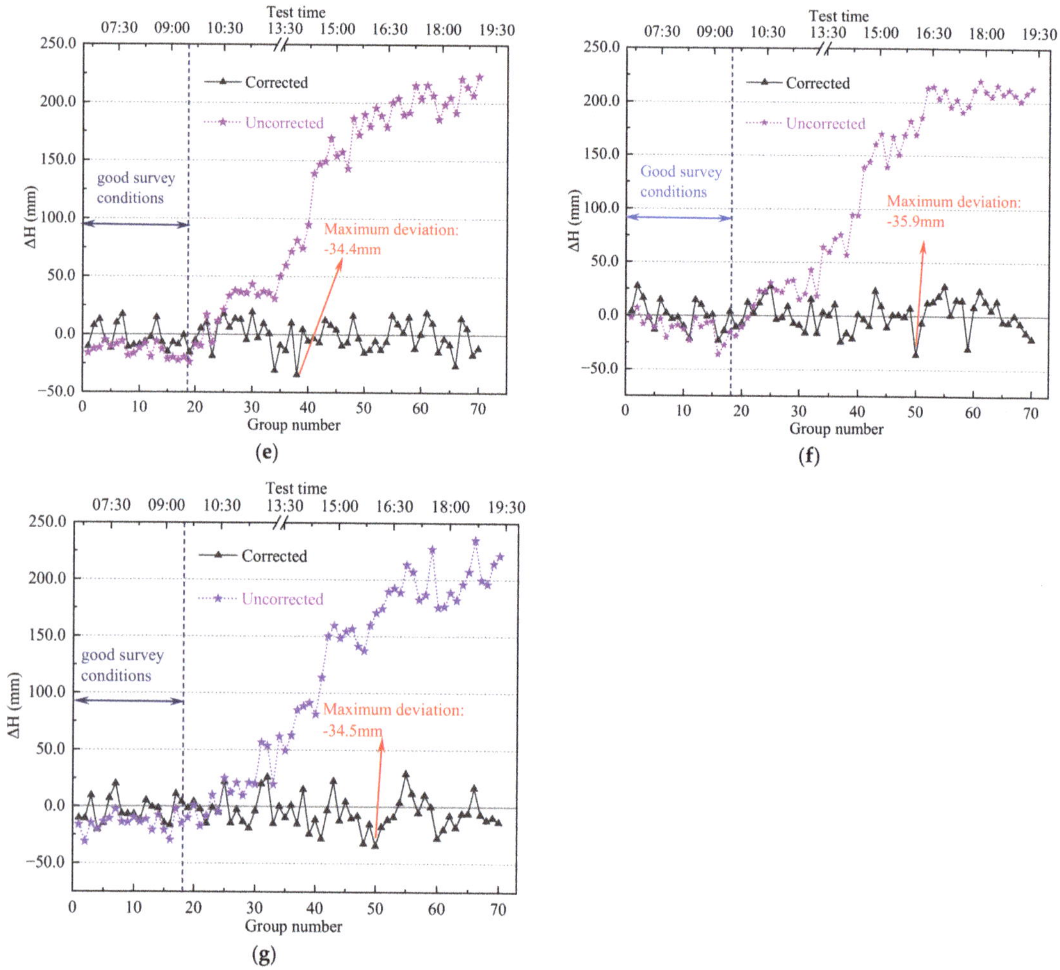

Figure 7. The curves of accuracy change: (**a**) DC-30; (**b**) DC-50; (**c**) DC-60; (**d**) DC-70; (**e**) DC-80; (**f**) DC-90; (**g**) DC-100.

Figures 8 and 9 show that:

a. With the increase in the distance between each test point and the datum point, the accuracy assurance rate generally shows a downward trend under the same accuracy requirement.

b. By correcting the survey results, within a 30 m horizontal distance from the datum point, the error range of 15 mm can reach a 97.2% assurance rate, and the error range of 20 mm can reach 100% assurance rate; within the transverse distance of 60 m from the datum point, the error range of 20 mm can reach about 90% assurance rate.

c. With the increase in the distance between each test point and the datum point, the ratio of different accuracy conditions shows different trends. The ratio of accuracy conditions of 0~5 mm and 5~10 mm shows a decreasing trend. The ratio of 10~15 mm accuracy conditions remains unchanged, while the ratio of greater than 15 mm accuracy conditions shows an upward trend.

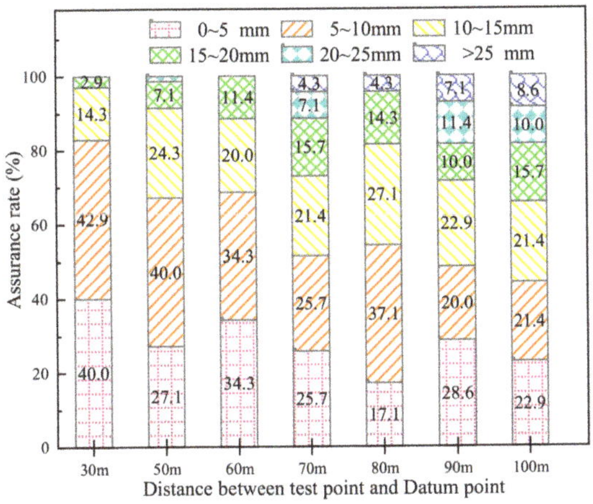

Figure 8. Accuracy assurance rate distribution.

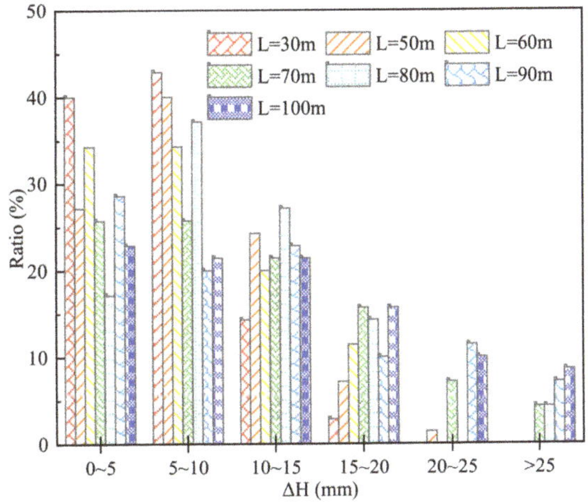

Figure 9. Accuracy ratio trend.

3.2.2. Data Dispersion

The mean value of the data sample reflects the centralized trend of the data. Table 3 shows the corrected mean value of multiple groups of survey results at all test points.

Table 3. The corrected mean value.

Test Point	Distance from Datum Point L/m	Corrected Deviation Range/mm	Corrected Deviation Mean/mm	Remarks
DC-30	30	−18.0~14.8	−1.9	The lowest temperature: is 21 °C; The highest temperature: is 37 °C; The data statistics of each test point: 70 groups.
DC-50	50	−19.8~22.0	−1.0	
DC-70	70	−19.3~19.8	0.2	
DC-60	60	−28.7~29.5	−1.1	
DC-80	80	−34.4~19.1	−1.5	
DC-90	90	−35.9~27.7	0.7	
DC-100	100	−34.5~29.1	−5.0	

It can be seen from the data in Table 3 that after the elevation survey method provided in this paper is used to correct the survey results, the survey results can better approximate the accurate value by taking the mean value of multiple groups of corrected elevation data and finally reach millimeter-level accuracy. Moreover, the mean value of the corrected data is less affected by the distance between the test point and the datum point and the change in the survey environment.

4. Discussion

Because of its great convenience, the use of a high-precision total station for elevation surveys has been widely concerned and used by the engineering community. In a long-distance elevation survey, its accuracy is significantly affected by atmospheric refraction, the Earth's curvature, temperature, and other factors, and the survey results may have a large deviation [10,11]. It is therefore very necessary to compensate for and correct the survey results.

A method of conducting an elevation survey for long distances based on dynamic compensation is proposed in this paper, which makes use of the principle that the compensation amount of two points that are close to each other shall be the same in a short time to correct the elevation survey results. Through the above experimental analysis, this method had a good correction effect on the elevation survey results within the experimental range, and its correction effect was affected by the distance between the test point and the datum point, thus showing the change in the corresponding accuracy assurance rate under different distances. In the actual project, the elevation survey method proposed in this paper can be reasonably selected according to the accuracy requirements of the project. In the above experiments, the uncorrected total station elevation survey results showed an obvious window period for the impact of environmental changes. In the morning, when the environmental conditions were suitable, the survey results were relatively stable, and the survey accuracy was relatively high. Therefore, it is recommended to adjust the long-distance elevation survey to the morning or night with good weather conditions, to reach a more ideal survey effect. Since the elevation survey method can maintain good adaptability under any environmental conditions, if combined with the ATR function of the total station and the corresponding data-processing program for the total station is developed, the method of taking the mean value after multiple surveys can also be used as a high-precision elevation survey method suitable for harsh environmental conditions.

To sum up, the new method proposed in this paper for elevation surveys under adverse conditions can provide sufficient technical support for elevation surveys in the construction of large-scale sea-crossing bridges due to its efficiency, high precision, and strong environmental adaptability. Limited by space, this paper only presents experimental research on the application of this method under the condition of a 1000 m range and does not involve research on the accuracy that this method can achieve under different range conditions. The relevant research will be carried out in the follow-up to clarify the change rules for the applicability, coverage area, and accuracy of measurement results with the distance of range.

5. Conclusions

By experimenting, this paper verified the feasibility of a total station elevation survey method based on dynamic compensation. At the same time, it made a quantitative study on the horizontal coverage of this method and further analyzed the practical situation of this method by counting the accuracy of the survey results under the condition of different distances from the datum point. The main conclusions are as follows:

(1) The analysis shows that the total station elevation survey method based on dynamic compensation proposed in this paper shows a good correction effect on the survey results within 100 m between the test point and the datum point, making the survey results fluctuate within a certain range near the accurate value, which proves that the survey method proposed in this paper is feasible.

(2) The accuracy of the corrected survey results is affected by the distance between the test point and the datum point. Through the correction of the survey results, the assurance rate of 97.2% can be reached in the error range of 15 mm and 100% in the error range of 20 mm within the range of 30 m horizontally from the datum point.
(3) By using this method to correct the survey results and combining it with the built-in ATR technology of the total station to obtain multiple groups of data, and then calculating the mean value, the elevation survey results can reach millimeter-level accuracy within the application range of 100 m from the datum point.
(4) The correction effect of the survey method provided in this paper on the elevation survey results is less affected by the environment. The method of taking the mean value after obtaining multiple groups of data can also be considered a high-precision elevation survey method suitable for harsh environmental conditions.

Author Contributions: Conceptualization, J.X. (Jun Xiao) and J.X. (Jianping Xian); methodology, J.X. (Jun Xiao), S.L. and S.Z.; validation, J.X. (Jun Xiao), J.X. (Jianping Xian) and S.L.; formal analysis, S.Z.; investigation, J.X. (Jun Xiao), S.L. and S.Z.; resources, J.X. (Jun Xiao); data curation, J.X. (Jun Xiao), J.X. (Jianping Xian) and S.Z.; writing—original draft preparation, J.X. (Jun Xiao), J.X. (Jianping Xian), S.L. and S.Z.; writing—review and editing, S.Z.; visualization, J.X. (Jun Xiao), J.X. (Jianping Xian) and S.L.; supervision, J.X. (Jun Xiao); project administration, S.Z.; funding acquisition, J.X. (Jun Xiao). All authors have read and agreed to the published version of the manuscript.

Funding: This research was funded by the Science and Technology Project of CCCC (grant number: No. 2020-ZJKJ-QNCX04) and the Science and Technology Special Major Project of CCCC (grant number: No. 2019-ZJKJ-07).

Institutional Review Board Statement: Not applicable.

Informed Consent Statement: Not applicable.

Data Availability Statement: Data are contained within the article.

Conflicts of Interest: The authors declare no conflict of interest.

References

1. Pan, G.R.; Liu, S.T.; Che, J.R.; Fan, W.P.; Zhan, Z.Y. Bridge engineering survey techniques—A case study of Hangzhou Bay Sea-crossing Bridge survey. *Bridge Constr.* **2009**, *1*, 33–36.
2. Zhong, J.; Yang, T.; Pang, Y.T.; Yuan, W.C. A novel structure-pulse coupled model for quantifying the column ductility demand under pulse-like GMs. *J. Earthq. Eng.* **2021**, 1–19. [CrossRef]
3. Zhong, J.; Ni, M.; Hu, H.M.; Yuan, W.C.; Yuan, H.P.; Pang, Y.T. Uncoupled multivariate power models for estimation performance-based seismic damage states of column curvature ductility. *Struct* **2022**, *36*, 752–764. [CrossRef]
4. Khedmatgozar Dolati, S.S.; Caluk, N.; Mehrabi, A.; Khedmatgozar Dolati, S.S. Non-destructive testing application for steel bridges. *Appl. Sci.* **2021**, *11*, 9757. [CrossRef]
5. Dolati, S.S.K.; Malla, P.; Mehrabi, A.; Polanco, J.O.; Nanni, A. Non-destructive testing applications for in-service FRP reinforced/strengthened concrete bridge elements. In *Nondestructive Characterization and Monitoring of Advanced Materials, Aerospace, Civil Infrastructure, and Transportation XVI*; SPIE: Bellingham, WA, USA, 2022; pp. 59–74.
6. Guo, B.J.; Xu, T.D. Leica TCA1800 High-precision ultra-long-distance sea-crossing height leveling. *Rail. Investig. Surv.* **2007**, *1*, 1–8.
7. Wu, D.J.; Xiong, W. Experiment and application of sea-crossing elevation transference based on trigonometric leveling. *Sci. Surv. Mapp.* **2016**, *41*, 130–140.
8. Xiang, F.; Liu, C.L.; Chen, Y.W.; Yang, X.F. Experiment and application of sea-crossing elevation measurement of Lingdingyang Bridge based on change rate of elevation anomaly. *Geomat. Spat. Inf. Technol.* **2022**, *45*, 51–54.
9. Wu, D.J.; Xiong, W. A method of sea-crossing trigonometric leveling by total station in the condition of survey station rocking. *Bull. Surv. Mapp.* **2016**, *6*, 87–90.
10. Yan, B.D.; Zhang, L.C.; Sun, C.J.; Wang, Y.P.; Sun, H.C. Precision analysis and application of trigonometric leveling with total station. *Geotech. Eng. Technol.* **2021**, *35*, 351–360.
11. Wu, D.J.; Xiong, W. An improved method of precise trigonometrical river crossing leveling. *Bull. Surv. Mapp.* **2010**, *3*, 4–20.
12. Yang, Q.; Wang, J.G.; Song, H.X. Using sea-crossing leveling measurement to implement elevation transmission between islands. *Hydrogr. Surv. Charting* **2018**, *38*, 75–78.
13. Xu, Y.M.; Wang, J.G.; Song, H.Y. Application of an improved trigonometrical elevation method in sea-crossing elevation transference. *Bull. Surv. Mapp.* **2013**, *S0*, 111–123.

14. Xu, Y.M.; Chen, Y.; Guan, X.; Zhou, J.Y. Approach to minimizing the influence of changeable observational environment on long-distance sea-crossing trigonometric leveling. *J. Surv. Eng.* **2020**, *146*, 04020018. [CrossRef]
15. Zhang, H.J.; Chen, G.H.; Jiang, H.J. Research on the improvement of a long-distance cross-sea triangulation elevation measurement method. *Geomat. Spat. Inf. Technol.* **2020**, *43*, 85–94.
16. Mai, J.K.; Chen, G.H.; Zhou, J.Y.; Zhou, B.Y. Application on extra-long distance and precision cross-sea elevation transfer. *Sci. Surv. Mapp.* **2021**, *46*, 59–64.
17. Wang, Z.P.; Zhou, J.Y.; Chen, G.H.; Dong, B.B.; Huang, R.Z. Data processing and application of super-long-distance cross-sea second-class leveling. *Hydrogr. Surv. Charting* **2021**, *41*, 40–48.
18. Zhang, H.; Hu, B.; Yuan, C.Z. Research and application of precise trigonometric leveling to replace second-class leveling achieving river-crossing leveling. *Bull. Surv. Mapp.* **2019**, *11*, 121–125.
19. Liu, M.B.; Fang, J.C.; Lu, E.F. Calculation method of refraction coefficient based on least square theory. *J. East China Univ. Sci. Technol.* **2016**, *39*, 163–165.
20. Li, J.Z.; Zhang, J.S.; Gao, H.; Liu, Y.; Li, L.P. Application of precise trigonometric leveling to the left and right bank elevation survey of Datengxia Hydro-junction Project. *J. Yangtze River Sci. Res.* **2022**, *39*, 160–164.
21. Li, J.Z.; Liu, Y.; Li, L.P.; Yang, S.M. Application of precise triangulation leveling system in Qinghai area. *Beijing Surv. Mapp.* **2021**, *35*, 529–533.
22. Yang, D.L.; Zou, J.G.; Shen, Y.W.; Zhu, H.B. Research and application of trigonometric leveling to replace precise leveling. *J. Surv. Eng.* **2021**, *147*, 04021012.
23. Tang, Y.J.; Kujawski, E.; Sztubecki, J. Improved leveling approach and its application in civil engineering. *J. Surv. Eng.* **2018**, *144*, 06018002. [CrossRef]
24. Guo, J.Y.; Chen, Y.N.; Liu, X.; Zhong, S.X.; Mai, Z.Q. Route height connection across the sea by using the vertical deflections and ellipsoidal height data. *China Ocean Eng.* **2013**, *27*, 99–100. [CrossRef]
25. Yang, P.B.; Lai, J.F.; Wu, Y.Q. Discussion on high precision elevation transfer method in wide sea of Gangzhu'ao Bridge. *Yangtze River* **2021**, *46*, 65–87.
26. Du, W.J.; Zhang, H.; Jin, S.Y. Research on precise trigonometric leveling instead of second order leveling. *Rail. Investig. Surv.* **2020**, *46*, 1–4.
27. Xu, G.H. A study of EDM precise trigonometric leveling. *Bull. Surv. Mapp.* **2002**, *10*, 22–24.

sustainability

MDPI

Article

Design and Working Performance Evaluation of a Combined Survey Platform under Strong Wave and Deep-Water Conditions

Jun Xiao [1,2,3], Jianping Xian [1,2], Shuai Zou [2,4,*], Song Li [1] and Yongshui Zhang [4]

1 CCCC Second Highway Engineering Co., Ltd., Xi'an 710065, China
2 Shaanxi Union Research Center of University and Enterprise for Bridge Intelligent Construction, Xi'an 710199, China
3 School of Highway, Chang'an University, Xi'an 710064, China
4 School of Civil Engineering, Chongqing Jiaotong University, Chongqing 400074, China
* Correspondence: zs2448602237@163.com; Tel.: +86-186-8169-2538

Abstract: To meet the engineering requirements of long-distance and high-precision elevation transfer under adverse sea conditions, a new structure for a combined survey platform based on the basic principle of vibration reduction and isolation is designed in this paper. The combined survey platform uses the barrier effect of the external support system on the wave-current load to reduce the influence of the wave-current load on the internal support system and the platform, so it can maintain good performance, even in harsh sea conditions, under the premise of no collision between the internal and external support systems. The expected working performance of the structure was verified by numerical simulation, and the influence of the layout, structure size, waves, and water flow on the working performance of the structure was quantitatively analyzed. The results show that: (1) the external support system can better realize the barrier effect of the wave-current load and significantly reduce the influence of the wave-current load on the internal support system and platform; (2) for the independent combined survey platform, when the pile diameters of the internal and external support systems are 0.8 m and 1.2 m and the wall thicknesses are 11.0 mm and 12.0 mm, respectively, the period of the wave is 8.0 s, the water depth is 20.0 m, the speed of water flow is 1.0 m/s, and the wave height is 4.0 m, then the maximum variation of the vertical angle of the platform is only 19.3″; (3) for the attached combined survey platform, the lateral stiffness of the external support system is increased and the displacement of the external support system is significantly reduced because the external support system is connected with the cushion cap through the attachments; therefore, the structure size of the survey platform can be greatly reduced.

Keywords: sea-crossing elevation; combined survey platform; numerical simulation; strong wave and deep-water conditions; wave-current load; working performance

Citation: Xiao, J.; Xian, J.; Zou, S.; Li, S.; Zhang, Y. Design and Working Performance Evaluation of a Combined Survey Platform under Strong Wave and Deep-Water Conditions. *Sustainability* **2022**, *14*, 14360. https://doi.org/10.3390/su142114360

Academic Editors: Kai Wei, Mingjin Zhang, Jian Zhong and Yutao Pang

Received: 14 September 2022
Accepted: 31 October 2022
Published: 2 November 2022

1. Introduction

During the construction of large-scale sea-crossing bridge projects, it is necessary to carry out a high-precision sea-crossing elevation transfer to unify the elevation data on both sides of the bridge and accurately transfer the elevation on both sides to the pier and main beam in the sea. Combined with the many sea-crossing bridges that have been built, the precision of the elevation control survey is generally required to meet national second-class survey standards [1]. The corresponding control parameters are that the accidental mean square error of elevation difference per kilometer is ±1 mm and the total mean square error of elevation difference per kilometer is ±2 mm. However, the span of sea-crossing bridges is generally long, as some of these bridges reach several kilometers or even tens of kilometers; therefore, it is extremely difficult to carry out a sea-crossing elevation transfer survey using the conventional leveling method [2,3]. In the project, the elevation control network is generally densified by setting a survey platform in the sea to shorten the survey distance. With the continuous progress of the project, the survey platform is often set on the foundation works, such as the test pile and some piers under priority construction, which greatly shortens the distance measurement [4], creating conditions for the sea-crossing elevation transfer survey using the total station. However, in the actual construction process, due to the influence of the wind, waves, tide, and other factors, the survey platform is always subject to slight shaking, resulting in the vertical angle change range of the total station installed on the survey platform reaching tens of seconds or even minutes [5,6], and the problem is particularly prominent for engineering projects located in strong wave and deep-water area.

At present, the main challenge to realize long-distance and high-precision elevation transfer on the sea is that, under complex sea conditions, the traditional survey platform will have micro-amplitude shaking, which makes the accuracy of elevation difficult to meet the high standard requirements of relevant specifications or standards.

In the previous engineering construction process, the main methods to reduce the impact of external wave-current load (the wave-current refers to waves and currents, which are independent of each other) on the survey were: (1) reducing the shaking amplitude of the survey platform through the designed structure, and (2) reducing the survey error by combining the method based on taking the average value of multiple repeated observations [5,7]. In order to keep the offshore survey platform stable under the impact of wave-current load, the most straightforward method is to increase the stiffness of the platform by increasing the geometric size of the platform. In references [5,8], the sea-crossing elevation survey under the influence of current and waves is carried out directly by increasing the structural stiffness of the survey platform and setting the forced centering observation stand. Practical tests show that the maximum shaking amplitude of the survey platform is still more than 100″ under the influence of sea conditions, such as waves and currents. At the same time, the construction cost of a survey platform designed with this concept is huge and even unacceptable when it adapts to the environment of strong waves and deep water. Although the survey error can also be reduced by taking the average value of multiple repeated observations, the method is different from the traditional single survey, and the survey methods and the data processing systems require special design, which is inefficient. In essence, this method assumes that the impact of waves in the survey window period is stable to ensure the statistical stability of the survey data. In fact, waves are mostly unsteady [9], and their impact on the survey platform and the survey reference point is also unsteady, so the statistical survey data itself changes with time. To ensure the accuracy of the survey data, it is necessary to observe for a long time or select an appropriate survey window period. The elevation survey is generally carried out at night [10], which causes obvious interference with the construction.

In view of this, to meet the engineering requirements of a long-distance high-precision elevation transfer under severe wave-current conditions, this paper proposes a combined survey platform suitable for strong wave and deep-water conditions based on the idea of vibration reduction and isolation. A reasonable design is carried out according to engineering practice to ensure the stability of the platform under harsh surge conditions. At the same time, through numerical simulation, the influence of different structural sizes and different sea conditions on the working performance of the combined survey platform is quantitatively studied. The main purpose of the combined survey platform proposed in this paper is to properly reduce the project investment. At the same time, it can make the offshore elevation survey free from the influence of the survey window period, and it can realize fast and accurate real-time offshore elevation monitoring [11]. It is of great engineering and practical significance.

2. Principles and Methods

2.1. Research Background

2.1.1. Project Case

This paper is based on the projects of the Peljesac Bridge in Croatia and the Shenzhen-Zhongshan Bridge in China.

The main bridge of the Peljesac Bridge is a six-tower central single cable plane steel box girder low tower cable-stayed bridge. The total length of the bridge is 2404 m, and its span group is (84 + 108 + 108 + 189.5 + 5 × 285 + 189.5 + 108 + 108 + 84) m. The bridge has 14 piers and abutments in total, including 4 piers and abutments arranged on the land on both banks and the remaining 10 piers and towers arranged in the water, as shown in Figure 1.

Figure 1. The layout of the Peljesac Bridge.

The Peljesac Bridge crosses Mariston Bay and the European Union Nature Reserve. It has strict control over the setting of temporary structures in the sea, and the average water depth of the bay at the bridge location is 27 m. The bridge adopts a high pile cap foundation with a soft foundation. During the construction of the bridge, no marine survey platform was set. The elevation transfer is carried out by arranging the survey station on the cushion cap, as shown in Figure 2.

Figure 2. The offshore survey layout of the Peljesac Bridge.

The Shenzhen-Zhongshan Bridge is located in the middle of the Pearl River in the core area. It is a bridge under construction that connects Shenzhen City and Zhongshan City in Guangdong Province, and it has a total length of 24 km. To meet the needs of the construction survey, two permanent survey platforms and two temporary survey platforms are set up, as shown in Figure 3.

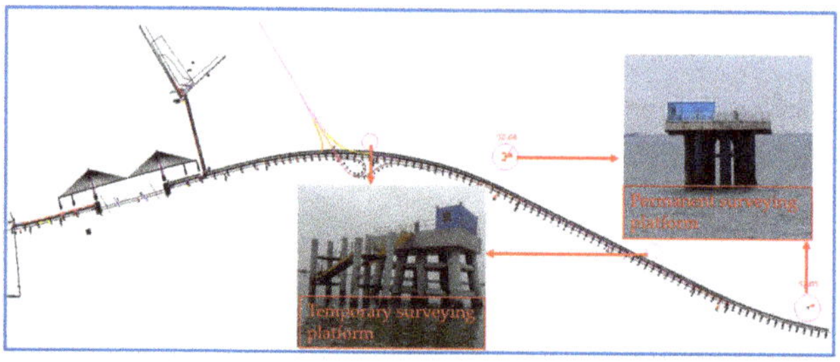

Figure 3. The survey platform layout of the Shenzhen-Zhongshan Bridge.

2.1.2. Working Performance Test

Due to the influence of the waves, the water flow, and the tide, the survey instruments placed on the pier, the cushion cap, and the separately set survey platform each shake greatly during the working period. Figures 4 and 5, respectively, show the changes in the instrument coordinates and the vertical angles of a total station set on a cushion cap in the sea that were caused by the sea conditions present during the construction of the Peljesac Bridge.

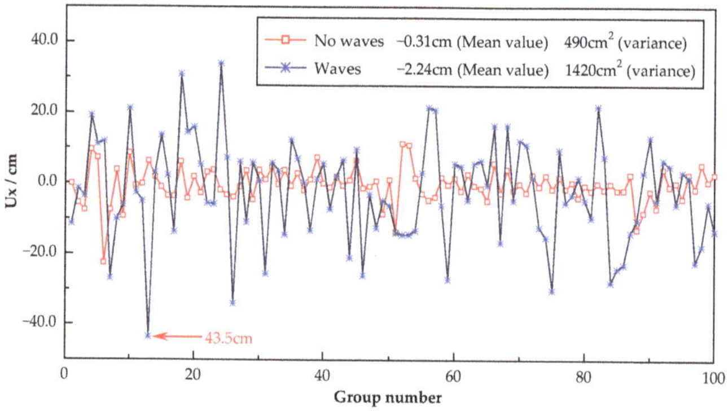

Figure 4. The coordinate changes in the survey instrument.

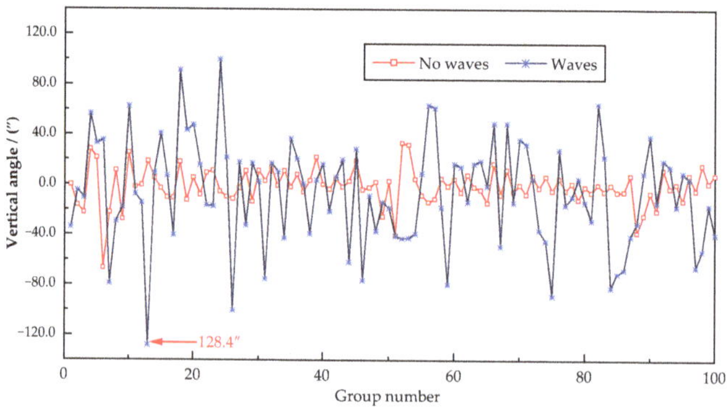

Figure 5. The vertical angle changes in the survey instrument.

The changes in the coordinates and the vertical angles of the measuring instrument under the conditions of waves and no waves show that:

(1) The shaking of the survey platform is relatively significant due to the impact of the wave current, and the maximum coordinate change in the survey instrument in the direction of wave propagation can reach about 43.5 cm, with a corresponding vertical angle change value of 128.4″.
(2) Compared with the condition of waves, the working performance of the survey platform is more stable under the condition of no waves, and the change in the coordinates and the vertical angle in the survey instrument is relatively small.

2.2. The Principle of Design

For the previous survey platform, the wave-current load on the platform is borne by the pile foundation. When the sea conditions are relatively bad, the survey platform will shake to some extent, which affects the survey accuracy of the total station erected on it.

Based on the basic principle of vibration reduction and isolation, this paper designs a new structure for a combined survey platform, and its principle of design is shown in Figure 6. The combined survey platform is composed of internal and external support systems and platforms, and the internal and external support systems are independent of each other in terms of load. The external support system directly bears the wave-current load, while the internal support system is not directly affected by the wave and water flow. It only bears the load imposed by the top platform and the load transmitted by the water and the soil between the internal and external support systems so the combined survey platform can maintain stability even under adverse sea conditions.

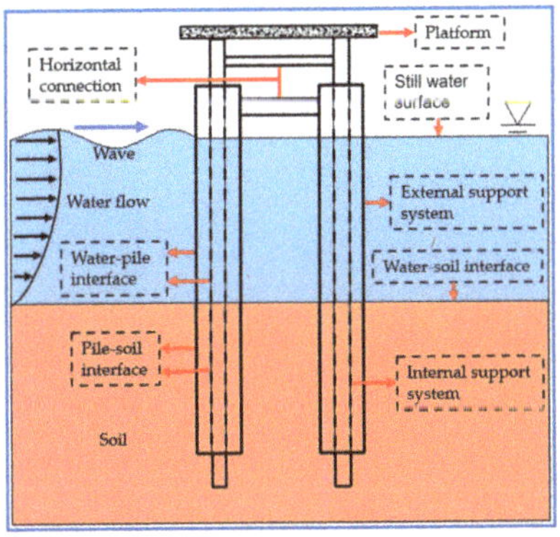

Figure 6. The design principle of the combined survey platform.

According to the different functions of the internal and external support systems, the corresponding structural dimensions need to be designed separately. The larger the inner diameter of the external support system is, the greater the rigidity. Additionally, the larger the gap between the external support system and the internal support system is, the less likely it is that the internal support system will touch the external support system, but this will increase the investment of the entire survey platform. At the same time, with the increase in the inner diameter of the external support system, the wave-current load on the external surface increases correspondingly. Selecting the appropriate stiffness and geometric size while meeting the needs of the working performance is an important measure to achieve the dual goals of applicability and economy.

2.3. Design of the Layout Scheme

According to the different layouts, the design of the combined survey platform can be divided into two schemes: independent and attached.

2.3.1. The Independently Set Combined Survey Platform

Since the construction of sea-crossing bridges is mostly in wide sea areas, the survey platform needs to work as an independent structure. Therefore, the loads caused by waves and water flow are all borne by the survey platform. The rigidity of the traditional survey platform should be strong enough to meet the high standard requirements of survey instruments for the stability of the survey platform. The direct problem is a large amount of investment, and the economy of engineering construction cannot be guaranteed. For the combined survey platform proposed in this paper, because its external wave-current load is completely loaded onto the external support system and the internal support system only bears the load generated by the upper platform and the load transmitted by the soil and the water between the internal and external support systems, its stability does not need to be ensured by greatly increasing the rigidity of the structure like the traditional survey platform, but the high standard working performance requirements of the survey platform are realized.

For the upper platform part of the independently set combined survey platform, two HN-600 load-bearing beams can be set on top of the internal support system. UB-25b distribution beams with a spacing of 488 mm are placed on the load-bearing beams. Fifteen mm thick steel plates are laid on the distribution beams. The design drawing is shown in Figure 7.

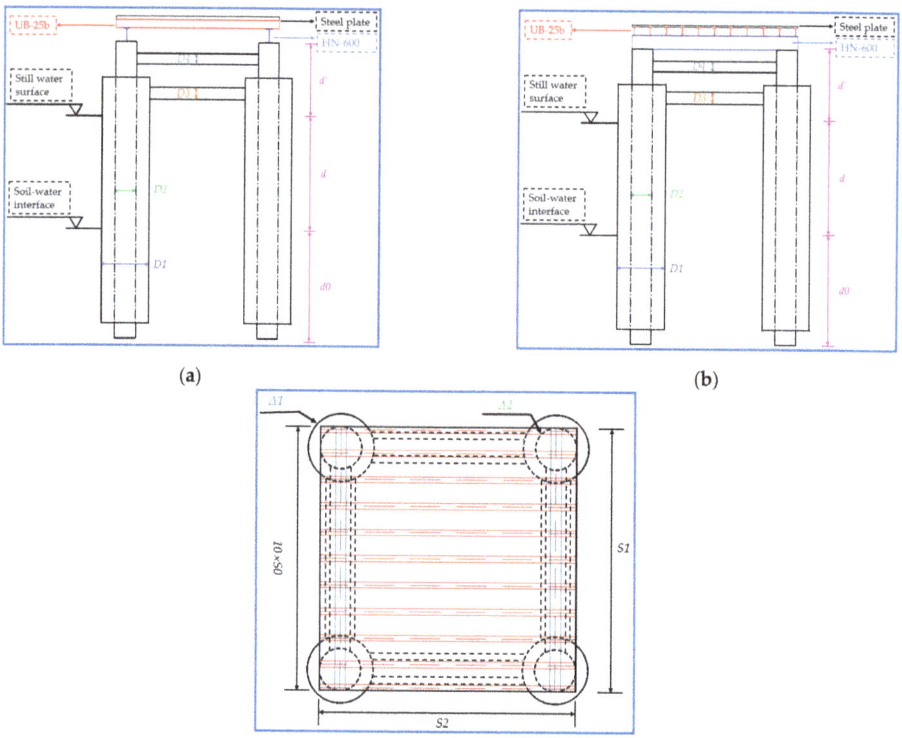

(a) (b) (c)

Figure 7. The design drawing of the independently set combined survey platform. (**a**) The elevation view of the survey platform. (**b**) The side view of the survey platform. (**c**) The vertical view of the survey platform.

D_1 and D_2 are the diameters of the external and internal support systems, respectively; D_3 and D_4 are the diameters of the external lateral connection and the internal lateral connection, which are 0.8 m and 0.4 m, respectively, and their thicknesses are 0.01 m; d_0 is the penetration depth of the internal and external support systems in the soil, which are 30 m and 25 m, respectively; d is the height from the still water surface to the seabed, taken as 20 m; d' is the distance from the top of the internal and external support systems to the still water surface, taken as 5 m and 4 m, respectively; S_0 is the spacing between the distribution beams, taken as 0.448 m; S_1 and S_2 are the length and the width of the platform, which are both 5 m; and Δ_1 and Δ_2 are the thicknesses of the external and internal support systems, respectively.

2.3.2. The Attached Combined Survey Platform

The survey and positioning accuracy of the pile foundation and the cushion cap is lower. In the early stage of the project, cross-water surveys can be carried out by GNSS and other methods. When the construction of the cushion cap is completed, the external support system of the combined survey platform can be attached to the cushion cap through attachments (a steel pipe or other structures with sufficient rigidity). The stiffness of the cushion cap can make up for the deficiency in the stiffness of the external support system so the external support system will not shake significantly under the wave-current load. The cushion cap will shake to a certain extent under the wave-current load, and if the internal support system is also connected with the cushion cap, the shaking will have a negative impact on the working performance of the independent combined survey platform; therefore, the internal support system will not be connected with the cushion cap. At the same time, the attached combined survey platform can be a single-pile platform. The gap between the external support system and the internal system can be properly reduced. The design drawing is shown in Figure 8.

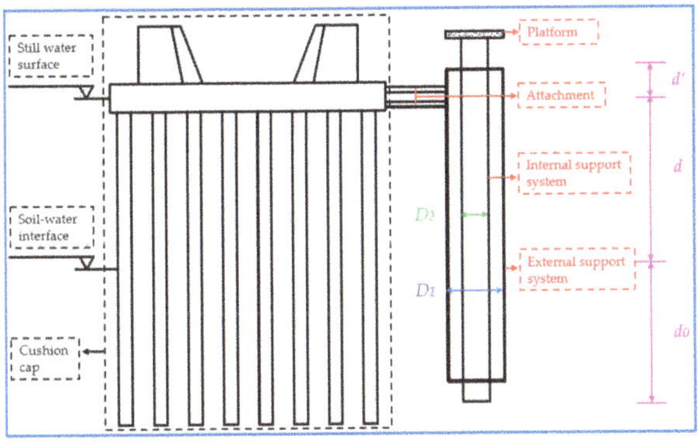

Figure 8. The design drawing of the attached combined survey platform.

Where the parameters in the figure have the same meanings as those in Figure 7.

2.3.3. Project Cost Comparison

In order to make an approximate evaluation of the project cost required for the combined survey platform designed in this paper, the materials required for the construction of the independently set combined survey platform and the permanent survey platform (as shown in Figure 9) in project case 2 are compared. The steel pipe piles of the platform shown in Figure 9 are all ⌀1500 × 12 mm, and the size of the concrete platform is 10 × 10 × 2 m. The structural dimensions of the external and internal support systems of the independently set combined survey platform are: $\Delta_1 = 12.0$ mm, $\Delta_2 = 11.0$ mm, $D_1 = 1.2$ m, $D_2 = 0.8$ m.

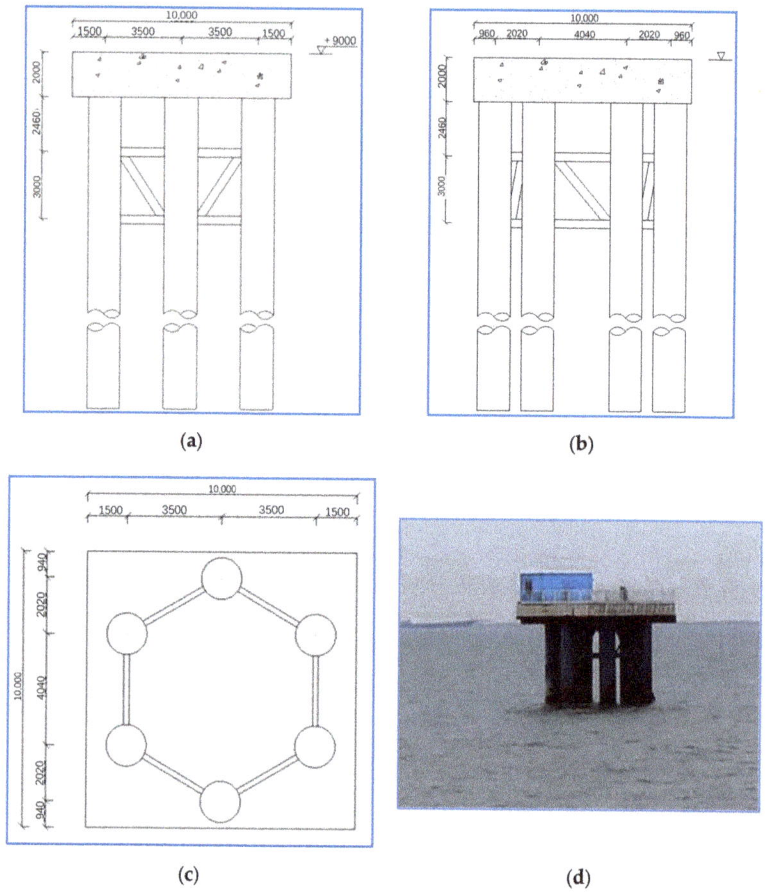

Figure 9. The permanent survey platform. (**a**) The elevation view. (**b**) The side view. (**c**) The vertical view. (**d**) The general arrangement plan. Note: All dimensions in the figure are in millimeters.

The comparison of material consumption is shown in Figure 10. It can be seen from the figure that the traditional survey platform needs to provide enough stiffness to resist the impact of the external wave current load, so its structure size is often large, and the required engineering investment is more than that of the combined survey platform proposed in this paper.

2.4. Dynamic Equations and Boundary Conditions

This section describes the dynamic equations and boundary conditions used in the numerical simulation in Section 2.6.

2.4.1. Governing Equation of Water

It is assumed that the fluid is a compressible, small-disturbance fluid without rotation and viscosity. The governing equation of water, represented by the velocity potential of water, φ, is

$$\frac{\partial^2 \varphi}{\partial x^2} + \frac{\partial^2 \varphi}{\partial y^2} + \frac{\partial^2 \varphi}{\partial z^2} = \frac{1}{c^2}\frac{\partial^2 \varphi}{\partial t^2} \quad (1)$$

where c is the speed of sound in water, $c = \sqrt{K/\rho_w}$, and $c \to \infty$ when the compressibility of the water is neglected; K is the bulk modulus of water [12]; ρ_w is the water density; and T is the time.

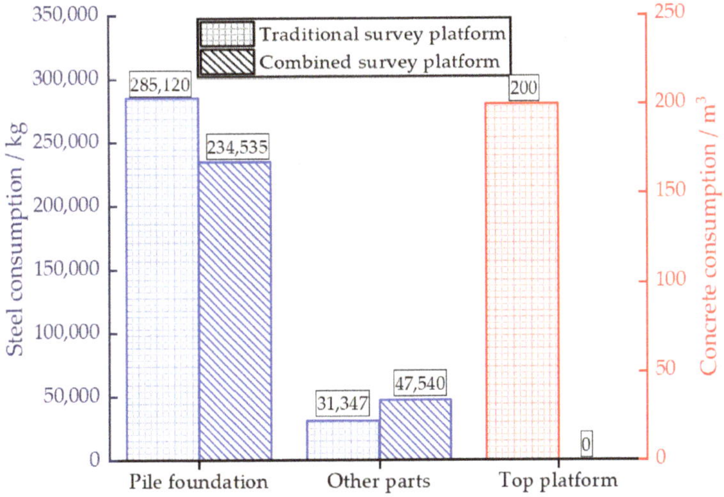

Figure 10. The comparison of material consumption.

2.4.2. Governing Equations of Soil and Pile

According to the basic equations of elastodynamics, the governing equations of soil and pile can be expressed as

$$(\lambda + G)\nabla\nabla\cdot u + G\nabla^2 u = \rho\ddot{u} \tag{2}$$

$$\nabla\nabla\cdot u = \begin{cases} \frac{\partial^2 u_x}{\partial x^2} + \frac{\partial^2 u_y}{\partial x \partial y} + \frac{\partial^2 u_z}{\partial x \partial z} \\ \frac{\partial^2 u_x}{\partial x \partial y} + \frac{\partial^2 u_y}{\partial y^2} + \frac{\partial^2 u_z}{\partial y \partial z} \\ \frac{\partial^2 u_x}{\partial x \partial z} + \frac{\partial^2 u_y}{\partial y \partial z} + \frac{\partial^2 u_z}{\partial z^2} \end{cases} \tag{3}$$

$$\nabla^2 u = \begin{cases} \nabla^2 u_x \\ \nabla^2 u_y \\ \nabla^2 u_z \end{cases} \tag{4}$$

where $\nabla = \left\{ \frac{\partial}{\partial x}\ \frac{\partial}{\partial y}\ \frac{\partial}{\partial z} \right\}^T$ is the vector differential operator of rectangular coordinates; λ and G are the lame constant; and u_x, u_y, and u_z are the displacements of soil and pile in x, y, and z directions respectively.

2.4.3. Surface Boundary Conditions of Water

The pressure on the free surface of still water is equal to the atmospheric pressure. For simplicity of the following calculation,

$$p_{z=H_L} = 0 \tag{5}$$

the hydrodynamic pressure at any time and space can be expressed as

$$p(x, y, z) = -\rho_w \frac{\partial \varphi}{\partial t} \tag{6}$$

where $p(x, y, z)$ is the hydrodynamic pressure of water.

If a small amplitude wave is formed on the free surface of the water, the gravity wave included in the analysis shall be considered through the linear surface wave condition [13]. The pressure on the surface can be expressed as

$$p_{z=H_L} = \rho_w g h \quad (7)$$

The above equation can be expressed as

$$\rho_w g h = -\rho_w \frac{\partial \varphi}{\partial t} \quad (8)$$

2.5. *Calculation of Wave-Current Load*

In addition to the load generated by the wave, the offshore structure is often acted upon by the current load generated by the water flow. Therefore, when analyzing the response of the structure under the action of the sea conditions, the wave load and the current load should be considered together. In this section, the wave parameter calculation model and the model of current velocity varying with water depth used in the numerical calculation in Section 2.6 are described.

2.5.1. Wave Parameter

Waves can be divided into regular and irregular waves according to their shapes. The waves that propagate freely after leaving the wind zone are close to regular waves, while the wind waves in the ocean are irregular. The regular linear wave theory is selected as the wave theory in this paper, and the expression of the wave surface equation is

$$\eta = \frac{H}{2}\cos(kx - \omega t) \quad (9)$$

where H is the wave height and K is the wave number.

The strong wave and deep-water conditions described in this paper are defined according to the international wave level table (as shown in Table 1) and the micro amplitude wave theory.

Table 1. The international wave level table.

Degree	Wave Height (m)	Classification	Description
0	-	Clam sea	Clam (glassy)
1	<0.1	Smooth sea	Clam (rippled)
2	0.1–0.4	Small sea	Smooth
3	0.5–1.2	Slight sea	Slight
4	1.3–2.4	Moderate sea	Moderate
5	2.5–3.9	Rough sea	Rough
6	4.0–5.9	Very rough sea	Very rough
7	6.0–8.9	High sea	High
8	9.0–13.9	Very high sea	Very high
9	>14.0	Precipitous sea	Phenomenal

According to the micro amplitude wave theory, wave types can be divided as follows. The deep-water wave:

$$\frac{d}{L} \geq 0.5 \quad (10)$$

The finite depth wave:

$$0.5 > \frac{d}{L} > 0.05 \quad (11)$$

The shallow water wave:

$$\frac{d}{L} \leq 0.05 \quad (12)$$

where L is the wavelength and d is the water depth. In engineering, d = 0.5 L is often used as the boundary between infinite water depth and finite water depth.

The dispersion equation of linear waves is

$$\mu = \frac{\mu_0}{\tanh(\mu)} \tag{13}$$

where $\mu = \frac{2\pi}{L}d$ and $\mu_0 = \frac{4\pi^2}{gT^2}d$.

After the period is determined, the wavelength can be solved through the above dispersion equation, and vice versa. Many scholars have put forward some approximate methods for a direct solution to the above equations. In this paper, the following calculation (Formula 14), proposed by Vatankhah and Aghashariatmadari [14] through numerical analysis, is used to solve them.

$$\mu = \frac{\mu_0}{\sqrt{\tanh(\mu_0)}}\left(1 + \mu_0 e^{-(1.835+1.225\mu_0^{1.35})}\right) \tag{14}$$

Under any water depth condition, the maximum relative error of the formula is less than 0.019%.

2.5.2. The Flowing Speed Model of Seawater

In the direction perpendicular to the sea level, the speed of the seawater flow changes with the depth of the seawater. Generally, the speed changes from the sea level to the seabed in a regular manner of "large up and small down." In this paper, the 1/7 exponential rate distribution model [15], proposed by Soulsby, is adopted. The model divides the vertical distribution of the flow speed into two layers for calculation. The upper layer adopts a linear distribution, and the lower layer adopts an exponential distribution. The schematic diagram of the model is shown in Figure 11. The vertical distribution of flow speed is given by

$$u(z) = \begin{cases} u_0\left(\frac{z}{0.32h}\right)^{\frac{1}{7}} & 0 \le \frac{z}{h} \le 0.5 \\ 1.07u_0 & 0.5 < \frac{z}{h} < 1 \end{cases} \tag{15}$$

where u_0 is the velocity of the seawater surface, m/s.

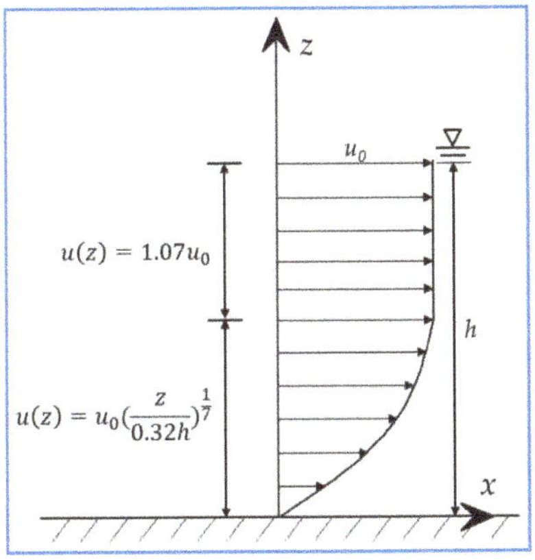

Figure 11. The flowing speed model of seawater in the direction of water depth.

2.5.3. Wave-Current Load

There are many methods used to calculate the wave-current load [16,17]. In this paper, the nonlinear wave force [18] is not considered. The Morison formula with current participation is used to calculate the wave-current load on the pile. The wave-current load is given by

$$p_H = \frac{1}{2}\frac{\gamma}{g}C_D D u_{wc}|u_{wc}| + \frac{\gamma}{g}C_M A\frac{\partial u}{\partial t} \quad (16)$$

$$u_{wc} = u_w + u \quad (17)$$

where p_H is the wave-current load on a unit length micro-segment at any height z of the pile, N/m; A is the sectional area of the pile, m^2; C_D is the drag coefficient; C_M is the inertia force coefficient; γ is the weight of the water, kN/m^3; $\frac{\partial u}{\partial t}$ is the horizontal acceleration of the orbital motion of the water particles, m/s^2; u_{wc} is the total horizontal velocity of the water particles, m/s; u_w is the flow speed of the water, m/s; and u is the flow speed of the water particles generated by waves, m/s.

2.5.4. Hydrodynamic Coefficient

The values of the two hydrodynamic coefficients in the Morison formula, namely the drag coefficient C_D and the inertia force coefficient C_M, are related to the accuracy of the final calculation results. For the values of these two coefficients, local scholars and scholars abroad have done a lot of experimental research, and they have given many recommended values in relevant papers, as well as specifications for reference in engineering practice. In general, if we want to get more accurate results, we must carry out hydrodynamic experiments in the actual water area to determine the value of the correlation coefficient. In this paper, the C_D and C_M values are determined according to references [19,20]. Finally, the C_D is taken as 1.2, and the C_M is taken as 2.0.

When calculating the wave-current load on the pile foundation of offshore structures, if the relative pile distance S/D (S is the pile center distance and D is the pile diameter) is small, the blocking effect of the pile array on the wave-current will increase the flow speed at the pile foundation, and the pile group effect cannot be ignored. When $S/D \geq 3.0$, the pile group effect is weak, and the impact can be ignored [21].

2.6. Numerical Simulation

2.6.1. Method

In this paper, the Coupled Acoustic-Structure approach (*CAS*) is used for numerical simulation analysis in the finite element software ABAQUS. The acoustic wave control equation expressed by the acoustic pressure p as an independent variable is also called the Helmholtz equation, which is given as

$$\frac{\partial^2 p}{\partial t^2} - c^2\nabla^2 p = 0 \quad (18)$$

where c is the speed of sound in the liquid, $c = \sqrt{K/\rho}$, and k is the bulk modulus of the liquid.

The advantage of using the *CAS* approach for modeling is that it is relatively simple and effective for numerical processing because it assumes no material flow, so no mesh deformation occurs. In addition, since the acoustic element has only pressure freedom at each node, it significantly reduces the computational time required for simulation [22].

The impedance boundary condition is defined on the free surface of the acoustic element using linear wave theory. The boundary impedance relates the pressure of the acoustic element to the normal motion at the acoustic structure interface. The speed ($\dot{u}_{out}$) of acoustic particles in the direction of the external normal of the fluid surface is related to the pressure and the rate of change in the pressure over time, which is given as

$$\dot{u}_{out} = \frac{1}{k_1}\dot{p} + \frac{1}{c_1}p \quad (19)$$

where p is the acoustic pressure; $\dot{p}$ is the change rate of sound pressure with time; $\frac{1}{k_1}$ is the ratio coefficient between the acoustic pressure and the displacement in the normal direction of the surface; and $\frac{1}{c_1}$ is the ratio coefficient between the acoustic pressure [23].

2.6.2. The Finite Element Model

In the modeling and numerical analysis of the combined survey platform shown in Figure 7, acoustic elements are used to simulate the internal and external support systems and the water between the support systems [24–26]. The internal and external support systems are simulated by shell elements, and the seabed is simulated by solid elements. The simulation of the water-pile and water-soil interfaces is realized by "tie" constraints, and the simulation of the pile-soil interface is realized by "embedded" constraints. The interaction between the platform and the distribution beam above the internal support system is defined by "tie" constraints in the software [27], and the analysis program uses dynamic implicit analysis. The parameters of pile and soil in the model are shown in Table 2.

Table 2. The parameters of the model.

Part	Poisson's Rati	Density/(kg/m^3)	Elastic Modul/(MPa)	Size/(m)	Bulk Modulus/(MPa)
Pile	0.2	2500	30,000	/	/
Soil	0.4	1800	100	$1.5d_0 + 20D_1$ *	/
Water	/	1024		/	206.78

* Note: To eliminate the boundary effect, the size of the seabed in the depth direction is 1.5 times the embedded depth of the pile, and it is 20 times the pile diameter in the direction perpendicular to the depth [28].

During the shaking process in the water between the internal and external support systems, gravity is essential for resilience, but it is not allowed to directly define the uniformly distributed gravity load in the acoustic elements. In order to solve this problem, the acoustic impedance is introduced [22]. In the "geostatic step" of the model, the hydrostatic pressure corresponding to the water depth should be applied to the soil, and the treatment of the stress balance should be carried out by introducing the stress field. The wave-current load caused by the wave and water flow on the external support system is loaded by using the time history load obtained by the calculation method described in Section 2.5.3. The modeling process is shown in Figure 12.

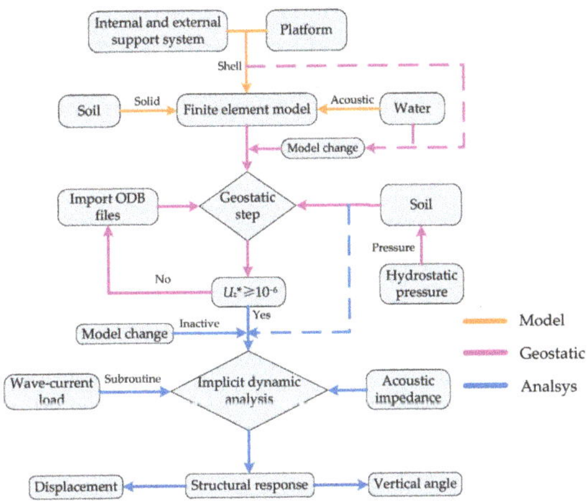

Figure 12. The flow chart of the numerical simulation modeling. * Note: U_z is the displacement in the depth direction of the soil in the geostatic step.

2.6.3. The Development and Verication of Subroutine

The "AQUA" module in ABAQUS can only calculate the wave-current load on small-scale structures, and the model element can only be calculated by the beam element. To load the combined survey platform proposed in this paper with the wave-current load, the load on the beam element is converted to the shell element by the principle applied in reference [29], and then the subroutine "Utracload" is compiled in the Fortran language to load the corresponding load.

The principle applied in this process is: assuming that the pressure P is uniformly distributed on the half circumference of the pile, which is the surface facing the direction of the wave, the resultant force F_x of the pile in the x direction is the total wave load on the pile, and the resultant force F_y of the pile in the y direction is 0 [29]. The calculation principle is shown in Figure 13. Reference [29] has studied the rationality of this method. The research results show that the method can be used to simplify the wave load on small-scale structures, and the accuracy of the calculation results fully meets the engineering requirements.

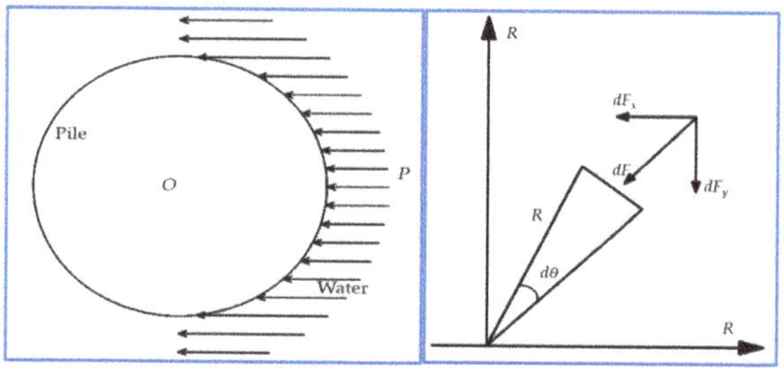

(a) (b)

Figure 13. The conversion principle of the wave-current load on the shell element.

The expression of the wave-current load on the shell element is given as

$$F_H = \int_{-\frac{\pi}{2}}^{\frac{\pi}{2}} dF_x = 2\int_0^{\frac{\pi}{2}} Pcos\theta ds = 2\int_0^{\frac{\pi}{2}} PRcos\theta d\theta = 2PR = PD \quad (20)$$

where θ represents the angle between the X axis and the line connected by the wave propagation direction and the center of the steel pipe. Its function is to decompose the wave force of the steel pipe pile along the X axis and the Y axis, respectively.

To verify the accuracy of the load subroutine, the beam element and the shell element are used to establish models for the steel pipe pile with a diameter of 1 m and a pile length of 20 m in ABAQUS, respectively, and contrast analysis of the two models in the wave height of 1 m, a period of 8 s wave conditions, and the steel pipe piles by the total difference of wave load. Among them, the beam element model uses the "AQUA" module to load the wave load, and the shell element model uses the above-mentioned load subroutine to load the wave load calculated by the theoretical formula. The total wave load received by the beam element model is the support reaction force received by the pile bottom in the wave propagation direction, while the shell element model is the sum of the support reaction forces received by each node at the pile bottom in the wave propagation direction. The comparison of the results is shown in Figure 14.

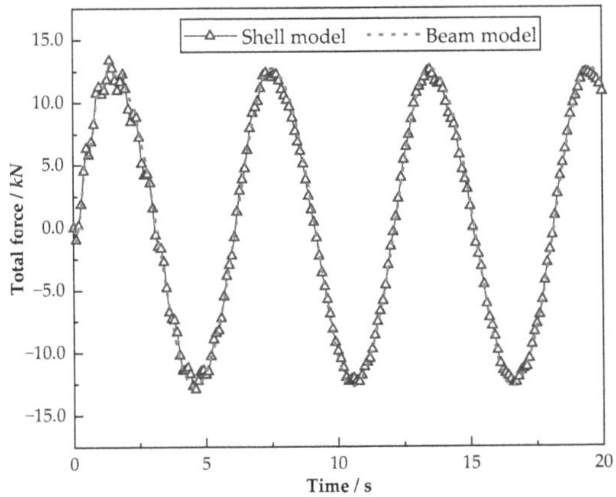

Figure 14. The comparison of the results.

It can be seen from the comparison of the two models shown in Figure 14 that the results obtained by the shell element and the beam element are the same. This shows that the principle of the method used to convert the wave load on the beam element into the shell element is reasonable, and it also verifies the correctness of the Fortran subroutine written in this paper.

3. Analysis of Working Performance

Based on the basic principle of vibration reduction and isolation, a new structure for a combined survey platform is proposed in this paper. Two kinds of layout designs are carried out according to the different positions of the survey platform: independent and attached. Quantitative research is carried out to understand the specific effects of the structural size and the sea conditions on the working performance of the combined survey platform.

For the convenience of expression, the pile diameter and the wall thickness mentioned below refer to the combination of the pile diameter and the wall thickness of the external support system, and the platform described is the survey platform on the top of the internal support system.

3.1. The Working Performance of the Independent Layout Scheme

According to the different roles played by the internal and external support systems in the process of quantitative research on the independently set combined survey platform, six combination schemes are designed according to the different structure sizes, as shown in Table 3.

Table 3. The schemes of size combinations.

Scheme	The External Support System		The Internal Support System	
	Δ1/mm	D1/m	Δ2/mm	D2/m
Scheme 1	8.0	1.0	11.0	0.8
Scheme 2	10.0	1.0	11.0	0.8
Scheme 3	12.0	1.0	11.0	0.8
Scheme 4	8.0	1.2	11.0	0.8
Scheme 5	10.0	1.2	11.0	0.8
Scheme 6	12.0	1.2	11.0	0.8

Note: In the process of construction, the verticality of steel pipe piles is often difficult to ensure. Therefore, during the driving of the steel pipe piles of the internal and external support systems, limit devices should be used to keep their inner wall surfaces parallel to the maximum extent and to provide more guarantees for avoiding collisions between the internal and external support systems under the wave-current load.

3.1.1. The Structural Response of the Independently Set Combined Survey Platform

In order to verify the working performance of the independently set combined survey platform under the wave-current load, the wave height, the wave period, the speed of the water flow, and the water depth are set as 1.0 m, 8.0 s, 1.0 m/s, and 20.0 m, respectively. The time history of the wave-current load is loaded on the survey platform mentioned in Scheme 2 in Table 3. Figure 15a shows the displacement variation time history of the free end of the internal and external support systems. As the figure shows, under the action of the wave-current load, the variation amplitude of the displacement of the external support system is large, with a maximum value of up to 65.1 mm, while the variation amplitude of the displacement of the internal support system is significantly reduced compared to the external support system, and the maximum displacement is only 5.4 mm.

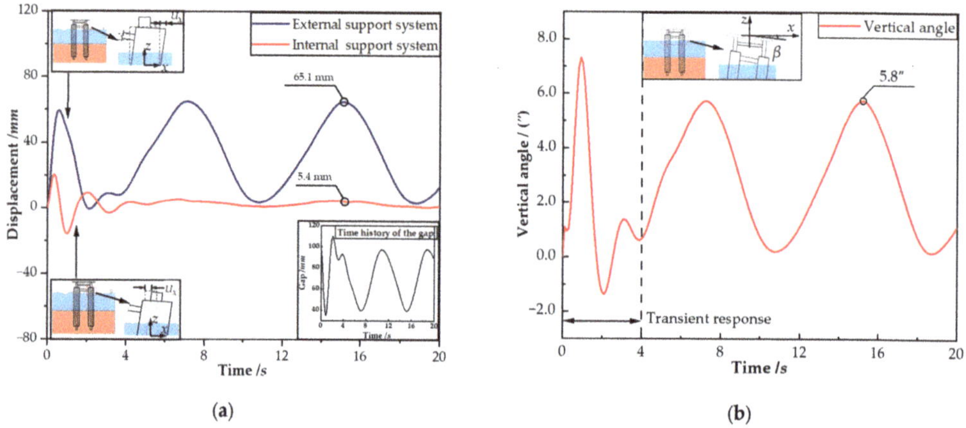

Figure 15. The time history of structural response. (**a**) The displacement of the internal and external support systems. (**b**) The vertical angle of the platform.

The gap between the internal and external support systems can be expressed as

$$\Delta U = (D_1 - D_2) \times 1000 - U_1 + U_2 \quad (21)$$

where ΔU is the gap between the internal and external support systems, mm; D_1 is the diameter of the external support systems, m; D_2 is the diameter of the internal support systems, m; U_1 is the displacement of the external support system, mm; and U_2 is the displacement of the internal support system, mm.

The subgraph in Figure 15a shows the time-history curve of the gap between the internal and external support systems. The large difference in displacement between the internal and external support systems indicates that: (1) the external support system achieves the expected barrier effect to the wave-current load, and the internal support system only bears the load transferred by soil and water quality between the internal and external support systems, which can significantly reduce the wave-current load on the influence of the internal support system; and (2) since the displacement of the internal support system is small, whether the internal and external support systems collide is mainly affected by the displacement of the external support system. Figure 15b shows the variation time history of the vertical angle of the independently set combined survey platform. As the figure displays, under the action of the wave-current load, the variation amplitude of the vertical angle of the platform is small, and the maximum value of the vertical angle is only 5.8″, which proves that the independently set combined survey platform designed in this paper still has a good working performance under the influence of the wave and the current.

The prerequisite for the good working performance of the combined survey platform proposed in this paper is that the internal and external support systems do not collide. Additionally, since the displacement of the internal support system is small compared to the displacement of the external support system, the quantitative research carried out in the following is evaluated by the displacement of the external support system and the vertical angle of the top platform.

3.1.2. The Influence of Sea Conditions

To study the influence of the wave-current period on the working performance of the independently set combined survey platform, the wave height, the velocity, the water depth, and the wave-current period are set as 1.0 m, 1.0 m/s, 20.0 m, and 4.0–10.0 s, respectively. Then, the wave-current load is loaded on the survey platform with different structure size combination schemes. Figure 16a,b shows the influence of the maximum displacement of the external support system and the maximum vertical angle of the platform by the wave-current period. As shown in the figure, for the same combined survey platform model, the maximum displacement of the external support system and the maximum vertical angle of the platform gradually decrease with the gradual increase of the wave-current period, but the decreasing trend is relatively gentle. In general, the variation of the wave-current period has little influence on the working performance of the independently set combined survey platform. When the wave-current load is the same, the model results with different structural sizes have certain differences. By increasing the pile diameter or the wall thickness, the displacement of the external support system can be reduced to a large extent, but the change in the vertical angle of the platform is small.

Due to the influence of tides, the water depth at the location of the survey platform will change constantly. The wave height, the speed of the water flow, the wave-current period, and the water depth are set as 1.0 m, 1.0 m/s, 8 s, and 17.0~23.0 m, respectively. The most unfavorable effects of the water depth on the working performance of the dependently set combined survey platform are shown in Figure 16c,d. The figure highlights that the maximum displacement of the external support system and the maximum vertical angle of the platform basically show a trend of linear increase with the increase in the water depth. An increase in the water depth from 17 m to 23 m results in a change in the maximum displacement of the external support system of about 45 mm and a change in the maximum vertical angle of the platform of about 2.7″. When the wave current load is the same, the displacement of the external support system can be reduced, to a large extent, by increasing the pile diameter or the wall thickness, but the magnitude of the change in the vertical angle of the platform is relatively small.

In order to quantify the influence of the water flow on the working performance of the independently set combined survey platform, the wave height, the wave-current period, the water depth, and the water flow speed are set as 1.0 m, 8.0 s, 20.0 m, and 0~2.5 m/s, respectively. Figure 16e,f respectively show the influence of the maximum displacement of the external support system and the maximum vertical angle of the platform by the water flow speed. It can be seen from the figure that the water flow speed has a great influence on the performance of the independently set combined survey platform. With the increase in water flow speed, the maximum displacement of the external support system and the maximum vertical angle of the platform increase significantly. Due to the large displacement of the external support system, the gap between the internal and external support systems should be increased accordingly to avoid a collision between the internal and external support systems. As shown in the figure, by increasing the pile diameter and the wall thickness of the external support system, the independently set combined survey platform can still work normally at a water flow speed of 2.0 m/s, and the maximum vertical angle of the platform is not more than 20″. It is thus clear that the independently set combined survey platform is sensitive to the influence of the water flow speed. It has a good effect by changing the structure size to reduce the displacement of the external support system and prevent collisions between the internal and external support systems under the influence of large water flow speeds, but the impact on the vertical angle is relatively small.

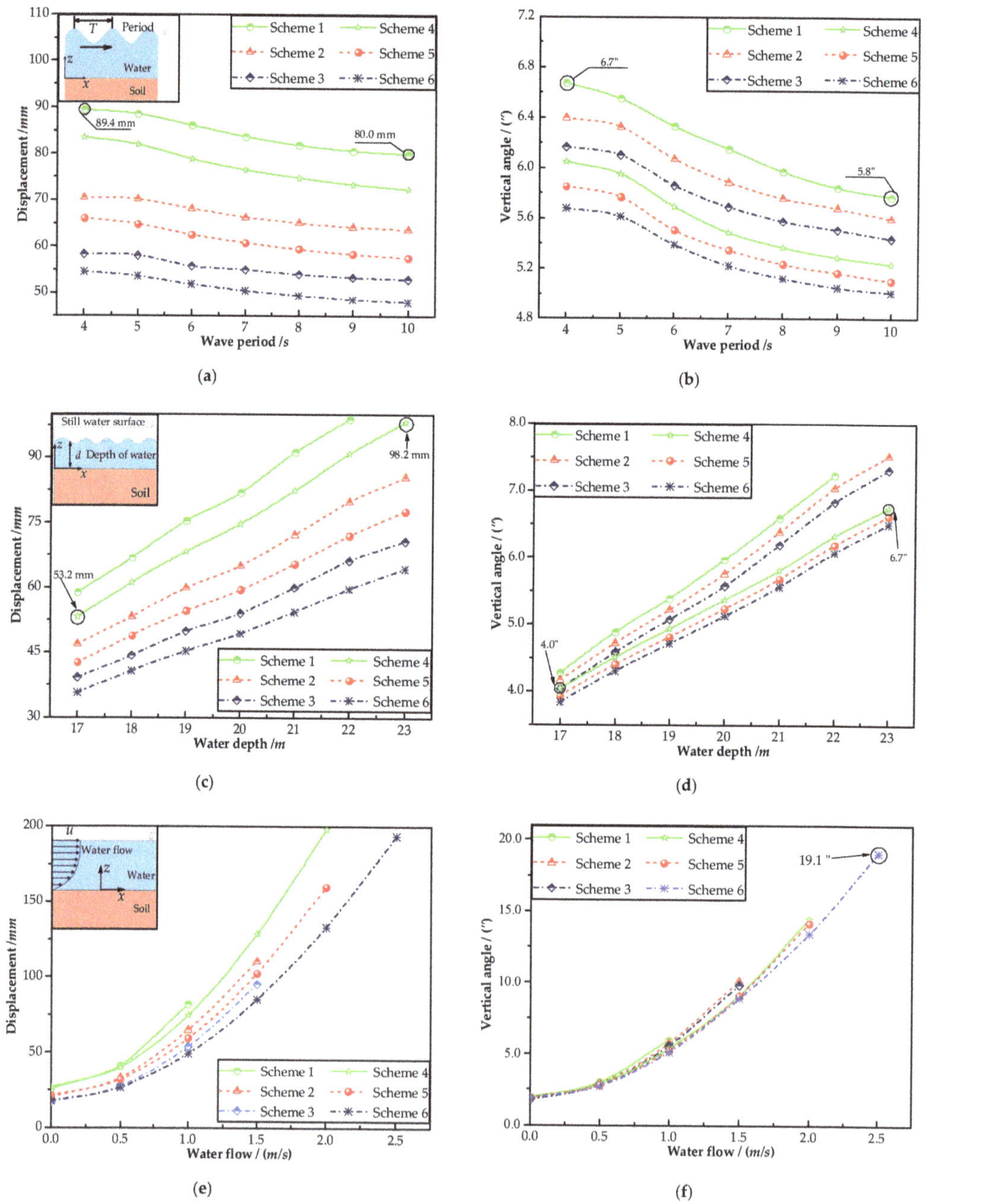

Figure 16. *Cont.*

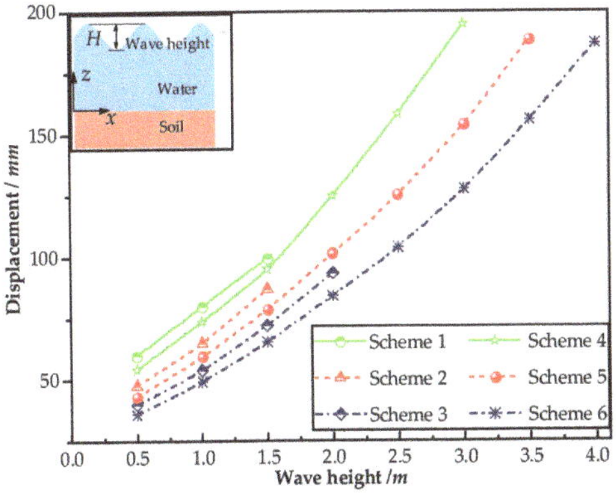

(**g**)

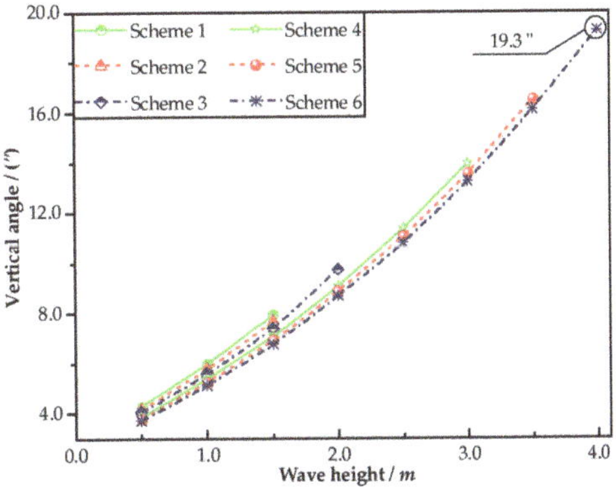

(**h**)

Figure 16. The most unfavorable effect of sea conditions on the working performance of the dependently set combined survey platform. (**a**,**c**,**e**,**g**) The maximum displacement of the external support system. (**b**,**d**,**f**,**h**) The maximum vertical angle of the platform.

To clarify the influence of the wave height on the working performance of the independently set combined survey platform, the period, the water depth, the water flow speed, and the wave height are set to 8 s, 20 m, 1.0 m/s, and 0.5~4.0 m, respectively. Figure 16g,h respectively show the influence of the maximum displacement of the external support system and the maximum vertical angle of the platform by the wave height. As the figure shows, due to the gradual increase in wave height, the maximum displacement of the external support system and the maximum vertical angle of the platform both increase significantly, and the possibility of collision between the internal and external support systems also increases accordingly. The internal and external support systems need to provide a larger gap to avoid a collision. As shown in the figure, by increasing the pile diameter and the wall thickness of the external support system, the displacement of the external support system can be significantly reduced, and the collision between the internal and external support systems can be effectively prevented, but it has little effect on the vertical angle of the platform.

3.2. The Working Performance of the Attached Layout Scheme

The displacement of the external support system caused by the wave-current load in the direction parallel to the attachments is small for the attached combined survey platform. Therefore, the influence of the wave-current load in the direction perpendicular to attachments on its working performance is studied in this paper. As shown in Table 4, according to different combinations of sizes, nine layout schemes are designed.

Table 4. The schemes of the size combinations.

Scheme	The External Support System		The Internal Support System	
	Δ1/mm	D1/m	Δ1/mm	D2/m
Scheme 1	14.0	2.2	14.0	2.0
Scheme 2	16.0	2.2	14.0	2.0
Scheme 3	18.0	2.2	14.0	2.0
Scheme 4	14.0	2.4	14.0	2.0
Scheme 5	16.0	2.4	14.0	2.0
Scheme 6	18.0	2.4	14.0	2.0
Scheme 7	14.0	2.6	14.0	2.0
Scheme 8	16.0	2.6	14.0	2.0
Scheme 9	18.0	2.6	14.0	2.0

3.2.1. Effect Verification of Attachment Layout Scheme

The attached layout scheme proposed in this paper is to attach the external support system of the combined survey platform to the cushion cap with attachments to improve the lateral stiffness of the external support system, reduce the structure size, and ensure the working performance of the survey platform. To verify the working performance of the attached combined survey platform under the influence of the wave-current load, the wave height, the wave period, the water flow speed, and the water depth are set as 1.0 m, 8.0 s, 1.0 m/s, and 20.0 m, respectively. The combined survey platform mentioned in Scheme 5 in Table 4 is respectively arranged in independent and attached layouts. Then, the wave-current load time history loading is carried out. Figure 17a,b show the time history of the displacement of the free end of the external support system and the vertical angle of the platform, respectively, for the single-pile combined survey platform with two types of arrangement. It can be seen from the figure that, under the influence of the wave-current load, the maximum displacement of the external support system of the independently set combined survey platform is 20.4 mm, and the maximum vertical angle of the platform is 32.4″. For the attached combined survey platform, the maximum displacement of the external support system is only 2.6 mm and the maximum vertical angle of the platform is 10.8″.

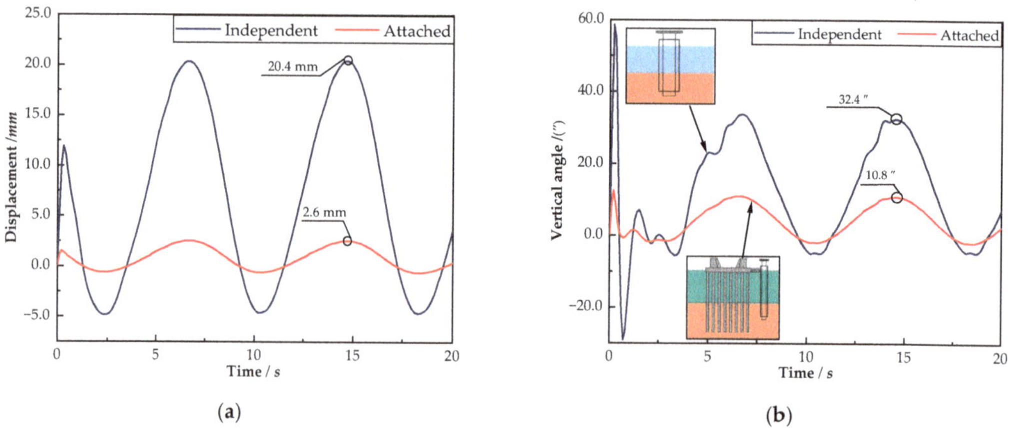

Figure 17. The time history of the structural response of the combined survey platform with two types of layouts. (**a**) The displacement of the external support system. (**b**) The vertical angle of the platform.

The stiffness of the external support system can be increased by attaching the external support system to the cushion cap, and the displacement of the external support system and the vertical angle of the platform can be significantly reduced, which verifies the expected effect of the attached layout scheme proposed in this paper.

3.2.2. The Influence of Sea Conditions

The wave height, the water flow speed, the water depth, and the wave-current period are set as 1.0 m, 1.0 m/s, 20.0 m, and 4.0–10.0 s, respectively. The wave-current load is loaded on the survey platform with different size combination schemes. Figure 18a,b, respectively, show the influence of the maximum displacement of the external support system and the maximum vertical angle of the platform by the wave-current period. As shown in the figure, due to the change in the wave-current period, the displacement of the external support system and the vertical angle of the platform are less affected—the maximum displacement of the external support system is only 8.1 mm, and the maximum vertical angle of the platform is 11.8″. With the period of the wave-current increasing gradually, the maximum displacement of the external support system decreases gradually and then flattens out, while the maximum vertical angle of the platform increases gradually and then flattens out.

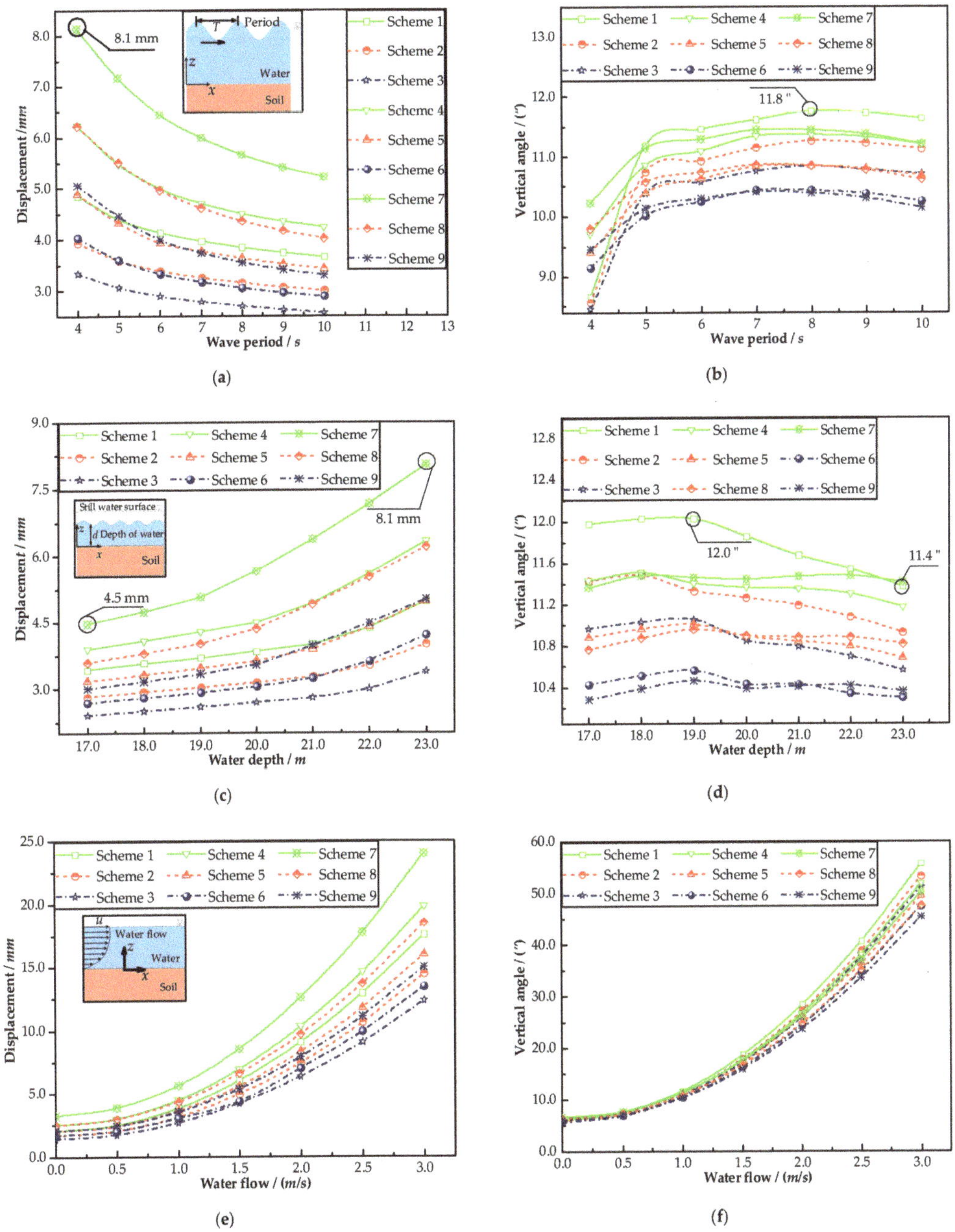

Figure 18. *Cont.*

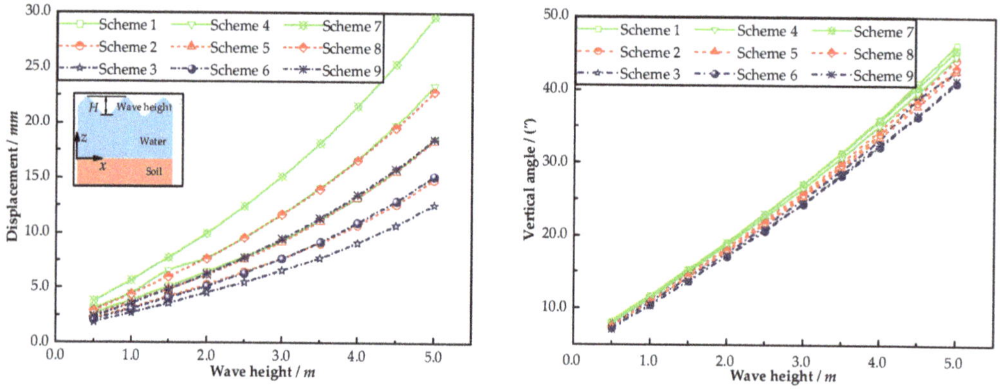

(g) (h)

Figure 18. The most unfavorable effect of sea conditions on the working performance of the attached combined survey platform. (**a**,**c**,**e**,**g**) The maximum displacement of the external support system. (**b**,**d**,**f**,**h**) The maximum vertical angle of the platform.

The wave height, the water flow speed, the wave-current period, and the water depth are set as 1.0 m, 1.0 m/s, 8.0 s, and 17.0~23.0 m, respectively. Figure 18c,d, respectively, show the influence of the maximum displacement of the external support system and the maximum vertical angle of the platform by the tide. As shown in the figure, the variation in the water depth caused by the tides has a low impact on the displacement of the external support system and the vertical angle of the attached combined survey platform. The maximum displacement of the external support system gradually increases with the increase in water depth. Within the selected water depth range, the maximum change in the maximum displacement of the external support system is about 3.6 mm. The vertical angle of the platform shows a trend of rising first and then decreasing, and the maximum change in the vertical angle within the selected water depth change range is about 0.6″.

The wave-current period, the water depth, the wave height, and the water flow speed are set as 8.0 s, 20.0 m, 1.0 m, and 0–3.0 m/s, respectively. Figure 18e,f, respectively, show the influence of the maximum displacement of the external support system and the maximum vertical angle of the platform by the water flow speed. The figure shows that the water flow speed has a significant influence on the performance of the attached combined survey platform. When the water flow speed increases, the maximum displacement of the external support system and the maximum vertical angle of the platform increase significantly. Changing the pile diameter and the wall thickness affects the displacement of the external support system to a certain extent, but the vertical angle of the platform is less affected. Increasing the pile diameter increases the displacement of the external support system, and increasing the wall thickness can reduce the displacement of the external support system, but the displacement of the external support system is generally small.

The wave-current period, the water depth, the water flow speed, and the wave height are set as 8.0 s, 20.0 m, 1.0 m/s, and 0.5~5.0 m, respectively. Figure 18g,h, respectively, show the influence of the maximum displacement of the external support system and the maximum vertical angle of the platform by the wave height. As shown in the figure, the maximum displacement of the external support system and the maximum variation of the vertical angle of the platform increase significantly with the increase in wave height. It has great influence on the displacement of the external support system by changing the structural size, but it has little influence on the vertical angle of the platform. With the same wave height, the maximum displacement of the external support system decreases with the increase in the wall thickness and increases with the increase in the pile diameter, and the difference between the different combinations of sizes increases significantly with the increase in the wave height.

4. Discussion

Due to wave-current load, the traditional survey platform has a large shaking amplitude, so it is difficult to meet the high standard requirements of long-distance and high-precision elevation transfer in the construction of sea-crossing bridges. This paper designs a new structure for a combined survey platform based on the basic principle of vibration reduction and isolation to better solve this problem. The internal support system is only subjected to loads transferred by soil and water quality between the internal and external support systems, which has little influence on the displacement of the internal support system and the vertical angle of the platform, so the new structure has excellent working performance.

Two kinds of layout designs are carried out according to the different positions of the survey platform: independent and attached. The wave-current load is loaded on the models of the two layouts. The time history response analysis of the structure shows that: (1) the external support system achieves the expected blocking effect on the wave-current load, significantly reduces the impact of the wave-current load on the internal support system, and provides a guarantee for stable work of the platform on the top of the internal support system; and (2) attaching the external support system to the cushion cap through attachments can increase the stiffness of the external support system, significantly reduce the displacement of the external support, and effectively prevent collision between the internal and external support systems.

Different sea areas have different sea conditions. Through the reasonable setting of sea conditions, the specific effects of the wave-current period, the wave height, the current velocity, and the tide on the two layouts are quantitatively analyzed. The analysis results show that: (1) tide and wave-current periods have little impact on the working performance of the combined survey platform; and (2) the increase in wave height and water flow speed has a significant impact on the working performance of the survey platform.

Through the quantitative analysis of the structural dimensions of the combined survey platforms of the two layouts, it is found that the structural stiffness of the external support system can be improved by increasing the wall thickness and the pile diameter of the external support system, but as the pile diameter increases, the wave-current load on the external support system will also increase. The above analysis shows that: (1) changing the pile diameter and the wall thickness of the external support system has a significant impact on the displacement of the external support system, but it has a small impact on the vertical angle of the platform; (2) the displacement of the external support system of the combined survey platform of the two layouts can be effectively reduced by increasing the wall thickness; and (3) for the independently set combined survey platform, increasing the pile diameter can improve the overall stiffness of the external support system and better reduce the displacement of the external support system. For the attached combined survey platform, increasing the pile diameter can increase the stiffness of the structure, but since the internal and external support systems are single-pile structures and the overall stiffness of the structure is small, the increase in the stiffness caused by the increase in the pile diameter is insufficient to resist the deformation caused by the increase in the wave-current load, and the displacement of the external support system will increase. The above analysis also shows that: (4) for the independent combined survey platform, the pile diameter can be appropriately increased to provide greater rigidity, and the gap between the internal and external support systems is larger to prevent collision between the internal and external support systems; and (5) for the attached combined survey platform, its external support system is attached to the cushion cap through attachments. The displacement of the external support system caused by the wave-current load is small, so the pile diameter and wall thickness of the external support system can be appropriately reduced to ensure the economy of engineering construction.

Different layouts have their characteristics. The structural size of the attached combined survey platform is small, so the construction cost is low. However, its internal support system is a single-pile structure with poor stiffness. It is sensitive to the load transmitted by the water and the soil between the internal and external support systems, and the vertical angle of the platform changes greatly. Moreover, the external support system needs to be attached to the cushion cap, so the construction of the survey platform can only be carried out after the construction of the cushion cap is completed, which limits the construction period. When the longitudinal bridge alignment of the sea-crossing bridge is a curve, the sight of the observation instrument erected on the survey platform is easily blocked, and the field of view is narrow. For the independently set combined survey platform, the wave-current load is borne by the external support system. The displacement of the external support system is large, the gap between the internal and external support systems must be high, and the structural size is large, so the construction cost is high. However, its internal support system is integrated by multiple single piles through lateral connections. It has a large overall stiffness, and the load transmitted by the soil and the water has little impact on the vertical angle of the platform. The position of the survey platform can be reasonably arranged according to the project requirements. Therefore, the field of view of the observation instruments erected on the survey platform is relatively wide.

In sum, the combined survey platform of the two layouts has advantages and disadvantages. In actual engineering, the survey platform of the two layout forms can be reasonably adopted according to the needs.

5. Conclusions

The main research conclusions are as follows:

a. Through the comparative analysis of the displacement time-history responses of the internal and external support systems under the influence of the wave-current load, the barrier effect of the external support system on the wave-current load is verified, and the influence of the wave-current load on the internal support system is significantly reduced.
b. The external support system of the independently set combined survey platform produces a large displacement. Increasing the pile diameter of the external support system can improve the stiffness of the structure, and the gap between the internal and external support systems can also increase, effectively preventing collision between the internal and external support systems.
c. For the independent combined survey platform, when the pile diameters of the internal and external support systems are 0.8 m and 1.2 m and the wall thicknesses are 11.0 mm and 12.0 mm, respectively, the period of the wave is 8.0 s, the water depth is 20.0 m, the speed of water flow is 1.0 m/s, and the wave height is 4.0 m, the maximum variation of the vertical angle of the platform is only 19.3″.
d. For the attached combined survey platform, the lateral stiffness of the external support system is increased, and the displacement of the external support system is significantly reduced, because the external support system is connected to the cushion cap through the attachments. When the pile diameters of the internal and external support systems are 2.0 m and 2.4 m, the wall thickness is 14.0 mm and 14.0 mm, and the period of the wave, the water depth, the water flow speed, and the wave height are set as 8.0 s, 20.0 m, 1.0 m/s, and 5.0 m, respectively, the maximum displacement of the outer support system is only 29.6 mm.

6. Patents

There is an invention patent resulting from the work reported in this manuscript. The invention relates to a combined survey platform suitable for strong wave conditions (application number is 202010461340. X).

Author Contributions: Conceptualization, J.X. (Jun Xiao) and J.X. (Jianping Xian); methodology, J.X. (Jun Xiao), J.X. (Jianping Xian) and S.Z.; software, S.Z.; validation, J.X. (Jun Xiao), J.X. (Jianping Xian) and Y.Z.; formal analysis, S.Z.; investigation, J.X. (Jun Xiao), S.L. and Y.Z.; resources, J.X. (Jun Xiao); data curation, S.Z.; writing—original draft preparation, J.X. (Jun Xiao), J.X. (Jianping Xian), S.Z., S.L. and Y.Z.; writing—review and editing, S.Z.; visualization, J.X. (Jun Xiao), J.X. (Jianping Xian) and S.L.; supervision, J.X. (Jun Xiao); project administration, S.Z.; funding acquisition, J.X. (Jun Xiao). All authors have read and agreed to the published version of the manuscript.

Funding: This research was funded by the Science and Technology Project of CCCC (grant number: No. 2020-ZJKJ-QNCX04) and the Science and Technology Special Major Project of CCCC (grant number: No. 2019-ZJKJ-07).

Institutional Review Board Statement: Not applicable.

Informed Consent Statement: Not applicable.

Data Availability Statement: Data are contained within the article.

Conflicts of Interest: The authors declare no conflict of interest.

References

1. *GB/T 12897-2006*; Specifications for the First and Second Order Leveling. Standards Press of China: Beijing, China, 2006.
2. Zhu, Y.H.; Xu, T.D. Survey scheme of Hangzhou Bay Sea-crossing Bridge. In *Proceedings of the National Conference on Bridge Engineering, China-2005, Ningbo, China, 12–15 October 2015*, 1st ed.; China Communication Press: Beijing, China, 2005; pp. 51–58.
3. Wu, D.J.; Xiong, W.; Zheng, Q. Research on methods and techniques of repetition surveying of the first order control network for Hong Kong-Zhuhai-Macao Bridge. *Geotech. Investig. Surv.* **2011**, *9*, 74–78.
4. Guo, B.J.; Xu, T.D. Leica TCA1800 High-precision ultra-long-distance sea-crossing height leveling. *Rail. Investig. Surv.* **2007**, *1*, 1–8.
5. Wu, D.J.; Xiong, W. A method of sea-crossing trigonometric leveling by total station in the condition of survey station rocking. *Bull. Surv. Mapp.* **2016**, *6*, 87–90. [CrossRef]
6. Xiang, F. Research on a New Method of Control Survey and Construction Survey of Lindingyang Bridge in Shenzhen-Zhongshan Bridge. Master's Thesis, Southwest Jiaotong University, Sichuan, China, 2021.
7. Zheng, Q.; Zhao, M.; Xiong, W. Technical research on sea-crossing elevation transference of Huangmaohai Link. *Geospatial. Inf.* **2019**, *4*, 178–180,188. [CrossRef]
8. Xu, Y.; Chen, Y.; Guan, X.; Zhou, J. Approach to Minimizing the Influence of Changeable Observational Environment on Long-Distance Sea-Crossing Trigonometric Leveling. *J. Surv. Eng.* **2020**, *146*, 04020018. [CrossRef]
9. Sang, S.; Xu, X.J. Overview of nonlinear motion response of floating structures. *Eng. J. Wuhan Univ.* **2010**, *43*, 608–612.
10. Wang, Z.P.; Zhou, J.Y.; Chen, G.H.; Dong, B.B.; Huang, R.Z. Data processing and application of super-long-distance cross-sea second-class leveling. *Hydro. Surv. Chart.* **2021**, *41*, 40–43,48.
11. Wang, X.; Zhang, C.; Deng, J.; Su, C.; Gao, Z. Analysis of Factors Influencing Miners' Unsafe Behaviors in Intelligent Mines using a Novel Hybrid MCDM Model. *Int. J. Environ. Res. Public Health* **2022**, *19*, 7368. [CrossRef] [PubMed]
12. Zienkiewicz, O.C.; Taylor, R.L. *The Finite Element Method, Volume 1: The Basis*, 5th ed.; Butterworth-Heinemann: Oxford, UK, 2000.
13. Haroun, M.A. Vibration studies and tests of liquid storage tanks. *Earthq. Eng. Struct. Dyn.* **1983**, *11*, 179–206. [CrossRef]
14. Vatankhah, A.R.; Aghashariatmadari, Z. Improved explicit approximation of liner dispersion relationship for gravity waves: A discussion. *Coast. Eng.* **2013**, *78*, 21–22. [CrossRef]
15. Soulsby, R. Tides, surges and mean sea-level: A handbook for engineers and scientists. *Mar. Geol.* **1990**, *91*, 345. [CrossRef]
16. Hu, Y.; Lei, L.P.; Yang, J.X. Study of wave force on foundation of sea-crossing bridges. *J. Waterway. Harbor.* **2012**, *33*, 101–105.
17. Yao, W.W. An Engineering Approach for Computation of Wave Loads on Pile-Slab Structures. Ph.D. Thesis, Shanghai Jiao Tong University, Shanghai China, 2009.
18. Fang, C.; Li, Y.L. Analysis of nonlinear wave load on small size substructure of sea-crossing bridge. *Rail. Stand. Des.* **2017**, *61*, 100–104.
19. Techet, A. 13.42 Design Principles for Ocean Vehicles. In *MIT Course Lecture Notes*; Massachusetts Institute of Technology: Cambridge, MA, USA, 2005.
20. *JST 145-2015*; Code of Hydrology for Harbor and Waterway. China Communication Press: Beijing, China, 2005.
21. Pan, L.; Zhu, B.; Zhang, J.W.; Kang, A.Z. Dynamic response of pile foundation of cross sea bridge to wave-current load in multiple soil layers. *Rail. Stand. Des.* **2020**, *64*, 76–82.
22. Rawat, A.; Mittal, V.; Chakraborty, T.; Matsagar, V. Earthquake induced sloshing and hydrodynamic pressures in rigid liquid storage tanks analyzed by coupled acoustic-structural and Euler-Lagrange methods. *Thin-Walled Struct.* **2019**, *134*, 333–346. [CrossRef]
23. Systemes, D. *Abaqus/Explicit Users' Manual, Version, 6.14*; Dassault Systemes Simulia Corporation: Johnston, RI, USA, 2014.
24. Qian, D.J.; Miao, X.H.; Wang, X.R. Sound Radiation of Underwater Structure Based on the Arithmetic of Coupled Acoustic-Structural Analysis with ABAQUS. *Appl. Mech. Mater.* **2012**, *226–228*, 2249–2252. [CrossRef]

25. Li, Y.P.; Li, J.B.; Lin, G. Effects of reservoir bottom wave absorption on dynamic response of gravity dams. *J. Hydroel. Eng.* **2021**, *40*, 145–154. [CrossRef]
26. Qiu, Y.X.; Wei, C.H.; Wu, Z.G.; Wang, J.T. Effects of reservoir water on simulations of arch dam dynamic characteristics. *J. Hydroel. Eng.* **2020**, *39*, 109–120. [CrossRef]
27. Huang, Y.M.; Zhao, M.; Wang, P.G.; Cao, Y.H.; Du, X.L. Simplified analysis of water-pile-soil interaction under dynamic loads. *J. Vib. Eng.* 2021; *in press*. Available online: https://kns.cnki.net/kcms/detail/32.1349.tb.20210727.1734.008.html(accessed on 27 August 2022).
28. Kong, D.S.; Liu, Y.; Deng, M.X.; Hou, D. Analysis of influencing factors of monopile foundation-soil interaction characteristics for offshore wind power. *Ocean. Eng.* **2021**, *39*, 100–111.
29. Dai, S.; Xie, D. An approach for computation of wave load on small dimension pile foundation based on Abaqus. *Chin. J. Solid. Mech.* **2011**, *32*, 288–295.

sustainability

Article

Load-Bearing Performance and Safety Assessment of Grid Pile Foundation

Rui Tang [1], Yongyi Wang [2], Weili Zhang [2,*] and Yuyong Jiao [2,*]

[1] Anhui Road & Bridge Engineering Group Co., Ltd., Hefei 230031, China; tangrui19920210@163.com
[2] Faculty of Engineering, China University of Geosciences, Wuhan 430074, China; yongyicomeon@163.com
* Correspondence: zwl@cug.edu.cn (W.Z.); yyjiao@cug.edu.cn (Y.J.)

Abstract: Group piles with cushion caps are a common structural form for deep-water bridge foundations. However, their application is limited by the challenges of complex construction, difficult recovery of the supporting large-scale temporary structure, and high engineering expenses. Therefore, we propose a new foundation form—grid pile foundation (GPF)—to improve the sustainability and reliability of foundations. In this study, the finite element software ABAQUS was used to investigate the mechanical properties and dimensional effects of the GPF. Subsequently, the Monte Carlo method was adopted to evaluate the safety under different geological conditions. The results demonstrated that along the depth, the inner frictional resistance of the GPF exhibits an exponential distribution, whereas the outer frictional resistance exhibits an approximate triangular distribution. In addition, the change in pile size has a non-negligible effect on the load-bearing capacity of the GPF. For the same work amount, the smaller pile and side lengths promoted the inner frictional resistance exertion of the GPF. Furthermore, the safety and reliability analysis suggested that the GPF proposed in this study can be used safely under complex geological conditions.

Keywords: grid pile foundation; ABAQUS; load-bearing performance; Monte Carlo method; safety assessment

check for updates

Citation: Tang, R.; Wang, Y.; Zhang, W.; Jiao, Y. Load-Bearing Performance and Safety Assessment of Grid Pile Foundation. *Sustainability* **2022**, *14*, 9477. https://doi.org/10.3390/su14159477

Academic Editor: Paolo Lonetti

Received: 3 June 2022
Accepted: 29 July 2022
Published: 2 August 2022

1. Introduction

Underground diaphragm walls (UGDWs) have been widely used in traditional foundation projects, such as foundation pits, water-block curtains, and bridge anchorages, and are gradually being applied in various engineering fields including bridges, high-rise buildings, and anti-seismic liquefaction [1]. A variety of ground improvement methods are in practice these days, including Maheshwari [2], Kumar et al. [3], Rashid et al. [4,5], etc. Group piles with cushion caps are a commonly used foundation form for long-span deep-water bridges. However, the construction of pile group foundations is complicated and requires significant amounts of concrete. In addition, cofferdams, which are used as temporary envelop enclosures during construction, are challenging to recover, thereby increasing the project cost. The aforementioned challenges can be overcome using the lattice (grid) foundation comprising UGDWs. The foundation form in which adjacent UGDWs are connected using rigid joints to form a planar closure with cushion caps is known as the lattice-shaped UGDW foundation. The grid foundation is significantly different from the single-width diaphragm wall foundation in terms of force mechanism and load-bearing performance [6]. Furthermore, it can conserve a significant amount of work in engineering projects compared with conventional pile-group foundations and can withstand complex loads such as horizontal seismic action, wave impact action, and high-speed vehicle braking, thus improving the sustainability and reliability of the entire engineering project [7].

The building block of the grid foundation is the UGDW whose load-bearing performance has been extensively studied. Ou et al. [8] obtained an analytical solution for the lateral displacement of a UGDW with internal support under an eccentric load using the energy method. Lei [9] determined the UGDW-caused ground settlement value using the Mindlin stress solution. Ba et al. [10] demonstrated that the increased UGDW flexibility leads to the redistribution of stresses in the soil and decreases the internal force of the foundation. Li et al. [11] used ABAQUS to simulate a closed UGDW, revealing that the bending and stiffness deformation is maximum at a third of the foundation height under lateral loads. Chen et al. [12] used FLAC 3D to demonstrate that the closed UGDW thickness under a vertical load is proportional to the outer friction and tip resistances, and that the soil core displacement under lateral loads is greater than the wall displacement. However, the construction of UGDW foundations is challenging under certain geological conditions, such as significantly soft silty soil and underwater operation. In addition, the challenges of non-alignment of adjacent wall segments and water leakage may arise. Therefore, a rigid connection of wall segments effectively helps overcome the limitations of water leakage and non-alignment during construction.

The existing grid foundations are predominantly rectangular, and their settlement calculations are primarily based on load-transfer and integral equation methods. Wu et al. [13] calculated the settlement values of simplified rectangular grid foundations using the load-transfer method; their values agreed well with the experimental measurements, thereby establishing the basis for the theoretical calculation for rectangular grid foundations. Wen [14] studied the pile–soil interaction and load-bearing performance of grid foundations; the indoor tests demonstrated that the soil resistance at the corner of the grid foundation end cap is greater than that at the boundary position, and the force at the middle of the boundary is the lowest. Panpan [15] proposed a nonlinear hyperbolic contact model for grid foundations and used numerical simulations to explain why a nonlinear contact model was selected over a linear contact model in grid foundations. In addition, Wu [16] numerically analyzed the "grid effect" and revealed that the ultimate bearing capacity of the grid UGDW does not increase linearly with an increasing number of grids under the same grid size; moreover, the bearing capacity of a single grid significantly weakens, and the stresses between grids overlap. Therefore, it is recommended to adopt the UGDW grid foundations with smaller grid sizes and fewer grids to enhance the bearing performance of the foundation as long as the bearing capacity requirement of the foundation is satisfied.

This study proposes a novel grid foundation with a hollow hexagonal cross-section–grid pile foundation (GPF), to solve the construction and expense challenges encountered while using the pile group foundation with a cushion cap, commonly used in large deep-water bridges. Subsequently, the numerical simulation software, ABAQUS, was used to conduct an in-depth analysis of the settlement, inner and outer frictional resistance, as well as soil pressure distribution laws of the GPF. In addition, the effect of dimensional changes of the GPF on its load-bearing capacity was explored. Finally, a safety assessment of the GPF was performed to support its further application.

2. Construction of the Numerical Model

2.1. Parameters of Numerical Models

In this study, ABAQUS is used to establish two types of numerical models, the single-wall model SWF and the grid pile model GPF. Model schematics are shown in Figure 1. In this paper, two SWF models of different sizes are established, SWF A and SWF B. The SWF A model adopts the dimensions and parameters of the underground diaphragm wall in the field test of Hou [17], and the calculation results of the model are compared with the test results to verify the accuracy of the model established in this paper. The parameters of the SWF B model are the same as those of the model SWF A, only the dimensions are changed, so the accuracy of the SWF B model is also guaranteed. GPF is a grid pile model composed of six SWFs B. The purpose is to compare and analyze the similarities and differences of the force and deformation laws of grid piles surrounded by a single wall and six single walls. In this paper, five GPF models with different sizes are established. The dimensions of all models are shown in Table 1.

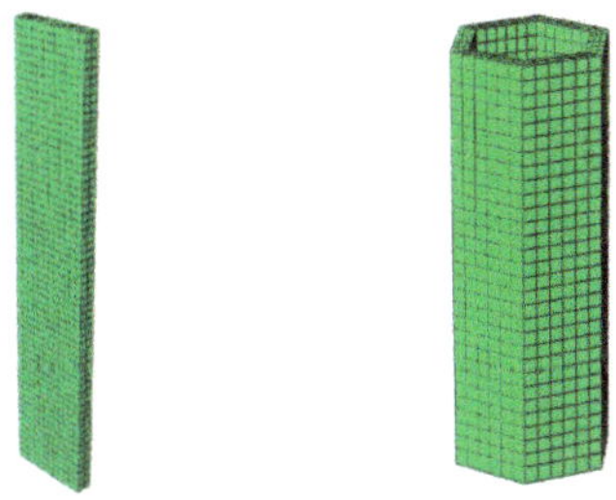

(a) (b)

Figure 1. Schematic of numerical models: (a) SWF, (b) GPF.

Table 1. Model dimensions.

Foundation Type	SWF		GPF				
	A	B	①	②	③	④	⑤
Length of side (m)	5	6	6	7.2	4.8	6	6
Thickness (m)	0.8	1.2	1.2	1.2	1.2	1.2	1.2
Length (m)	27	30	30	30	30	33	27

In order to compare the in situ loading test results of a UGDW of Hou [17], the material parameters of the numerical model are the same as in the literature [17]. The detailed physical parameters are shown in Table 2. The friction coefficient μ between the pile and soil is selected as 0.23, which is based on the recommended value of the friction coefficient between the pile foundation and the soil in the *Technical Code for Building Pile Foundations* (JGJ94-2008) [18] and combined with the literature [17].

Table 2. Pile and soil parameters.

	Density (kg/m^3)	Elasticity Modulus (E/MPa)	Internal Friction Angle φ/(°)	Cohesive Force (c/kPa)	Poisson's Ratio (v)	Initial Void Ratio e_0	Coefficient of Lateral Pressure k_0
Pile	2500	30,000	-	-	0.2	-	-
Soil layer 1	1860	60	27.5	10	0.3	0.71	0.54
Soil layer 2	1720	30	18	18	0.3	0.76	0.69

In ABAQUS, the pile is assumed to be a linear elastic material, and the soil adopts the Mohr-Coulomb elastic-plastic model. The soil is divided into two layers: Soil layers 1 and 2. The tangential contact between the pile and soil was calculated using a hard contact–penalty calculation model, whereas the normal contact was a "hard" form to limit the penetration of the pile into the soil. The lateral pressure coefficient K_0 between the pile and soil is calculated from the internal friction angle according to Equation (1).

$$k_0 = 1 - \sin(\varphi) \quad (1)$$

The C3D8R three-dimensional stress element is used for the pile and soil, respectively. According to the division method of the sweep, the GPF ① model has a total of 65,419 units.

2.2. Numerical Model for SWF

The settlement curve of SWF A is shown in Figure 2. A distinct inflection point is observed in the settlement curve, and the position of the inflection point is generally considered the ultimate load-bearing capacity of the foundation. Before the SWF reaches the ultimate load-bearing capacity, the load has a roughly linear relationship with the settlement; in addition, the settlement increases rapidly after the ultimate load-bearing capacity is reached. Numerical simulation calculations indicate that the ultimate load-bearing capacity of SWF is 14.8 MN, wherein the settlement is 32.9 mm.

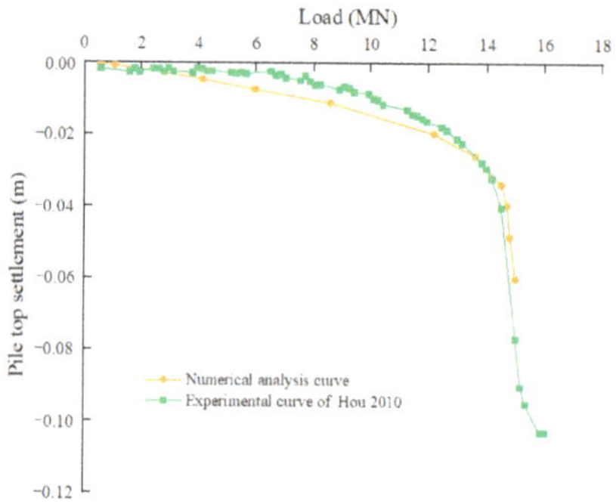

Figure 2. Settlement curve of SWF A [17].

The settlement curve derived from the field test by Hou [17] is shown in Figure 2. The settlement curve indicates that the ultimate load-bearing capacity of the single-width UGDW is 14 MN (corresponding to a settlement value of 28 mm), and the settlement exhibits a steep-drop pattern. Before reaching the ultimate load-bearing capacity, the settlement curve scales linearly. Subsequently, we compared the settlement curve obtained from the numerical model calculation with that of the field test data of Hou [17] (Figure 2), indicating a high degree of agreement and demonstrating the accuracy of the established numerical model for SWF.

2.3. Numerical Model for GPF

The dimensions of SWF A were changed to build SWF B. Subsequently, six SWFs B were used to build the GPF model ①. Then, we loaded both models to 60 MN and obtained the stress nephogram, as shown in Figure 3. S33 is the stress in the vertical direction. The stress distribution of SWF B and GPF under identical loads reveals that the soil pressure on both sides of the SWF is higher, whereas the pile core soil of the GPF exhibits high stress only near the pile end. Therefore, the area enclosed by six SWFs considerably decreases the stresses in the soil, particularly in the pile core soil.

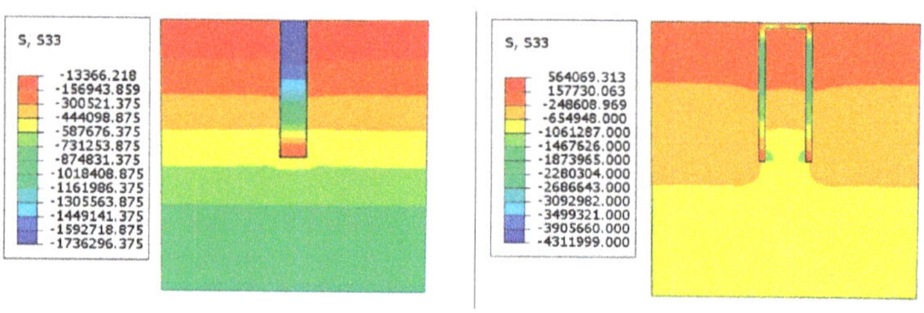

(**a**) (**b**)

Figure 3. Stress nephogram of numerical models: (**a**) The stress nephogram of single-width UGDW, (**b**) the stress nephogram of GPF.

3. Numerical Calculation and Analysis of GPF

3.1. Settlement Curve Analysis

Figure 4 shows that the settlement curve of the GPF changes more gradually. According to the *Code for Design of Building Foundation* (GB50007-2001) [19], when the settlement curve has no distinct inflection point, the ultimate load-bearing capacity is the corresponding load-bearing capacity when the foundation settlement value reaches 0.04 m. Therefore, the ultimate load-bearing capacity of the GPF is 45.7 MN, which is greater than that of an SWF of an identical size by a factor of 3.11. Although a GPF comprises six SWFs, the load-bearing capacity of the GPF is not greater than that of the SWF by a factor of six. In other words, the load-bearing capacity of the GPF is not equal to the sum of the load-bearing capacity of all the SWFs.

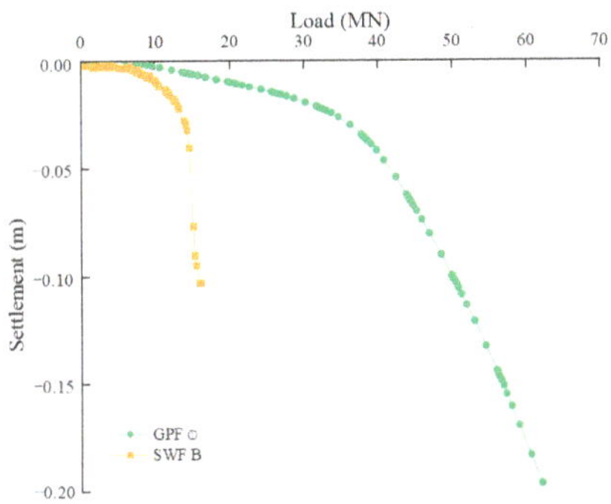

Figure 4. Settlement curves.

The vertical load-bearing capacity of a GPF is primarily composed of three parts, outer frictional resistance, inner frictional resistance, and tip resistance. During the beginning of loading, the outer frictional resistance is gradually exerted; with increasing load, the relative displacement between the pile body and outer soil body occurs, leading to the gradual exertion of the outer frictional resistance. Subsequently, relative displacement occurs between the pile core soil and pile body, at which point the inner frictional resistance and end resistance are exerted until both their limits are reached. Then, the pile settles faster and cannot maintain stability for long periods, and the load-bearing capacity at this time is the ultimate load-bearing capacity. This phenomenon of the load-bearing capacity of the GPF being less than the sum of the load-bearing capacities of all SWFs is known as the "group wall effect", which is similar to the "group pile effect" of the pile group foundation that reduces the load-bearing capacity of the pile group [20].

3.2. Stress Analysis of Pile and Soil

Figure 5 shows that the stress value of the GPF is slightly higher than that of the SWF when approaching their respective load ultimate bearing capacity. The pile stress curves of the SWF and GPF are of similar shapes. In addition, the stress decreases from approximately −1800 kPa at the top of the pile to approximately −200 kPa at the bottom; moreover, the stress in the upper third of the pile is higher and decreases gradually downward; in contrast, the stress in the lower third of the pile rapidly decreases downward. Therefore, the strength of the upper third of the pile should be strengthened while designing the pile foundation.

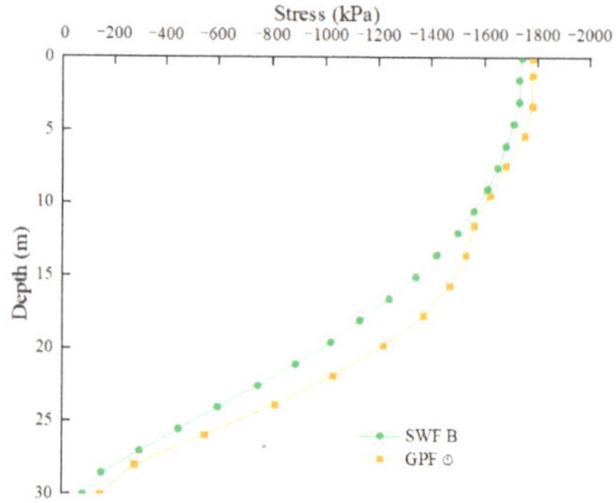

Figure 5. Stress curves of pile.

Figure 6 shows that the vertical stress in the soil increases with increasing depth. With increasing pile top load, the increase in the vertical stress in the pile core soil is slightly higher than that of the soil outside the pile. In addition, the stress variation of the pile core soil is in the lower two-thirds region of the soil core, whereas the stress variation of the soil outside the pile is in the lower one-third region of the pile body. This is because of the following: The interaction between the outer part of the pile body and the soil body distributes the load applied to the soil body at a certain diffusion angle, which is the same as the soil stress distribution law of solid piles. After the pile core soil is stressed, the lower portion of the soil core is continuously squeezed inward owing to the squeezing effect, consequently resulting in the soil core's tendency to move toward the upper portion of the pile; however, the cushion cap at the pile top restricts the upward movement of the soil core, resulting in increasing the stress in the soil core to greater than that in the outer soil body. With increasing load, the pile core soil is affected by squeezing in an increasing upward range, thereby resulting in the continuous bottom-to-top expansion of the pile core soil stress.

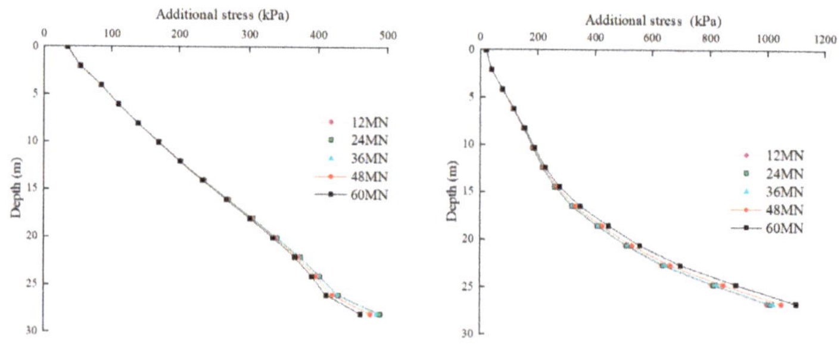

(**a**) (**b**)

Figure 6. Additional stress curves of soil inside and outside GPF: (**a**) Additional stress curves of soil outside pile, (**b**) additional stress curves of pile core soil.

Numerical calculations indicate that the inner frictional resistance accounts for a significant proportion of the pile resistance at the later loading stage, indicating its significant contribution to the load-bearing performance. When the ultimate load is reached, the load-bearing capacity of the soil core remains not fully exerted. Therefore, the continuous optimization of the GPF dimensions to determine the optimal cross-section size can enable earlier and fuller exertion of the inner frictional resistance in the load-bearing phase.

3.3. *Analysis of Lateral Frictional Resistance of Pile*

Figure 7 shows the variation curves of the outer and inner frictional resistance of the GPF with an increasing load. A comparison of Figure 7a,b reveals that the inner frictional resistance contributed by the soil core in the pre-loading period is significantly lower than the outer frictional resistance of the pile, whereas the inner frictional resistance is less than the outer frictional resistance. According to the aforementioned mechanism of pile core soil stress generation, the core soil is confined within the pile and sinks along with the pile body. The low relative displacement of the pile core soil and the pile body significantly decreases the inner pile soil frictional resistance. With an increasing load, the soil at the bottom of the pile is gradually extruded from the bottom into the pile, thereby resulting in the relative displacement of the lower portion of the soil core and pile body. Therefore, the inner frictional resistance is predominantly concentrated at the bottom of the pile and expands upward continuously with an increasing load. The inner frictional resistance is exponentially distributed along the pile body, and the percentage of total frictional resistance increases with increasing load, which is consistent with the observations of Liu [21].

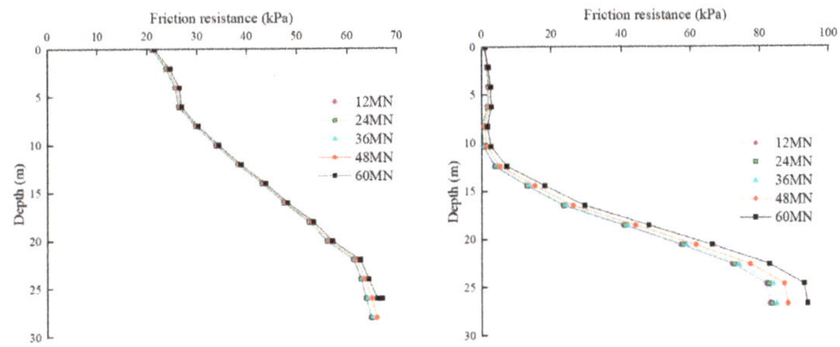

(**a**) (**b**)

Figure 7. Distribution curves of lateral frictional resistance of GPF: (**a**) Distribution curves of outer friction resistance, (**b**) distribution curves of inner friction resistance.

4. Impact of Change in GPF Size on Its Load-Bearing Performance

To study the impact of the side and pile lengths of the GPF on its load-bearing performance, five GPF models with different sizes were established, as shown in Table 2. Then, we calculated the five GPF models separately and analyzed the effect of pile and side length variations on its load-bearing performance based on the calculation results.

4.1. *Impact of Pile Length on Load-Bearing Performance*

The settlement curves of piles ①, ④, and ⑤ shown in Figure 8 characterize the impact of pile length on the ultimate load-bearing capacity and settlement value of the GPF. According to the *Code for Design of Building Foundation* [19], the maximum settlement value of the foundation must not exceed 0.04 m. The settlement curves of piles ①, ④, and ⑤ in Figure 8 have no inflection point before the settlement value reaches 0.04 m, and the corresponding load-bearing capacity values are 40.2, 49.6, and 38.72 MN, respectively, which meets the design requirements of the settlement value. Therefore, increasing the pile length can enhance the load-bearing capacity of the GPF. However, this is effort-intensive, consequently increasing the engineering cost.

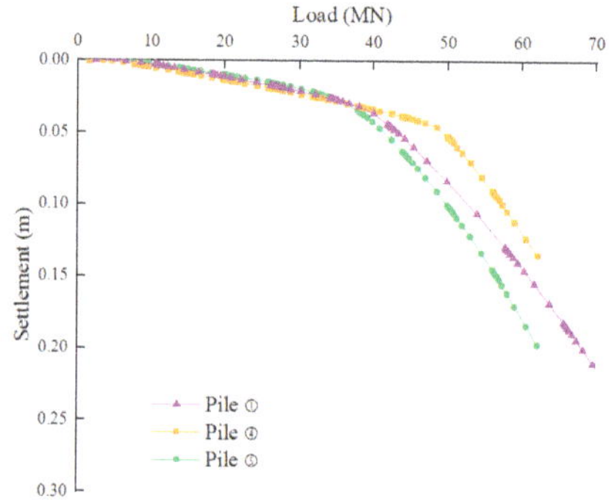

Figure 8. Settlement curves of piles ①, ④, and ⑤.

The axial force curves of piles ①, ④, and ⑤ shown in Figure 9 characterize the impact of pile length variation on the axial force. Pile ① is set as the reference pile. Comparing the axial force of piles ④ and ⑤ indicates that the pile length ④ increases compared with that of pile ①, and the area with the higher axial force in the upper part of pile ④ is significantly greater than that of pile ①. In addition, the tip axial force of pile ④ is greater than that of pile ①, indicating an increase in tip resistance. However, compared with that of pile ①, the length of pile ⑤ decreases, and the area with the higher axial force at the upper portion of pile ⑤ is significantly smaller than that of pile ①, with the axial force in the pile tip tending to zero. With increasing pile length, the influence range of the load extends downward. Figures 5 and 7 show that the variations of the stress and frictional resistance of the upper part of the pile are less along the depth, indicating that the frictional resistance of the GPF is not exerted in the upper part of the pile. Therefore, appropriately reducing the pile length can increase the frictional resistance and decrease the tip resistance.

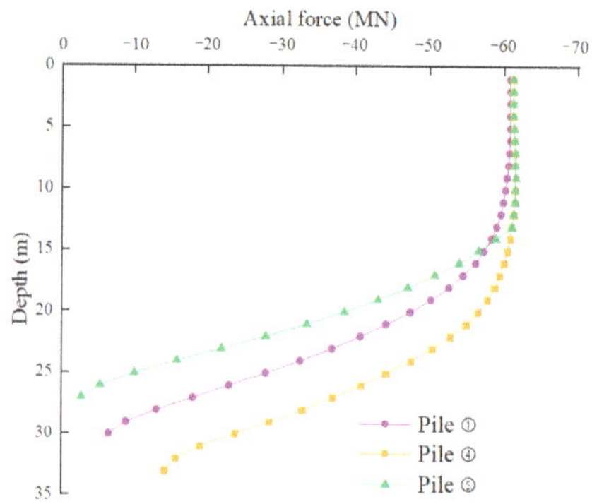

Figure 9. Axial force curves of piles ①, ④, and ⑤.

The frictional resistance distribution of piles ①, ④, and ⑤ in Figure 10 characterizes the impact of pile length on the inner and outer frictional resistance of the GPF. Figure 10a shows that the outer frictional resistance of the GPF has a trapezoidal distribution, and the outer frictional resistance increases approximately linearly along the pile body, with small variations in the three curves. Therefore, the impact of the outer frictional resistance caused by the varying pile length is significantly low and can be neglected. Figure 10b shows the inner frictional resistance of piles ①, ④, and ⑤. The distribution of the inner frictional resistance of piles ④ and ⑤ is virtually identical (exponential distribution). The inner frictional resistance of pile ⑤ is significantly higher than those of piles ① and ④, and its distribution along the length of the pile is comparable to that of a triangle. The inner frictional resistance in the middle and upper portions of pile ⑤ is also significantly higher than those in piles ① and ④. Therefore, pile length variations significantly impact inner frictional resistance. Furthermore, an increase in pile length inhibits the exertion of inner frictional resistance in the upper part of the pile and negligibly affects the outer frictional resistance.

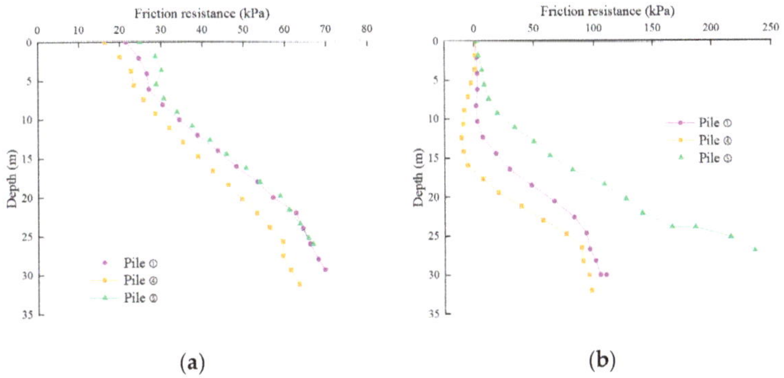

Figure 10. Lateral friction resistance of piles ①, ④, and ⑤: (**a**) The outer frictional resistance of piles ①, ④, and ⑤, (**b**) the inner frictional resistance of piles ①, ④, and ⑤.

4.2. Impact of Pile Side Length on Load-Bearing Performance

Figure 11 shows the settlement curves of piles ①, ②, and ③. The inflection points in the curves indicate that the ultimate load-bearing capacities of piles ③, ①, and ② are 30, 40, and 46 MN, respectively. Therefore, increasing the side length can improve the ultimate load-bearing capacity of the foundation. However, an increasing side length decreases the ultimate load-bearing capacity improvement rate. In this case, an increase in the load-bearing capacity only arises from the increase in outer frictional resistance and pile tip resistance.

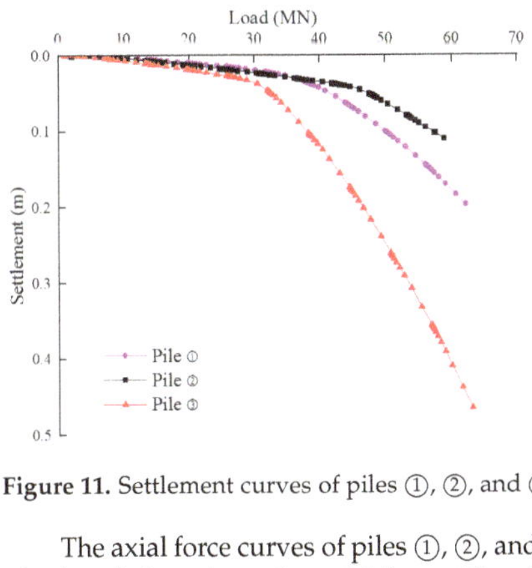

Figure 11. Settlement curves of piles ①, ②, and ③.

The axial force curves of piles ①, ②, and ③ shown in Figure 12 characterize the impact of pile side length on the axial force. The side length of pile ② is larger than that of pile ①, and the axial force at the upper part of pile ② increases, whereas the axial force at the lower part of the pile slightly changes. In addition, the side length of pile ③ is smaller than that of pile ①, and the overall reduction of axial force causes a reduction in the axial force at the tip of the pile, subsequently tending to zero.

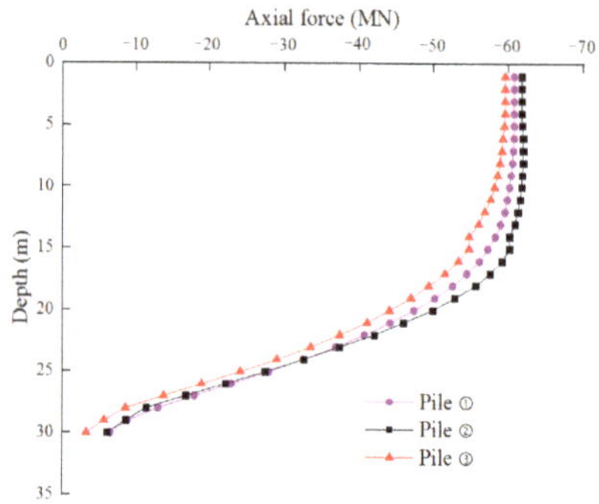

Figure 12. Axial force curves of piles ①, ②, and ③.

Figure 13a shows the distribution of the outer friction resistance of piles ①, ②, and ③. We inferred that the outer frictional resistance is not significantly affected by the change in pile side length. Figure 13b shows the distribution of the inner frictional resistance. The increase in the side lengths of piles ① and ② does not significantly change the frictional resistance. However, a decrease in the pile ③ side length and the relative decrease in the soil core diameter causes a larger inner frictional resistance posed by the soil core in the middle and lower portions of the pile. When the side length is small, the small diameter of the soil core causes a significant squeezing effect at the bottom of the pile, resulting in a gradual increase in the inner frictional resistance and its exertion toward the pile tip. Therefore, the inner frictional resistance contributes more to the increase in load-bearing capacity. The diameter of the soil core increases continuously with increasing side length, consequently resulting in a moderately weaker squeezing effect of the pile. Hence, the increase rate in the inner frictional resistance provided by the soil core decreases.

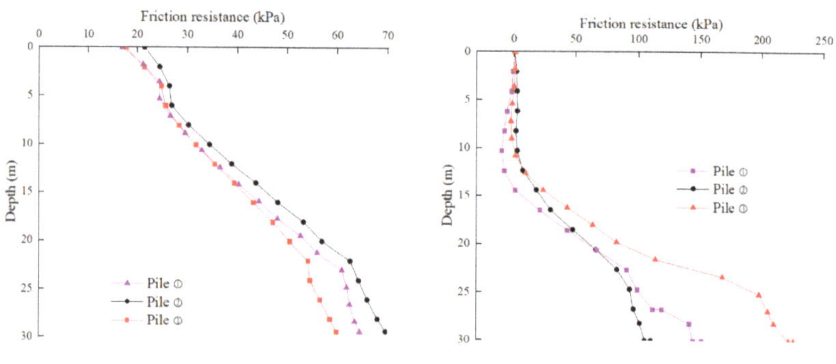

(a) (b)

Figure 13. Lateral friction resistance of piles ①, ②, and ③: (**a**) The outer frictional resistance of piles ①, ②, and ③, (**b**) the inner frictional resistance of piles ①, ②, and ③.

The above analyses indicate that the variations in the pile and side lengths have less effect on the outer frictional resistance. However, decreasing the pile and side lengths can enhance the exertion of the inner frictional resistance. In addition, decreasing the tip resistance can improve the load-bearing performance of the GPF.

5. Simplification Algorithm and Safety Evaluation Method for GPF Settlement

5.1. Simplification Algorithm for GPF Settlement Based on Mindlin Stress Solution

Definite theoretical research on the settlement calculation method of the GPF has not been reported yet. Therefore, we calculated the additional stresses in the foundation based on the settlement calculation method of tubular piles, which is similar to the GPF structure, based on the semi-infinite space mechanics model established by Mindlin's stress solution [22]. In addition, we derived the foundation settlement values using the layerwise summation method.

According to Zhu and Wang [23,24], the upper load on the silo pile is shared by the tip, outer frictional, and inner frictional resistance; in addition, additional stresses are generated in the foundation when the three resistances are exerted. Liu [25] calculated the additional stresses in the foundation according to Mindlin's method of deriving stresses, modified the integration area, and set the circular bottom surface of the tubular pile as well as the inner and outer sides as the integration areas of the end, inner frictional, and outer frictional resistances, respectively, to obtain the additional stresses in the foundation at different depths, subsequently deriving the total settlement using the splitting summation method. In this study, the method available in the literature [25] is adopted, and the resistances in the upper part of the GPF are classified as the tip, inner frictional, and outer frictional resistances. However, the additional stress integration region of the GPF changes, and the bottom surface is a positive hexagonal ring. Furthermore, the inner and outer sides are composed of six rectangles. Therefore, we determined the additional stresses generated in the foundation by end resistance, inner side frictional resistance, and outer side frictional resistance based on the calculation method of Mindlin's stress solution and superimpose them, as shown in Equation (2).

$$S = \psi_{\mathrm{P}} \frac{P_{\mathrm{P}}}{L^2} \sum_{j=1}^{m} \sum_{i=1}^{n_j} \frac{\Delta \mathrm{h}_{j,i}}{E_{\mathrm{s}j,i}} \sum_{k=1}^{n} \left[\alpha I_{\mathrm{P_A},\mathrm{k}} + \beta I_{\mathrm{P_i},\mathrm{k}} + (1-\alpha-\beta) I_{\mathrm{P_s},\mathrm{k}} \right], \tag{2}$$

where S is the final deformation amount of the foundation (mm), ψ_{P} is the empirical coefficient of settlement calculation (0.5), $E_{\mathrm{s}j,i}$ is the compression modulus of the ith layer of soil on the foundation floor (MPa), and α and β are the load distribution coefficients of the outer and inner frictional resistances, respectively.

$$I_{\mathrm{P_A},k} = \frac{L^2}{P_{\mathrm{P}}} \sigma_{z,P_A}, \tag{3}$$

$$I_{\mathrm{P_1},k} = \frac{L^2}{P_{\mathrm{P}}} \sigma_{z,Pi}, \tag{4}$$

$$I_{\mathrm{P_s},k} = \frac{L^2}{P_{\mathrm{P}}} \sigma_{z,Ps}, \tag{5}$$

where $I_{\mathrm{P_A},k}$, $I_{\mathrm{P_i},k}$, and $I_{\mathrm{P_s},k}$ are the respective stress influence coefficients generated by each part of the load for the settlement calculation.

The total settlement amount was calculated using Equation (2). In addition, the settlement curve generated using the same soil parameters as those listed in Table 1 is shown in Figure 14. Moreover, the settlement curve obtained from the numerical calculation includes the failure stage. Because the Mindlin stress solution used in Equation (2) is based on the elastic semi-infinite space, the settlement curve obtained only characterizes the elastic stage.

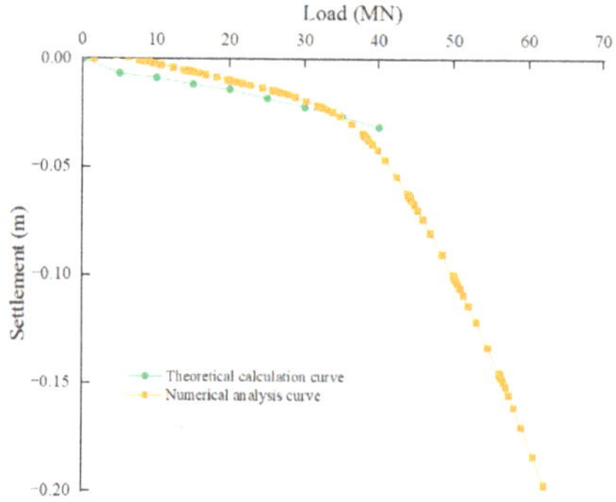

Figure 14. GPF settlement curves.

Numerical calculation results indicate that the GPF settlement curve has a distinct inflection point. When the inflection point is reached, the settlement value rapidly increases. Before reaching the inflection point, the theoretically calculated settlement value is slightly higher than the numerically calculated value, and the two curves exhibit a similar variation law and are in good agreement with each other. Therefore, Equation (2) is applicable to calculate the settlement of the GPF, with conservative prediction results.

5.2. Monte Carlo Method-Based Settlement Reliability Analysis of GPF

In Section 4.1, we assumed that the foundation soil is homogeneous for the GPF settlement value calculation. However, diverse soil layers and complex geological conditions are present in the actual projects. Therefore, it is required to determine whether the pile foundation fails under complex geological conditions. Herein, we adopted a simplified algorithm for the Monte Carlo-based settlement evaluation of the GPF, focusing on the effects of different geological conditions on the values tested to establish a more intuitive and comprehensive safety index of the GPF.

The Monte Carlo method is based on the concept of generating numerous subsamples by random sampling based on the distribution patterns of random variables [26–28]. We set the number of subsamples to N and introduced the subsample values to the structure work function to determine the failure and count the number of failures (n_f). According to the law of large numbers, the failure probability (P_r), which can be unbiasedly estimated via the failure frequency, is expressed as follows [29]:

$$P_r = \frac{n_f}{N}. \tag{6}$$

Let the structure work function be as follows:

$$Z = gX. \tag{7}$$

Then, the structural failure probability can be expressed as follows:

$$P_r = P[gX < 0]. \tag{8}$$

The reliability β can be expressed as follows:

$$\beta = \phi^{-1}1 - P_r, \tag{9}$$

where $\phi^{-1}(x)$ denotes the inverse function of a certain parameter.

The *Technical Code for Building Pile Foundations* [18] contains clear provisions for the settlement values of pile foundations. Generally, the settlement value of the pile foundation should be controlled to within 0.04 m. Therefore, pile foundations with settlement values that exceed this value can be considered supplementary to the required settlement to be used (i.e., load-bearing failure). We set Equation (2) as the work function. In addition, it is suggested that the random variables include the elasticity modulus, E, Poisson's ratio, ν, and pile top load, P [30,31]. The statistics of the random variables obtained based on the GPF dimensions described in this study are listed in Table 3.

Table 3. Statistics of random variables.

Random Variables	Distribution Pattern	Upper Limit	Lower Limit
E/MPa	Uniform distribution	60	40
ν	Uniform distribution	0.4	0.2
P/MN	Uniform distribution	60	50

The sample size, N, in this study is 1 million to ensure the accuracy of the estimated reliability [32–34]. The calculated failure probability (P_r) and reliability (β) are 8.7×10^{-5} and 3.754, respectively. The reliability of this study is greater than the threshold value (3.5) specified in the *Specifications of the National Transportation Association of the USA* [35]. Therefore, the GPF proposed in this study can be safely used under complex geological conditions.

6. Conclusions

(1) Under identical grid sizes, the soil stress and frictional resistance of the GPF change less with the increasing load, thereby resulting in significant changes in the inner frictional resistance in the lower portion of the pile.
(2) Along the depth, the inner frictional resistance of the GPF exhibits an exponential distribution pattern, whereas the outer frictional resistance shows an approximately triangular or trapezoidal distribution pattern.
(3) The stress in the upper third of the pile length is higher. Therefore, the strength of the upper third of the pile should be strengthened during pile foundation designing.
(4) The changes in the pile and side lengths have a small effect on the outer frictional resistance, and their reductions can improve the exertion of inner frictional resistance and reduce the tip resistance. Therefore, the GPF load-bearing performance can be enhanced.
(5) The safety and reliability analyses indicate that the GPF proposed in this study can be safely used under complex geological conditions.

Author Contributions: Conceptualization, R.T.; Formal analysis, R.T.; Funding acquisition, Y.J.; Investigation, Y.W.; Methodology, W.Z.; Project administration, Y.J.; Software, Y.W.; Supervision, W.Z.; Writing—review & editing, Y.J. All authors have read and agreed to the published version of the manuscript.

Funding: This research was funded by the state key international cooperation research project of the National Natural Science Foundation of China (Grant No. 41920104007).

Institutional Review Board Statement: Not applicable.

Informed Consent Statement: Not applicable.

Data Availability Statement: Not applicable.

Conflicts of Interest: The authors declare no conflict of interest.

References

1. Wu, J.; Cheng, Q.; Wen, H. Overview of the variety of diaphragm wall foundation developed in Japan in recent years. *Ind. Constr.* **2013**, *43*, 1.
2. Maheshwari, P. Analysis of combined footings on extensible geosyntheticstone column improved ground. *J. Civ. Eng. Sci. Technol.* **2017**, *8*, 57–71. [CrossRef]
3. Kumar, P.; Kumar, M.; Chandaluri, V.K.; Sawant, V.A. Uplift capacity of single and group of granular anchor pile system. *J. Civ. Eng. Sci. Technol.* **2018**, *9*, 34–40. [CrossRef]

4. A Rashid, A.S.; Black, J.; Kueh, A.B.H.; Mohamad, H.; Md Noor, N. Bearing capacity charts of soft soil reinforced by deep mixing. *Inst. Civ. Eng.* **2017**, *170*, 12–25. [CrossRef]
5. A Rashid, A.S.; Kueh, A.B.H.; Mohamad, H. Behaviour of soft soil improved by floating soil–cement columns. *Inst. Civ. Eng.* **2018**, *18*, 95–116. [CrossRef]
6. Wu, J.; Cheng, Q.; Wen, H.; Cao, J.-L. Vertical bearing behaviors of lattice shaped diaphragm walls and group piles as bridge foundations in soft soils. *Chin. J. Geotech. Eng.* **2014**, *36*, 1733–1744.
7. Wu, J.; Cheng, Q.; Wen, H.; Li, Y.; Zhang, J.; Wang, L. Comparison on the Horizontal Behaviors of Lattice-Shaped Diaphragm Wall and Pile Group under Static and Seismic Loads. *Shock Vib.* **2016**, *2016*, 1289375. [CrossRef]
8. Ou, Q.; Zhang, L.; Zhao, M.; Wang, Y. Lateral Displacement and Internal Force in Diaphragm Walls Based on Principle of Minimum Potential Energy. *Int. J. Geomech.* **2019**, *19*, 04019055. [CrossRef]
9. Lei, G.H.; Sun, H.S.; Ng, C.W.W. An Approximate Analytical Solution for Calculating Ground Surface Settlements Due to Diaphragm Walling. *Comput. Geotech* **2014**, *61*, 108–115. [CrossRef]
10. Yajnheswaran, B.; Akshay, P.R.; Rajasekaran, C.; Rao, S. Effect of stiffness on performance of diaphragm wall. *Procedia Eng.* **2015**, *116*, 343–349.
11. Liu, L.; Dai, G.; Gong, W. Numerical analysis about bearing characteristics of closed diaphragm wall under horizontal load. In Proceedings of the International Symposium on Innovation and Sustainability of Structures in Civil Engineering (ISISS 2011), Xiamen, China, 28–30 October 2011.
12. Chen, Q.; Wu, J.; Song, Z.; Wen, H. The behavior of a rectangular closed diaphragm wall when used as a bridge foundation. *Front. Struct. Civ. Eng.* **2012**, *6*, 398–420.
13. Wu, J.; Chen, Q.; Wen, H.; Wang, L.; Li, Y.; Zhang, J. A load transfer approach to rectangular closed diaphragm walls. *Inst. Civ. Eng.* **2016**, *169*, 509–526. [CrossRef]
14. Wen, H.; Chen, Q.; Meng, F.; Chen, X. Diaphragm wall-soil-cap interaction in rectangular-close diaphragm-wall bridge foundations. *Front. Archit. Civ. Eng.* **2009**, *3*, 93–100. [CrossRef]
15. Guo, P.; Gong, X.; Wang, Y. Research on performance of cellular diaphragm wall in deep excavation considering nonlinear contact between wall and soil. *Chin. J. Geotech. Eng.* **2021**, *43*, 1201–1209. (In Chinese)
16. Wu, J.; Chen, Q.; Tang, W. Numerical Analysis of the Chamber Effect of Lattice Shaped Diaphragm Wall as Bridge Foundation in Sand. *Constr. Technol.* **2016**, *45*, 60–63. (In Chinese)
17. Hou, Y. Study on Retaining Structure Behavior of Lattice Diaphragm Wall in Soft Soil Stratum. Ph.D. Thesis, Shanghai Jiaotong University, Shanghai, China, 2010. (In Chinese).
18. *(JGJ94-2008)*; Technical Code for Building Pile Foundations. China Architecture and Building Press: Beijing, China, 2008. (In Chinese)
19. *(GB50007—2011)*; Code for design of building foundation. China Architecture and Building Press: Beijing, China,, 2012. (In Chinese)
20. Ding, X.; Zhang, T.; Li, P.; Cheng, K. A Theoretical Analysis of the Bearing Performance of Vertically Loaded Large-Diameter Pipe Pile Groups. *J. Ocean Univ. China* **2016**, *15*, 57–68. [CrossRef]
21. Liu, H.; Fei, K.; Zhou, Y.; Gao, Y. Numerical simulation of inner frictional resistance of cast-in-situ concrete thin-wall pipe pile. *Rock Soil Mech.* **2004**, *25*, 211–216. (In Chinese)
22. Mindlin, R.D. Force at a point in the interior of a semi-infinite solid. *Physics* **1936**, *7*, 195–202. [CrossRef]
23. Zhu, M.; Wang, J.; Zhu, X. Performance analysis of cast-in-place tubular pile composite ground under embankment load. *J. Zhejiang Univ. (Eng. Sci.)* **2006**, *40*, 2186–2190. (In Chinese)
24. Wang, Z.; Zhang, Y. Analytical analysis of vertical load-transfer of large-diameter cast-in-situ concrete tubular piles. *Chin. J. Geotech. Eng.* **2007**, *29*, 1488–1492. (In Chinese)
25. Liu, H.; Wu, C.; Yu, F.; Xia, T. Dynasty Improved algorithm for settlement of cylindrical pile composite foundation under embankment load. *J. Geotech. Eng.* **2013**, *35*, 638–642. (In Chinese)
26. Pang, Y.; Meng, R.; Li, C.; Li, C. A probabilistic approach for performance-based assessment of highway bridges under post-earthquake induced landslides. *Soil Dyn. Earthq. Eng.* **2022**, *155*, 107207. [CrossRef]
27. Pang, Y.; Wei, K.; He, H.; Wang, W. Assessment of lifetime seismic resilience of a long-span cable-stayed bridge exposed to structural corrosion. *Soil Dyn. Earthq. Eng.* **2022**, *157*, 107275. [CrossRef]
28. Yang, H.; Pang, Y.; Tian, S.; Dang, X.; Yuan, W. Case study of the seismic response of an extra-dosed cable-stayed bridge with cable-sliding friction aseismic bearing using shake table tests. *Struct. Des. Tall Spec. Build.* **2017**, *26*, e1398. [CrossRef]
29. Shen, Z.; Wei, K. Stochastic model of tropical cyclones along China coast including the effects of spatial heterogeneity and ocean feedback. *Reliab. Eng. Syst. Saf.* **2021**, *216*, 108000. [CrossRef]
30. Pang, Y.; Wang, X. Cloud-IDA-MSA conversion of fragility curves for efficient and high-fidelity resilience assessment. *J. Struct. Eng.* **2021**, *147*, 04021049. [CrossRef]
31. Pang, Y.; Wang, X. Enhanced endurance-time-method (EETM) for efficient seismic fragility, risk and resilience assessment of structures. *Soil Dyn. Earthq. Eng.* **2021**, *147*, 106731. [CrossRef]
32. Zhong, J.; Ni, M.; Hu, H.; Yuan, W.; Yuan, H.; Pang, Y. Uncoupled multivariate power models for estimating performance-based seismic damage states of column curvature ductility. *Structures* **2022**, *36*, 752–764. [CrossRef]

33. Wei, K.; Arwade, S.R.; Myers, A.T.; Valamanesh, V. Directional effects on the reliability of non-axisymmetric support structures for offshore wind turbines under extreme wind and wave loadings. *Eng. Struct.* **2016**, *106*, 68–79. [CrossRef]
34. Zhong, J.; Yang, T.; Pang, Y.; Yuan, W. A novel structure-pulse coupled model for quantifying the column ductility demand under pulse-like GMs. *J. Earthq. Eng.* **2021**, 1–19. [CrossRef]
35. American Association of State Highway and Transportation Officials. *LRFD Bridge Design Specifications*, 8th ed.; American Association of State Highway and Transportation Officials: Washington, DC, USA, 2017.

sustainability

Article

Simulation Study on Sunshine Temperature Field of a Concrete Box Girder of the Cable-Stayed Bridge

Qiusheng Wang [1], Jianping Xian [1,2], Jun Xiao [1,2,*] and Shuai Zou [2,3]

1 CCCC Second Highway Engineering Co., Ltd., Xi'an 710065, China; qswang@vip.163.com (Q.W.); xianjianping@vip.163.com (J.X.)
2 Shaanxi Union Research Center of University and Enterprise for Bridge Intelligent Construction, Xi'an 710199, China; zs2448602237@163.com
3 School of Civil Engineering, Chongqing Jiaotong University, Chongqing 400074, China
* Correspondence: jun.xiao-cccc@hotmail.com; Tel.: +86-13668031328

Abstract: This paper investigates the distribution of the sunshine temperature field in bridge structures. To implement thermodynamic boundary conditions on the structure under the influence of sunshine, this study utilized the FILM and DFLUX subroutines provided by ABAQUS. Based on this method, the sunshine temperature field of the concrete box girder of a cable-stayed bridge was analyzed. The results showed that the simulated temperature values were in good agreement with the measured values. The temperature difference between the internal and external surfaces of the box girder under the influence of sunshine was significant, with the maximum negative temperature difference appearing around 6:00 a.m. and the maximum positive temperature difference appearing around 2:00 p.m. The temperature gradient of the box girder section calculated by the method presented a C-shaped distribution pattern, which differs from the double-line distribution pattern specified in the current "General Specifications for Design of Highway Bridges and Culverts" in China (JTG D60-2015). Furthermore, a sensitivity analysis of thermal parameters using the proposed simulation method for the sunshine temperature field of the concrete box girder was conducted, and the results indicated that the solar radiation absorption coefficient had a significant impact on the temperature field. A 30% increase or decrease in the solar radiation absorption coefficient caused the maximum temperature change on the surface of the structure to exceed 10 °C. This paper provides an accurate simulation of the sunshine temperature field of the concrete box girder of a cable-stayed bridge, and the research results are significant for controlling bridge alignment and stress state during the construction period, ensuring the reasonable initial operating state of the bridge, and enhancing the sustainability of the structure.

Keywords: bridge engineering; temperature gradient; numerical simulation; parameter sensitivity

Citation: Wang, Q.; Xian, J.; Xiao, J.; Zou, S. Simulation Study on Sunshine Temperature Field of a Concrete Box Girder of the Cable-Stayed Bridge. *Sustainability* **2023**, *15*, 7541. https://doi.org/10.3390/su15097541

Academic Editor: Gianluca Mazzucco

Received: 9 April 2023
Revised: 2 May 2023
Accepted: 3 May 2023
Published: 4 May 2023

1. Introduction

The temperature field of a bridge structure in the natural environment is constantly changing due to passive or active heat transfer between the structure and the external environment, such as solar radiation, convective heat transfer, and radiative heat transfer. The causes of temperature change in the bridge structure can be divided into the daily temperature cycle, the sudden temperature change, and the seasonal temperature cycle. Among these, the daily temperature cycle and the seasonal temperature cycle are the two key aspects of the bridge temperature field. The former is mainly generated by the effects of daily temperature variation, solar radiation, and sunshine shadow occlusion, while the latter is generated by the effects of seasonal climate change and annual temperature amplitude. This paper focuses on simulating and discussing the daily temperature cycle of the concrete box girder of a cable-stayed bridge. The concrete girder of a cable-stayed bridge is always exposed to the atmospheric environment, resulting in temperature differences between the surfaces of the structure under the influence of sunlight and environmental factors. These differences include an overall temperature difference between the external surfaces and the local temperature difference between the internal and external surfaces in the thickness, which form a non-uniform temperature field within the structure. The temperature difference between the external surfaces is caused by the difference in solar radiation received by the sunny surface and the shaded surface, leading to changes in the alignment of the concrete girder. The direction and magnitude of the alignment change will vary with the position of the sun and the angle of light exposure, making it challenging to monitor the alignment of the concrete girder during the construction phase [1]. Moreover, this temperature difference can also lead to changes in the stress state of the structure. The local temperature difference in the direction of the wall thickness is caused by the inconsistent environmental boundary conditions between the internal and external surfaces of the structure. The temperature on the external surface of the structure changes significantly with changes in atmospheric temperature and solar radiation, while the air in the inner cavity of the structure does not flow and the thermal conductivity of the concrete is poor, resulting in a much smaller temperature change on the internal surface of the structure. Research has shown that when the temperature difference between the internal and external surfaces is too large, the resulting temperature stresses can reach live load levels [2–6], which may cause concrete cracking and affect the safety and stability of the structure and, therefore, the sustainability of bridge operations. Therefore, conducting in-depth research on the sunshine temperature field of the concrete box girder of cable-stayed bridges is essential for ensuring the reasonable initial operation state of the bridge and enhancing the sustainable performance of the structure. Many scholars have carried out extensive research on the sunshine temperature field of the concrete box girder. Zuk [7] studied the measured temperature field of a bridge and found that the temperature difference between the internal and external surfaces of the concrete structure is the highest in summer, with a maximum value of up to 22 °C. The temperature of the external surface of a steel-concrete composite beam located near the Hardware River was tested. The test results show that the temperature difference between the upper and lower surfaces of the concrete bridge deck can reach up to 19° during the day, and the temperature difference at night is small. For steel beams, the internal vertical temperature difference is not obvious; the influence of atmospheric temperature, wind speed, solar radiation, and other related environmental factors on the temperature field of the structure is discussed for the first time, and the temperature difference equation of the composite concrete beam bridge deck is summarized in his paper. Hambly and Rao [8,9] monitored the temperature field of a real bridge under direct solar radiation and summarized the temperature gradient distribution mode of the concrete box girder. Churchward [10] analyzed the influence of atmospheric temperature, wind speed, solar radiation, and other related environmental factors on the temperature field of the structure and derived the temperature distribution function of the vertical section of the box girder under the combined action of the DTMAX and sunshine parameters. Hoffman [11] found that the temperature variation in the longitudinal direction of the bridge is small, and variations in solar radiation and atmospheric temperature cause a non-linear temperature gradient along the vertical height direction in the main girder section. Xia et al. [12] combined field measurements with finite element simulations to calculate the temperature distribution of different components of the bridge structure and analyze the displacement and strain response of the structure. The analysis results showed that the displacement of the bridge in the longitudinal and transverse directions was consistent with the monitoring data. Liu et al. [13] studied the influence of temperature on the elevation of towers and girder sections of a large-span concrete cable-stayed bridge and summarized the influence of temperature change on the elevation of the key positions of the girder and tower. Wu et al. [14] monitored the temperature field of a steel-concrete combined box girder bridge and found that the concrete and steel box girders had different forms of temperature gradient functions. The results were compared with the gradient patterns specified in the code, which showed that the temperature gradient distribution patterns specified in the code differed from the actual distribution patterns to some extent.

To summarize, research on the sunshine temperature field of concrete box girders has primarily relied on the analysis of measured temperature data from actual engineering or test models. Although this method can collect real temperature data at specific measuring point positions, there are limitations due to the number of measurement points and monitoring conditions, which often result in significant data errors and may not fully reflect the actual temperature field distribution of the structure. In contrast, the numerical simulation method is not limited by these conditions and can achieve a rapid calculation of the structural temperature field. Consequently, this method has been favored by many researchers and engineers and has become one of the mainstream methods for calculating and researching the structural temperature field of bridges.

In the daily temperature cycle, the bridge structure heats up during the day due to solar radiation and cools down at night when the ambient temperature decreases and the structure dissipates heat into the environment. The sunshine temperature field of the bridge structure is strongly time-varying due to the changes in solar radiation intensity and ambient temperature. Therefore, to achieve a refined simulation of the sunshine temperature field, accurate thermodynamic boundary conditions must be applied to the calculation model of the structure. This involves reflecting the heat transfer and heat exchange between the structure and the external environment through appropriate parameter models to accurately describe the changes in ambient temperature and solar radiation intensity at different times. In this study, the thermodynamic boundary conditions of the structure were imposed using the FILM and DFLUX subroutines provided by ABAQUS. This was based on previous research on the boundary conditions of the sunshine temperature field of bridge structures. This provided a simple and efficient method for the rapid analysis and prediction of the temperature field of the concrete box girders of cable-stayed bridges. The research results of this study are significant for controlling bridge alignment and stress state during the construction period, ensuring a reasonable initial operating condition of the bridge, and enhancing the sustainability of the structure.

2. Heat Transfer Boundary Condition

The heat transfer of the object should satisfy Fourier's Law [15,16], which states that the rate of heat transfer is proportional to the temperature gradient in the object. The differential equation of heat transfer without an internal heat source can be expressed as follows:

$$\rho c \frac{\partial T}{\partial \tau} = \frac{\partial}{\partial x}\left(k_x \frac{\partial T}{\partial x}\right) + \frac{\partial}{\partial y}\left(k_y \frac{\partial T}{\partial y}\right) + \frac{\partial}{\partial z}\left(k_z \frac{\partial T}{\partial z}\right) \tag{1}$$

here, ρ is the density of the material, kg/m^3; c is the specific heat capacity of the material, $J/(kg \cdot °C)$; k_x, k_y, and k_z are the thermal conductivity of the structure in x, y, and z directions, respectively, $W/(m \cdot °C)$.

Heat exchange between the bridge structure and the external environment, under the influence of sunshine, mainly includes solar radiation, convective heat transfer, and radiation heat transfer, as shown in Figure 1.

$$q = q_s + q_c + q_r \tag{2}$$

here, q is the heat flux density for comprehensive heat transfer on the structural surface, W/m^2; q_s is the solar radiation heat flux density absorbed by the structural surface, W/m^2; q_c is the heat flux density for convective heat transfer between the structural surface and the external environment, W/m^2; and q_r is the heat flux density for radiation heat transfer between the structural surface and the external environment, W/m^2.

Researchers have proposed several solar radiation models to consider the influence of the atmosphere on solar radiation [17–22], building upon existing research in the field of meteorology. To keep the input parameters for the model as simple as possible, the Duffie model [19] has been selected in this study for the calculation of solar radiation.

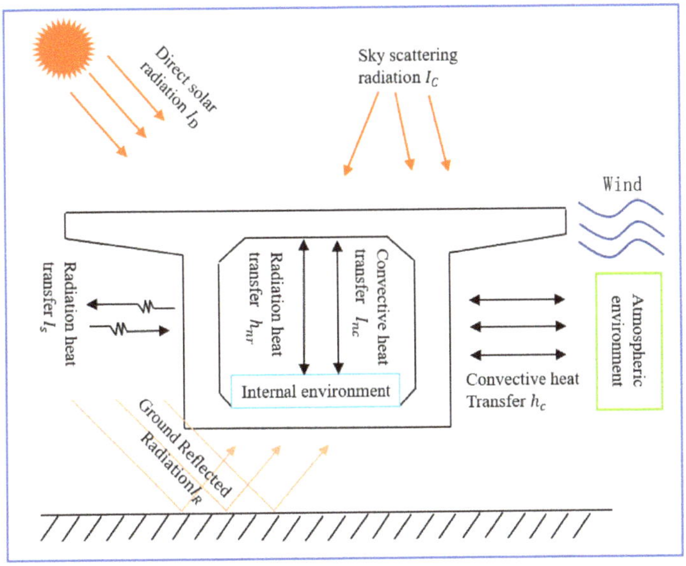

Figure 1. Heat exchange diagram between the structure and the external environment.

2.1. Solar Radiation Effect

Solar radiation on the external surface of the structure mainly comprises direct solar radiation (I_D), sky radiation (I_C), and ground reflected radiation (I_R), as shown in Figure 2. Thermal radiation on the surface of the bridge can be classified into long-wave radiation and short-wave radiation, with solar radiation being a type of short-wave radiation.

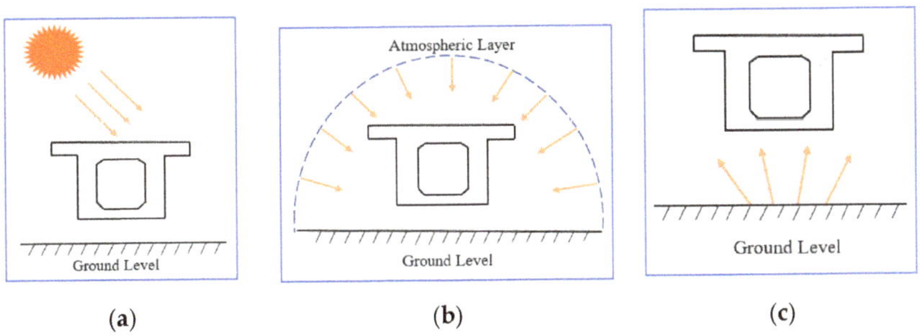

Figure 2. The solar radiation effect. (**a**) The direct solar radiation; (**b**) the sky radiation; (**c**) the ground reflected radiation.

The total solar radiation heat flux q_s of the total solar radiation on the surface of the structure is given by the following equation [23,24]:

$$q_s = \varepsilon I_{SOR} \tag{3}$$

here, I_{SOR} is the total intensity of solar radiation on the structure's surface, W/m^2; ε is the short-wave radiation absorption coefficient of the concrete surface.

2.1.1. Calculation of Sun Position Parameters

The declination angle δ, the elevation angle α_s and the azimuth angle γ_s, which are the relative position parameters between the sun and ground level, can be calculated [25] based on the laws of solar operation in astronomical knowledge.

The solar declination, also referred to as the solar inclination, is the angle between the solar ray and the Earth's equator at noon. This parameter is primarily used to describe the different seasons and can be approximately calculated according to FOR (4):

$$\delta = 23.45^\circ \sin\left[\frac{360^\circ}{365}(284 + N)\right] \quad (4)$$

here, N is the annual accumulated day, which is the total number of days from January 1 to the current date. The solar declination ranges from $-23.45^\circ \sim 23.45^\circ$, with a value of $+23.5^\circ$ during the summer solstice, zero during the spring and autumn equinoxes, and -23.5° during the winter solstice.

$$\sin(\alpha_s) = \cos(\varphi)\cos(\delta)\cos(\tau) + \sin(\varphi)\sin(\delta) \quad (5)$$

$$\gamma_s = \mathrm{sign}(\tau) \mid \cos^{-1}\left[\frac{\sin(\alpha_s)\sin(\varphi) - \sin(\delta)}{\cos(\alpha_s)\cos(\varphi)}\right] \mid \quad (6)$$

In the above formula, α_s is the solar elevation angle, which is 0° at sunrise and sunset. φ is the geographical latitude, which is positive in the Northern Hemisphere and negative in the Southern Hemisphere, and the value range is $-90^\circ \sim 90^\circ$; τ is the solar hour angle, and the solar hour angle at noon is exactly 0°, $\tau = (t - 12) \times 15^\circ$, where t is the true solar hour; the γ_s is positive to the west and negative to the east, and the value range is $-180^\circ \sim 180^\circ$; the sign is a symbol function.

The relative position relationship between the structural plane and the sun is shown in Figure 3. The incident angle i of the solar ray, the angle between the solar ray and the external normal n of the structural surface, is calculated by the following FOR (7) [26]:

$$\cos i = \sin\alpha_s \cos\beta + \cos\alpha_s \sin\beta\cos(\gamma_s - \gamma) \quad (7)$$

where β is the inclination angle of the structural surface relative to the horizontal plane, and the value range is $0^\circ \sim 180^\circ$. When it is greater than 90°, the surface is downward; γ is the angle between the exterior normal n of the structural surface and the positive south direction, ranging from -180° to 180°. The direction is positive to the west and negative to the east.

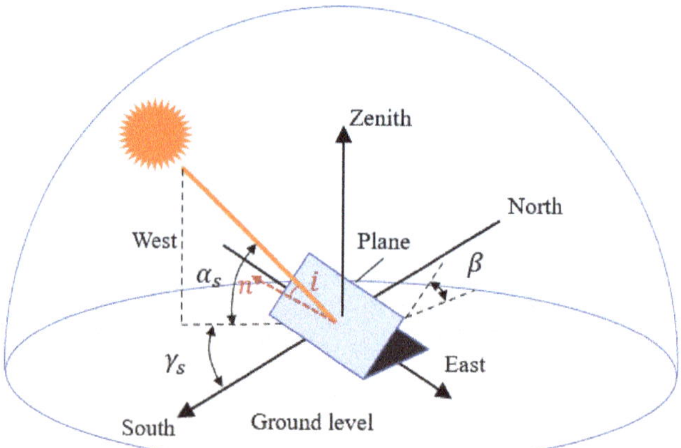

Figure 3. The relative position relationship between the sun and the inclined plane.

2.1.2. Intensity of Solar Radiation

Direct solar radiation is a component of the radiation intensity that reaches the earth's surface after undergoing multiple attenuations, such as absorption and scattering. In engineering calculations, Bouguer's Formula (8) is commonly used to estimate the intensity of direct solar radiation reaching the surface of the earth [27]:

$$I_{D0} = I_0 P^m \tag{8}$$

where I_0 is the solar constant, which represents the radiation intensity of the upper bound of the atmosphere projected onto the Earth's surface in unit time, W/m^2; m is the atmospheric optical quality; P is the composite atmospheric transparency coefficient [26].

$$I_0 = 1367\left[1 + 0.033\cos\left(360^\circ N/365\right)\right] \tag{9}$$

$$m = \frac{1}{\sin(\alpha_s)} \tag{10}$$

$$P = 0.9^{t_u k_a} \tag{11}$$

$$t_u = A_{tu} - B_{tu}\cos\left(\frac{360^\circ N}{365}\right) \tag{12}$$

where t_u is the Linke turbidity coefficient, which is related to time and geographical location; k_a is the relative atmospheric pressure at different altitudes, and its value varies with altitude [25]. Both A_{tu} and B_{tu} are empirical parameters representing the annual average value and variation range of the Linke turbidity coefficient t_u under different atmospheric conditions. The values of A_{tu} and B_{tu} can be determined from Table 1 [25].

Table 1. The Linke turbidity coefficient.

Region	A_{tu}	B_{tu}
Mountain area	2.2	0.5
Village	2.8	0.6
City	3.7	0.5
Industrial area	3.8	0.6

Solar radiation is scattered by air molecules, aerosol molecules, and other particles in the atmosphere before reaching the surface of the bridge structure. The part of solar radiation that can reach the surface of the bridge structure after being subjected to this effect is called solar scattered radiation. The intensity of solar scattered radiation I_{C0} (W/m^2) on the horizontal surface of a bridge can be calculated according to Equation (13) [28]:

$$I_{C0} = 0.5 I_0 \frac{1 - P^m}{1 - 1.4\ln(P)} \sin(\alpha_s) \tag{13}$$

As the Earth's surface absorbs and reflects solar radiation, a portion of solar radiation reaching the surface can be reflected on the surface of the bridge structure. This portion of the solar radiation is called ground-reflected radiation.

The horizontal ground reflected radiation intensity I_R (W/m^2) can be calculated using FOR (14) [29,30]:

$$I_{R0} = R_e[I_D \sin(\alpha_s) + I_C] \tag{14}$$

where R_e is the short-wave emissivity of the surface or water surface, typically taken at 0.2 for the ground [31].

2.1.3. Solar Radiation Intensity on Arbitrary Surface

The direct solar radiation intensity I_D, the sky scattered radiation intensity I_C, and the ground reflected radiation intensity I_R on the structural surface with an inclination angle of β are:

$$\begin{cases} I_D = I_{D0}\cos(\theta) \\ I_C = I_{C0}\dfrac{1+\cos(\beta)}{2} \\ I_R = I_{R0}\dfrac{1-\cos(\beta)}{2} \end{cases} \quad (15)$$

Under the influence of sunshine, there are different temperature differences on the surface of the bridge structure due to varying inclination angles, azimuth angles, and heights from the ground of the bridge components. However, the different components of the structure or different parts of the same component will be projected onto the surfaces of other components under the influence of sunshine, causing solar rays' shading. Three-dimensional shading will result in large temperature gradients on the surface of the bridge. The shadow occlusion relationship between the component and the component is complicated under the action of three-dimensional sunshine.

Based on existing research, the shading caused by solar ray irradiation can be categorized into no occlusion, self-occlusion, mutual occlusion, and permanent occlusion [32], as illustrated in Figure 4.

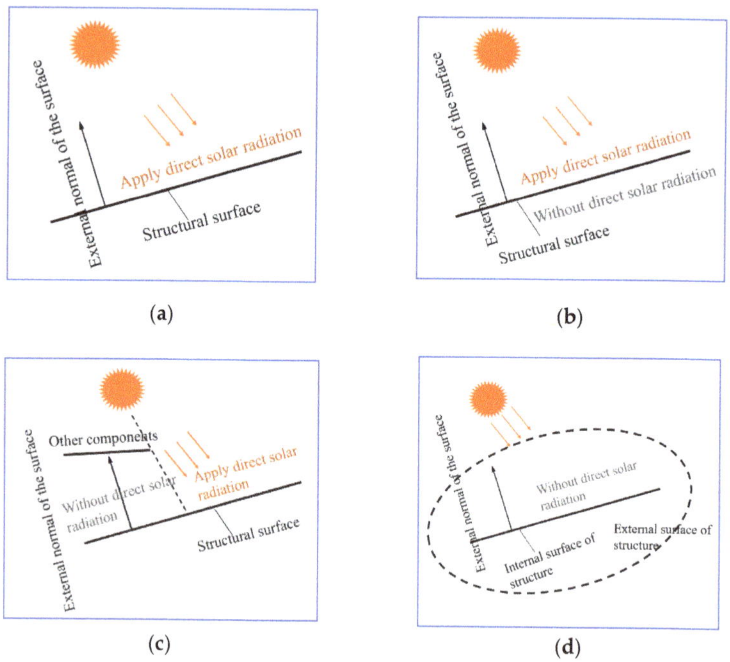

Figure 4. The solar radiation effect. (**a**) No occlusion; (**b**) self-occlusion; (**c**) mutual occlusion; (**d**) permanent occlusion.

The total amount of solar radiation I_{SOR} received by any surface should satisfy FOR (16) according to the shadow occlusion relationship of the structural surface shown in Figure 4.

$$I_{SOR} = \begin{cases} I_D + I_C + I_R & \text{Sunlight area} \\ I_C + I_R & \text{Shadow area} \end{cases} \quad (16)$$

As shown in Figure 5, the sun's trajectory varies in different seasons, and the intensity of solar radiation received on the surface of the same structure varies significantly. Figure 5b shows the variation of solar radiation intensity on the same horizontal surface at different times of the year, as calculated using the above solar radiation model for the 40° north latitude.

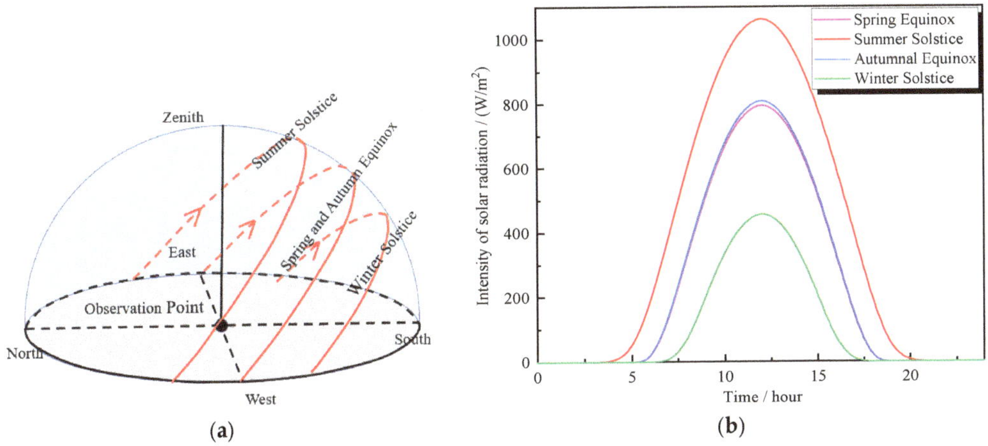

Figure 5. The variation of solar radiation intensity. (**a**) The schematic diagram of the solar trajectory changes in four seasons. (**b**) The variation of solar radiation intensity on the same horizontal surface at different times of the year.

2.1.4. Validation of the Solar Radiation Model

Compare the measured data with the theoretical values calculated by the above theoretical formula to verify the accuracy of the solar radiation model.

The measured data were obtained by monitoring the solar radiation intensity on a sunny day in March in the 40° north latitude area, as reported by Song et al. [33]. The comparison between the calculated value of the model and the measured values is shown in Figure 6. It can be observed from the figure that the simulated values of the radiation model used in this article agree well with the measured values, indicating that the radiation model is accurate and can be used to simulate the solar radiation intensity for the calculation of the structural sunshine temperature field.

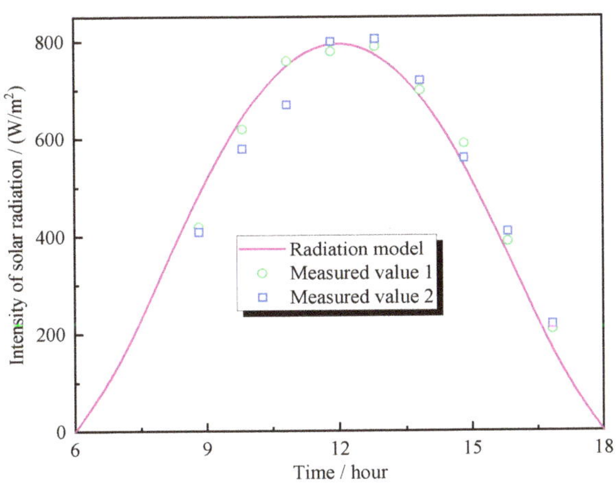

Figure 6. The comparison and verification of solar radiation intensity.

2.2. *Convective Heat Transfer Effect*

Convective heat transfer refers to the process of mutual heat transfer between the structural surface and the surrounding environment due to the temperature difference. The heat flux density q generated by convective heat transfer between the structural surface and the atmospheric environment q_c (W/m^2) can be calculated by Newton's law of cooling [34]:

$$q_c = h_c(T_a - T) \quad (17)$$

where T_a is the ambient temperature, °C; T is the surface temperature of the bridge structure, °C; and h_c is the convective heat transfer coefficient, $W/(m^2 \cdot °C)$.

The convective heat transfer coefficient h is related to the shape of the structure, wind speed, ambient temperature, and other factors. The FOR (18) is usually used to calculate the temperature field of a bridge structure [35,36].

$$h_c = 2.5\left(\sqrt[4]{T_a - T} + 1.54v\right) \quad (18)$$

where v is wind speed, m/s.

2.3. *Radiation Heat Transfer Effect*

The nature of radiation heat transfer is the emission of electromagnetic waves from a heat source. Radiation heat transfer occurs continuously between the surface of the bridge structure and the surrounding environment. The surface of the structure not only absorbs radiation from the atmosphere and the ground but also emits radiation to the surrounding environment. Unlike solar short-wave radiation, the thermal radiation generated by radiation heat transfer is long-wave radiation.

The radiation heat transfer between the structural surface and the external environment in this article is calculated using FOR (19) [25,37].

$$q_r = h_r(T_a - T) - q_{rn} \quad (19)$$

$$h_r = C_0\varepsilon\left[(T + 273.15)^2 + (T_a + 273.15)^2\right](T + T_a + 546.3) \quad (20)$$

$$q_{rn} = (1 - \varepsilon_a)\frac{1 + \cos(\beta)}{2}\varepsilon C_0(T_a + 273.15)^4 \quad (21)$$

where h_r is the radiation heat transfer coefficient, $W/(m^2 \cdot °C)$; C_0 is the Stefan–Boltzmann constant, taking $5.67 \times 10^{-8} W/\left(m^2 \cdot K^4\right)$; ε_a is the atmospheric radiation coefficient, its value range is $0.74 \sim 0.95$, generally taken as 0.82; ε is the radiation emissivity of the structure, generally $0.85 \sim 0.95$; and q_{rn} is the heat flux density of the inclined plane caused by the sky radiation effect, W/m^2.

2.4. *Daily Temperature Model*

Ambient temperature changes are mainly influenced by solar radiation on an ideal sunny day (without clouds or rain). The daily temperature starts to gradually rise after sunrise and typically reaches its peak around 14:00~15:00, after which it begins to fall gradually. To simulate the sunshine temperature field of the bridge structure, it is reasonable to assume that the atmospheric temperature varies as a sinusoidal function throughout the day. In this study, the segmental sinusoidal function model shown in Figure 7 is used to simulate the changes in temperature throughout the day with the following equations [38,39]:

$$\begin{cases} T_1 = 0.5\left[T_{sum} + \Delta T \sin\left(\frac{\pi(t+30)}{24}\right)\right] & 0 \le t < 6 \\ T_1 = 0.5\left[T_{sum} + \Delta T \sin\left(\frac{\pi(t-10.5)}{9}\right)\right] & 6 \le t < 15 \\ T_1 = 0.5\left[T_{sum} + \Delta T \sin\left(\frac{\pi(t-9)}{12}\right)\right] & 15 \le t \le 24 \end{cases} \quad (22)$$

where t is the moment, hour; $T_{sum} = T_{max} + T_{min}$; $\Delta T = T_{max} - T_{min}$; T_{max} is the daily maximum temperature, taking the temperature at 3:00 p.m., °C; and T_{min} is the daily minimum temperature, taking the temperature at 6:00 a.m., °C.

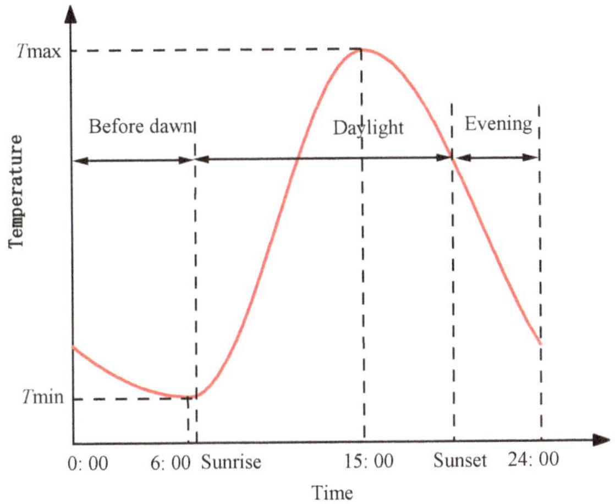

Figure 7. The daily temperature change model.

Since the solar altitude angle is zero at sunrise and sunset, the sunrise moment t_c and the sunset moment t_c can be calculated using FOR (23) [25]:

$$\begin{cases} t_c = 12 - \frac{1}{15}\cos^{-1}(-\tan\delta\tan\phi) \\ t_s = 12 + \frac{1}{15}\cos^{-1}(-\tan\delta\tan\phi) \end{cases} \tag{23}$$

where δ is the solar declination; ϕ is the geographical latitude.

The accuracy of the daily temperature change model was verified by comparing the simulated values with the measured temperature data. Figure 8 displays the comparison between the simulated and measured ambient temperatures on 17 May 2019, in the area of 116°20′ east longitude and 39°56′ north latitude.

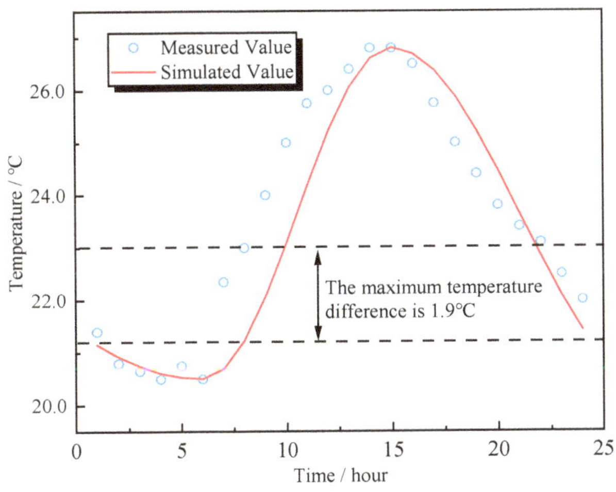

Figure 8. Verification of the daily temperature model.

As shown in Figure 8, the simulated temperature changes are in good agreement with the measured temperature changes, and the maximum temperature difference is only 1.9 °C. Therefore, it can be used to accurately simulate the temperature changes of environmental objects in engineering calculations.

3. Methods

3.1. Methodology for Simulation of Sunshine Temperature Fields

In this article, the DFLUX and FILM subroutines are provided by the ABAQUS platform for the redevelopment of the software to calculate the sunshine temperature field of the structure. Figure 9 shows the secondary development interface of the subroutine. The DFLUX subroutine can apply the surface heat source controlled by the custom heat source equation to the surface of the selected structure [40]. This is used to load the solar radiation heat load onto the surface of the structure. At the same time, the sunshine shadow recognition algorithm is introduced into the DFLUX subroutine to distinguish the sunshine shadow surface of the structure. The FILM subroutine is used to define the heat transfer coefficient related to ambient temperature and model parameters [40]. Hence, the convective heat transfer and radiation heat transfer-generated heat transfer coefficients between the structural surface and the surrounding environment are loaded by the FILM subroutine. Figure 10 shows the specific loading situation.

```
SUBROUTINE DFLUX(FLUX, SOL, KSTEP, KINC, TIME, NOEL, NPT, COORDS,
     1 JLTYP, TEMP, PRESS, SNAME)
C
      INCLUDE 'ABA_PARAM.INC'
C
      DIMENSION FLUX(2), TIME(2), COORDS(3)
      CHARACTER*80 SNAME


      user coding to define FLUX(1) and FLUX(2)


      RETURN
      END
```

(**a**)

```
      SUBROUTINE FILM(H, SINK, TEMP, KSTEP, KINC, TIME, NOEL, NPT,
     1 COORDS, JLTYP, FIELD, NFIELD, SNAME, NODE, AREA)
C
      INCLUDE 'ABA_PARAM.INC'
C
      DIMENSION H(2),TIME(2),COORDS(3), FIELD(NFIELD)
      CHARACTER*80 SNAME

      user coding to define H(1), H(2), and SINK

      RETURN
      END
```

(**b**)

Figure 9. Interface for secondary development of subroutines. (**a**) The interface of the DFLUX subroutine. (**b**) The interface of the FILM subroutine.

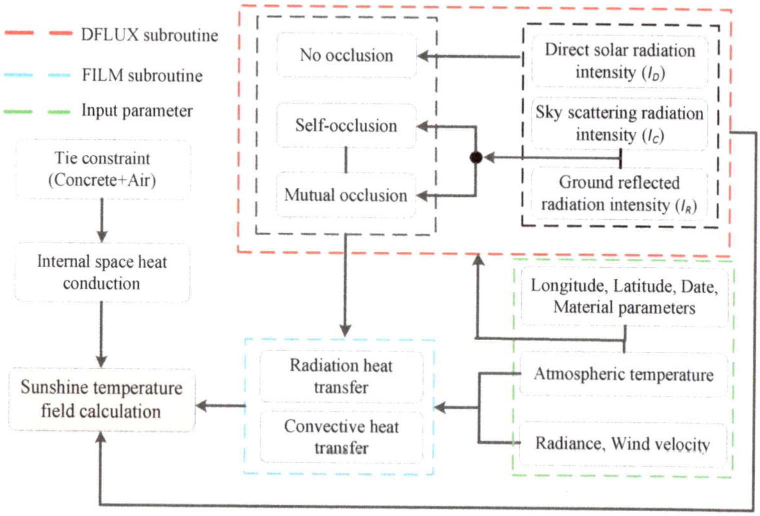

Figure 10. The simulation process of the sunshine temperature field.

3.2. Validation of the Simulation Method

The experimental model in Reference [41] is taken as a case, and the measured temperature value in the literature is compared with the simulated temperature value in this paper to verify the accuracy of the above temperature field simulation method.

3.2.1. Brief Description of the Case

The experimental model is a concrete-curve box girder with a total span of 10 m along the centerline of the bridge, a radius of curvature of 12 m, and a central angle of 48°. The main girder is a single-cell box section with a roof width of 1.7 m, a floor width of 0.62 m, a minimum vertical web thickness of 0.1 m, and a box girder height of 0.36 m. The section of the box girder is depicted in Figure 11. The test model bridge is located at 118°38′ east longitude and 32°05′ north latitude, and it exhibits an east-west orientation.

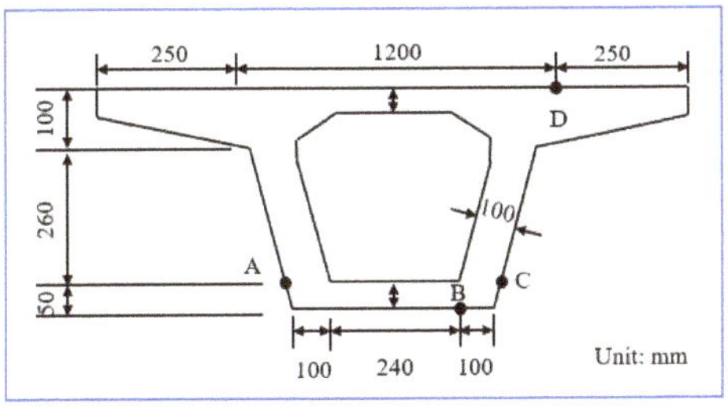

Figure 11. Section size and temperature measuring point layout diagram.

The simulation values of temperature at measuring points A, B, C, and D in the mid-span section of the experimental model were compared with the measured temperature values to verify the accuracy of the above sunshine temperature field simulation method. The arrangement of the measuring points is shown in Figure 11.

3.2.2. Setting of Model and Material Parameters

The concrete box girder is a hollow and closed structure, which makes it difficult to determine boundary conditions such as convective and radiative heat transfer between the internal surface of the box girder and the air inside. In the absence of directly measured data on the inner air temperature, this study uses heat conduction to simulate the heat transfer process between the internal surface of the box girder and the air in the inner cavity [42,43]. The finite element model takes into account the continuous changes in temperature and heat flux density at the interface between concrete and air. The eight-node linear heat transfer hexahedral solid element "DC3D8" provided in ABAQUS is used to simulate the concrete box girder and the air inside, and the two are connected using the "TIE" function in the program. The finite element model and meshing are shown in Figures 12 and 13, respectively.

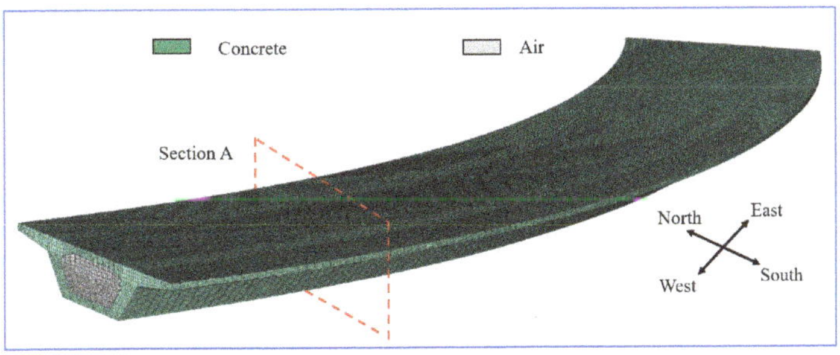

Figure 12. Schematic diagram of the finite element model.

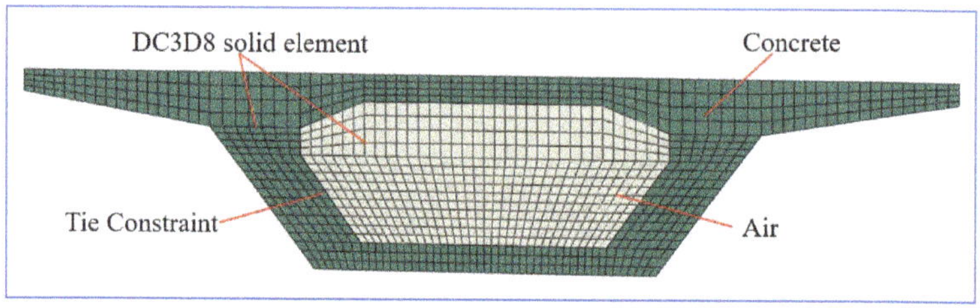

Figure 13. The mesh division diagram of section A.

Since the temperature change caused by sunshine has little effect on the thermal parameters of concrete structures, they can be considered fixed values [42]. Table 2 shows the thermal parameters of the materials used in the model.

Table 2. Thermal parameters of materials.

Thermal Parameter	Concrete	Air
Density (kg/m^3)	2400	1.29
Thermal Conductivity (W/(m·°C))	2.5	0.025
Specific Heat Capacity (J/(kg·°C))	900	1010
Radiation Absorption Rate	0.5	/
Longwave radiative emissivity	0.9	/
Longwave radiation absorptivity	0.9	/

3.2.3. Results Analysis

The accuracy of the proposed sunshine temperature field simulation method was verified by comparing the simulated temperature field values from 0 h on 13 August 2013 to 24 h on 14 August 2013 with the measured values of the temperature field during the corresponding period. The initial temperature field was obtained by introducing the temperature field calculation results of the first 3 days into the calculation model to eliminate the influence of the initial temperature of the model on the simulation results. The thermal parameters of the materials used in the model are shown in Table 2. The atmospheric temperature data for the calculation date are shown in Table 3.

Table 3. Atmospheric temperature.

Date	Weather Conditions	Maximum Temperature/°C	Minimum Temperature/°C
10 August 2013	Clear	39.0	29.0
11 August 2013	Clear	39.0	30.0
12 August 2013	Clear	39.0	29.0
13 August 2013	From sunny to cloudy	39.0	28.0
14 August 2013	Clear	38.0	27.0

Figures 14 and 15 show the contours of the calculated temperature field, and Figure 16 presents a comparative analysis of the simulated and measured temperature field values obtained using the sunshine temperature field simulation method proposed in this paper. The trend of the simulated and measured temperature field values is similar. Although the simulation results for the roof differ significantly from the measured results during the period from 10 a.m. to 5 p.m. on 13 August, the two are more consistent at other locations. The maximum errors between the simulated values and the measured values at points A, B, C, and D are 2.9 °C, 2.5 °C, 2.8 °C, and 2.8 °C, respectively. The weather changed from sunny to cloudy during the period from 10 a.m. to 5 p.m. on 13 August, and the direct solar radiation weakened, resulting in a large difference between the simulation results and the measured results for the roof. Table 4 shows that the mean deviations between the simulated values and the measured values of the temperature at points A, B, C, and D are 1.5 °C, 1.9 °C, 1.4 °C, and 1.5 °C, respectively.

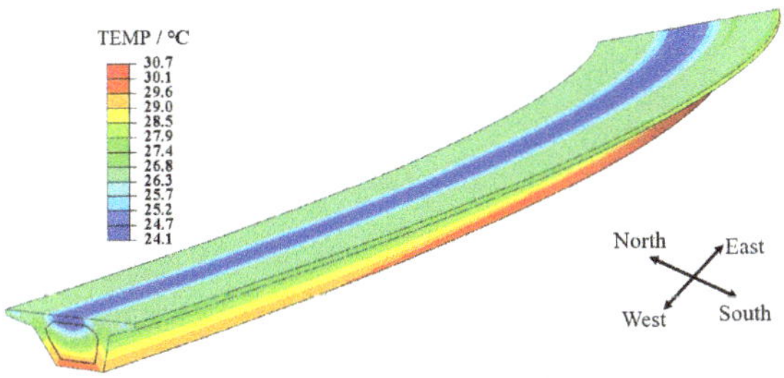

Figure 14. Temperature field calculation cloud map at 7:00 a.m. on 13 August.

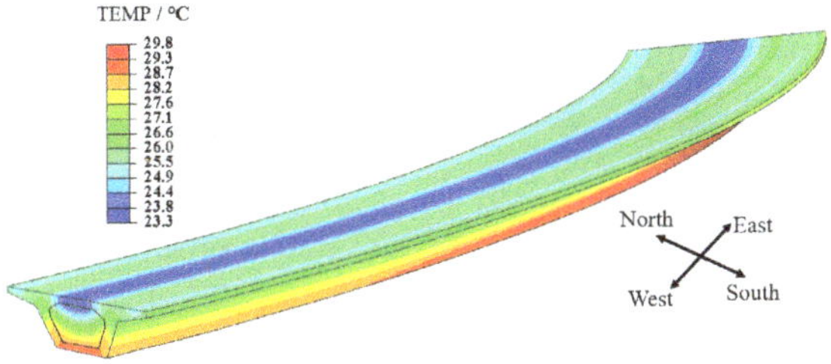

Figure 15. Temperature field calculation cloud map at 7:00 a.m. on 14 August.

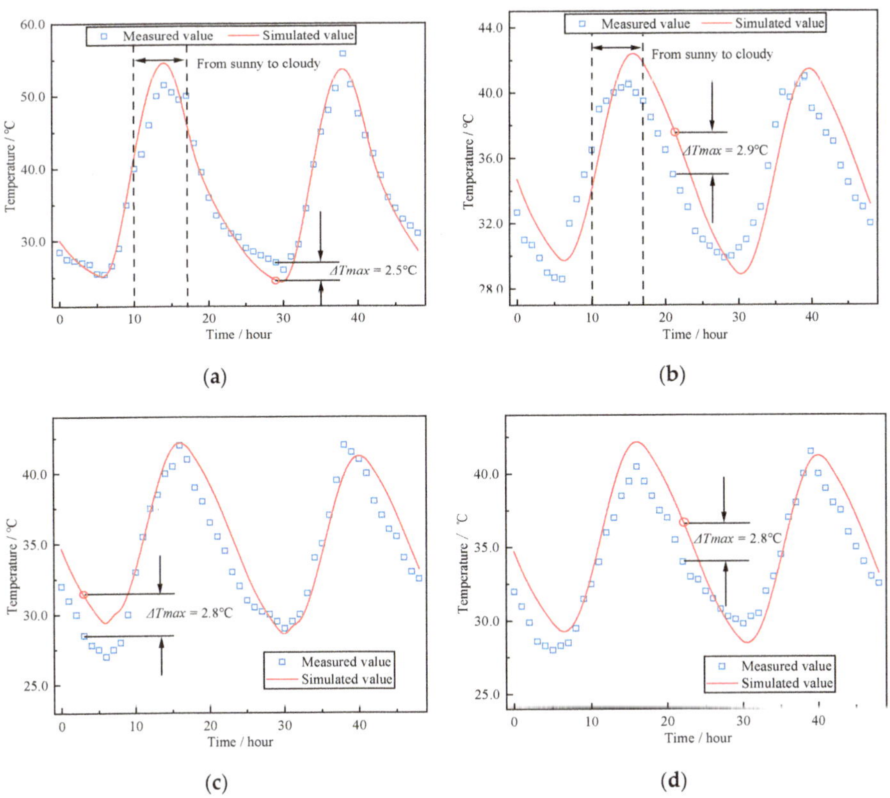

Figure 16. Comparison of temperature changes. (**a**) Comparison of temperature values at point A; (**b**) comparison of temperature values at point B; (**c**) comparison of temperature values at point C; (**d**) comparison of temperature values at point D.

Table 4. The deviation between the simulated and measured results.

Measuring Point	Maximum Deviation/°C	Mean Value of Deviation/°C
A	2.5	1.5
B	2.9	1.9
C	2.8	1.4
D	2.8	1.5

In summary, the proposed simulation method for the concrete box girder temperature field in this article provides results that are in good agreement with the actual situation, indicating that it can be used to accurately simulate the sunshine temperature field of bridge structures and analyze their temperature effects.

4. Results

In the daily temperature cycle of a bridge structure, solar radiation heats up the structure during the day, while the ambient temperature decreases at night and the structure dissipates heat into the environment to cool down. The sunshine temperature field of the bridge structure exhibits strong time-varying characteristics due to the changes in solar radiation intensity and ambient temperature. This section focuses on simulating and discussing the daily temperature cycle of the concrete box girder of a cable-stayed bridge.

4.1. Example Introduction

A single-box, three-cell concrete box girder with a length of 400 m (ignoring the diaphragm plate) is selected as an example to calculate the temperature field to analyze the sunshine temperature field of the closed concrete box girder, including the tuyere, which is 38.5 m. The vertical thickness of the upper and lower floors and the inclined webs on both sides is 0.55 m, and the thickness of the middle web is 0.8 m. The detailed dimensions of the box girder section can be seen in Figure 17. The bridge is located at $116^{\circ}20'$ east longitude and $39^{\circ}56'$ north latitude, with a north-south orientation. The meshing of the model section is shown in Figure 18.

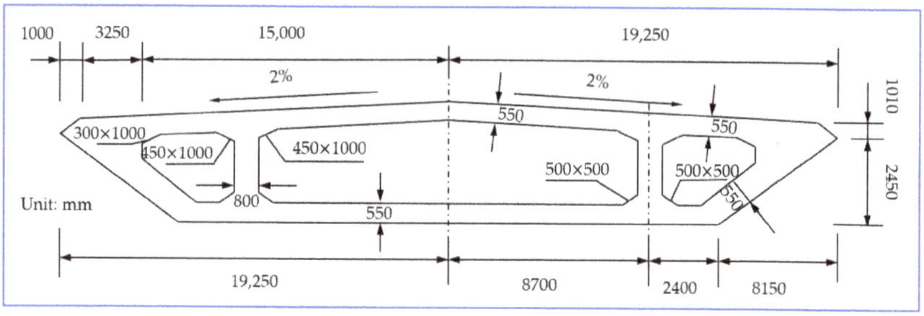

Figure 17. Detail size of the girder section.

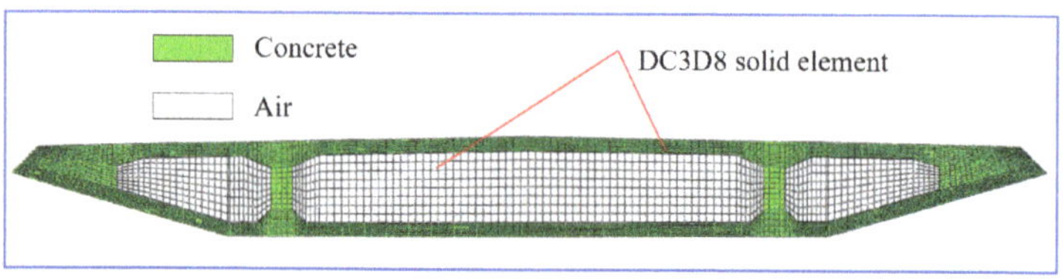

Figure 18. Schematic diagram of the mesh division.

The roof has no shading effect during the day; the bottom plate has been self-shielding; and the two outer inclined webs are exposed to the sun only for a short time during the day. Therefore, the T_1~T_8 measuring points in Figure 19 are selected to analyze the time-varying temperature of the internal and external surfaces of the roof, the bottom plate, and the outer inclined web. Among them, T_1, T_3, T_5, and T_7 are the temperature measuring points on the external surface, and T_2, T_4, T_6, and T_8 are the temperature measuring points on the internal surface.

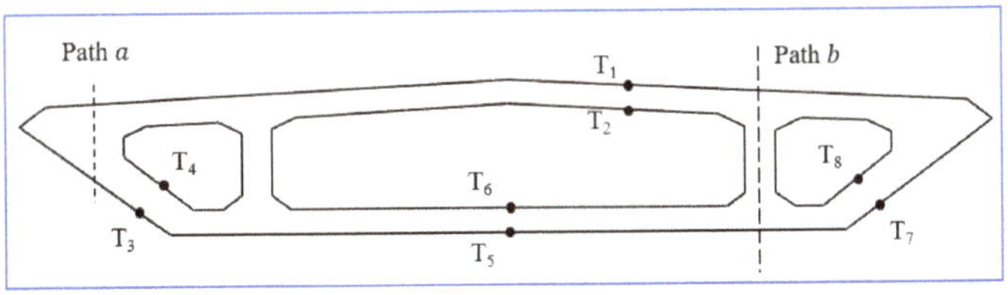

Figure 19. The setting of temperature measurement points and temperature extraction paths.

The influence of wind speed on the structural temperature is mainly reflected in the effect of the convective heat transfer coefficient on the surface of the structure; therefore, it cannot be ignored in the calculation of the convective heat transfer coefficient. In the absence of actual measurement data, many scholars often assume the wind speed as 1.0 m/s when analyzing the temperature field of bridge structures [44,45]. Therefore, this article also assumes the same value for the wind speed when analyzing the sunshine temperature field of the concrete box girders. The material parameters and environmental parameters required for analysis are shown in Tables 5 and 6. The results of the continuous calculation of the temperature field in the first three days are used as the initial temperature field for subsequent calculations to eliminate the influence of the initial temperature field on the calculation results.

Table 5. Thermal parameters of materials.

Thermal Parameter	Concrete	Air
Density (kg/m^3)	2400	1.29
Thermal Conductivity (W/(m·°C))	2.5	0.025
Specific Heat Capacity (J/(kg·°C))	900	1010
Radiation Absorption Rate	0.6	/
Long-wave radiative emissivity	0.9	/
longwave radiation absorptivity	0.9	/

Table 6. Weather conditions.

Date	Weather Conditions	Maximum Temperature/°C	Minimum Temperature/°C
5 September 2022	Clear	28.0	14.0
6 September 2022	Clear	29.0	15.0
7 September 2022	Clear	32.0	17.0
8 September 2022	Clear	32.0	19.0

4.2. Time-Varying Analysis of Temperature Field

Figure 20 shows the two-dimensional temperature distribution cloud map of the example at a typical time of the day. Figure 21 shows the time-varying curves of the surface temperature of the roof, floor, and inclined web of the concrete box girder. It can be observed that the temperature on the external surfaces of the roof, floor, and inclined web of the concrete box girder fluctuates greatly during the day, while the temperature on the internal surface changes slightly. The minimum temperature of each surface appears at around 6 a.m. The maximum temperature of the external surface of the roof appears at around

2 p.m., reaching about 44.3 °C, and the temperature difference between its internal and external surfaces can reach 16.5 °C. The change range of the external surface temperature of the floor and the inclined web is smaller than that of the roof. This is because the floor is always in a self-occlusion state and not exposed to the sun, while the two inclined webs are also exposed to the sunshine for a shorter period during the day and receive less direct solar radiation. As a result, the magnitude of the corresponding surface temperature variation is smaller compared to the surface of the top slab, which is always exposed to the sun. The maximum value of the temperature on the floor and the inclined web occurred at around 3 p.m., with a maximum value of approximately 31.0 °C, and the maximum temperature difference between the external surface and the internal surface is approximately 6.5 °C.

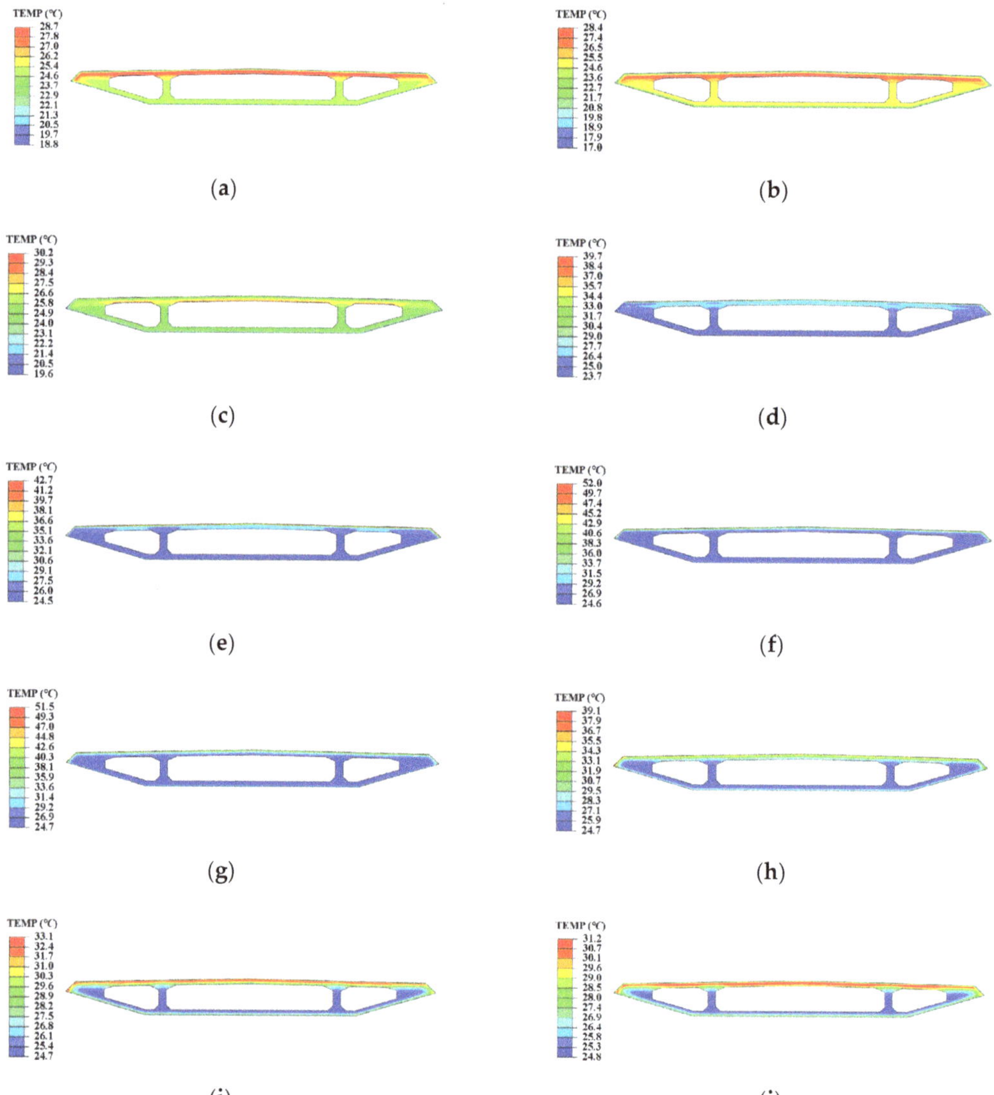

Figure 20. Two-dimensional temperature field distribution cloud chart at typical moments in a day. (**a**) 2:00; (**b**) 6:00; (**c**) 8:00; (**d**) 10:00; (**e**) 12:00; (**f**) 14:00; (**g**) 16:00; (**h**) 18:00; (**i**) 20:00; (**j**) 22:00.

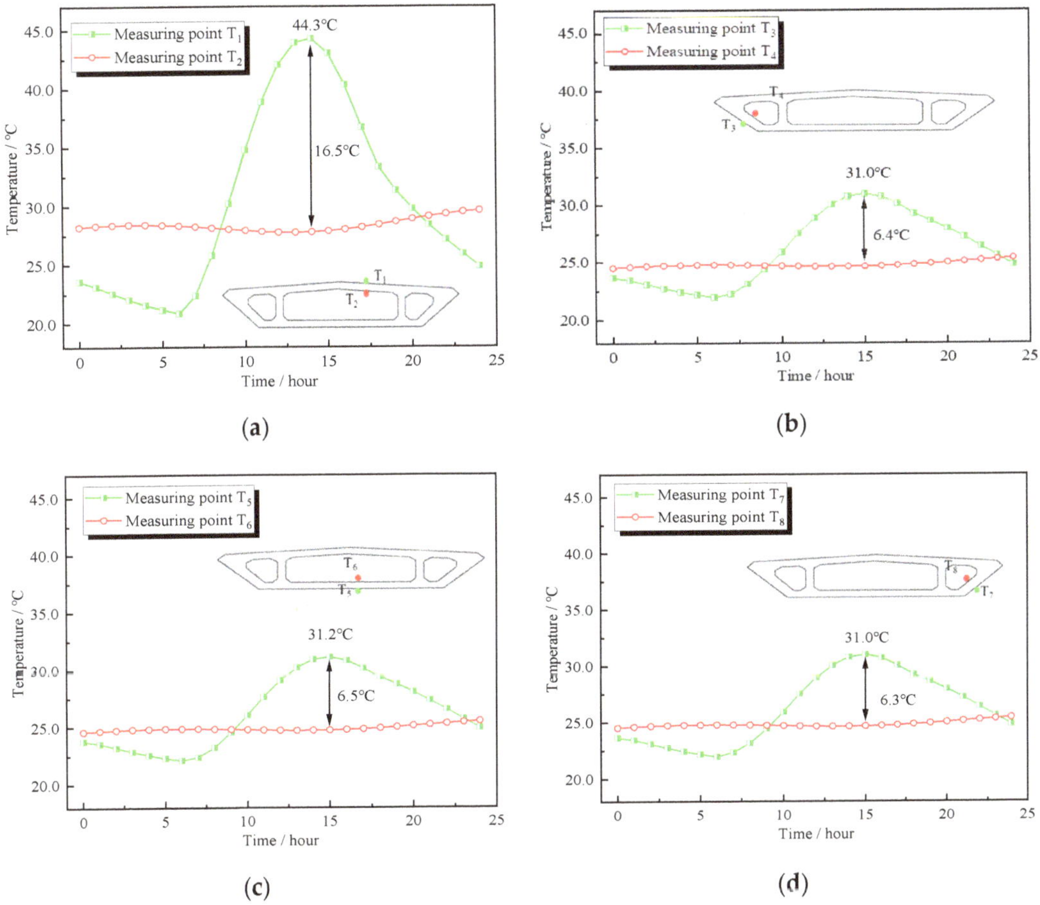

Figure 21. Time-varying curve of the internal and external surface temperatures of the concrete box girder. (**a**) Temperature time-varying curve of the roof; (**b**) temperature time-varying curve of the floor; (**c**) temperature time-varying curve of the left inclined web; (**d**) temperature time-varying curve of the left inclined web.

4.3. Vertical Temperature Gradient

A large temperature gradient is often generated in the direction of the beam height due to the low thermal conductivity of concrete. If the temperature of the roof is higher than the temperature of the web, it is referred to as the vertical positive temperature difference distribution mode, also known as the heating gradient. Conversely, when the temperature of the web is higher than the temperature of the roof, it is referred to as the vertical negative temperature difference distribution mode, also known as the cooling gradient. The difference between the maximum and minimum temperatures in the height direction of the beam represents the maximum gradient temperature difference.

The concrete box girder of the cable-stayed bridge is generally a multi-box structure. The outer web is affected by external factors and the temperature in the box, while the middle web is only affected by the temperature in the box. The time-varying analysis of the temperature field of the concrete box girder shows that the temperature of the top plate changes the most, and the temperature of the bottom plate and the outer web surface on both sides is the same. The temperature distribution of path a and path b in Figure 19 at different times is selected to analyze the vertical temperature gradient of the concrete box girder.

4.3.1. Temperature Distribution of the Path a

Figures 22–24 show the temperature changes of path a at different times of the day. According to the figures, the most unfavorable vertical negative temperature difference occurs at 6 a.m., which is 6.6 °C, while the most unfavorable vertical positive temperature difference appears at 2 p.m., which is about 17.3 °C.

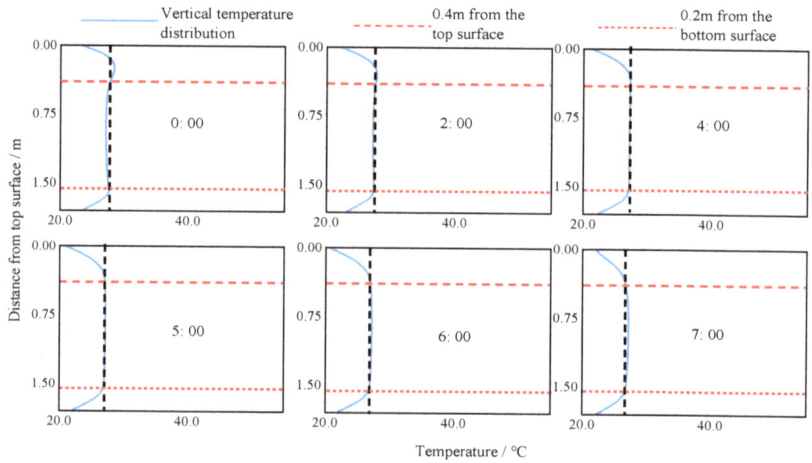

Figure 22. The vertical temperature distribution of path a in the early morning.

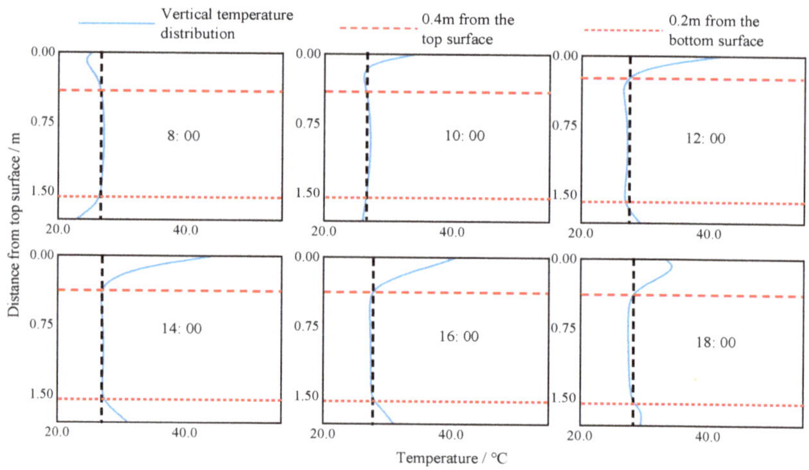

Figure 23. The vertical temperature distribution of path a in the daytime.

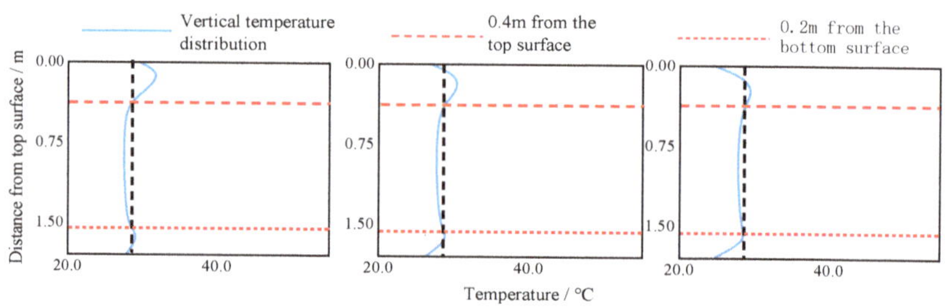

Figure 24. The vertical temperature distribution of path a at night.

4.3.2. Temperature Distribution of the Path b

Figures 25–27 show the temperature change of path b at different times of the day. According to the figure, the most unfavorable vertical negative temperature difference in a day occurs at 6 a.m., which is 6.7 °C; the most unfavorable vertical positive temperature difference appears at 2 p.m., about 17.2 °C.

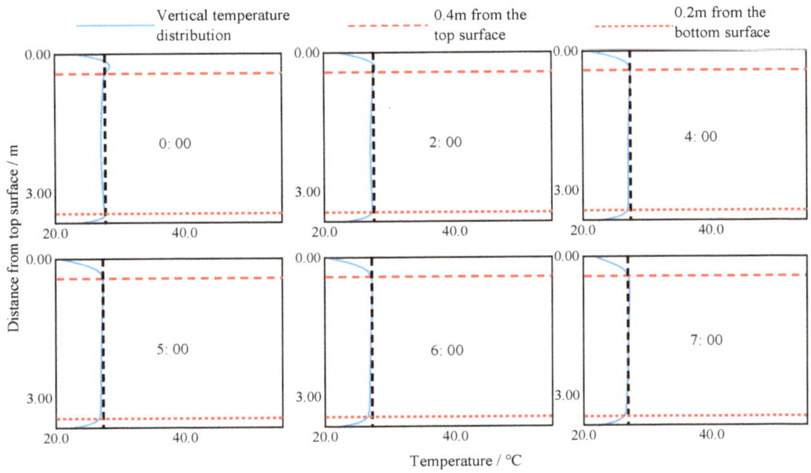

Figure 25. The vertical temperature distribution of path b in the early morning.

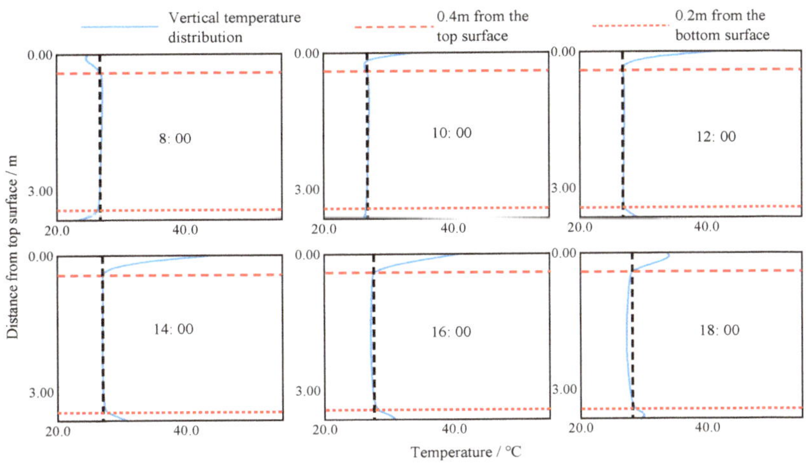

Figure 26. The vertical temperature distribution of path b in the daytime.

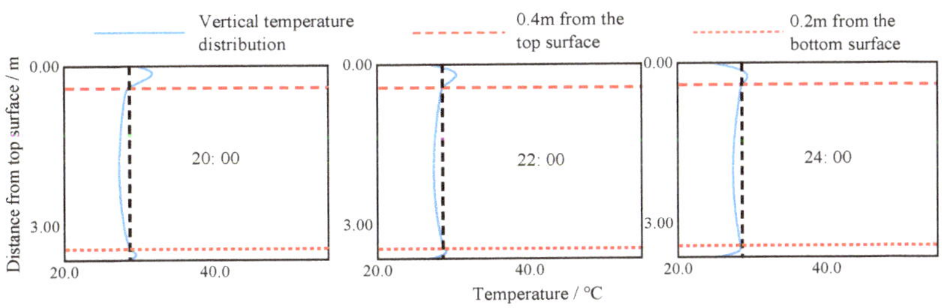

Figure 27. The vertical temperature distribution of path b at night.

4.3.3. Most Unfavorable Temperature Gradient

The above analysis shows that the most unfavorable heating gradient and the most unfavorable cooling gradient of paths a and b appear at the same time and both show a C-shaped distribution. Among them, the most unfavorable vertical negative temperature difference occurred at 6 a.m., about 6.7 °C, and the most unfavorable vertical positive temperature difference appeared at 2 p.m., about 17.2 °C.

There are significant differences in the regulations for the vertical temperature gradient of concrete box girders among different national bridge specifications. To quantify the differences, the most unfavorable temperature gradient distribution pattern calculated in this article is compared with the most unfavorable temperature gradient pattern recommended in several national specifications. Table 7 shows the most unfavorable vertical temperature gradient values in this article and several national specifications. The data in the table shows that the difference between the most unfavorable vertical temperature gradient value in this article and the most unfavorable vertical temperature gradient value in the General Specifications for Design of Highway Bridges and Culverts (JTG D60-2015) is the largest, with a difference of 7.8 °C and 5.8 °C for the heating gradient and cooling gradient, respectively.

Table 7. The most unfavorable vertical temperature gradient value.

Temperature Gradient Mode	Heating Gradient/°C	Cooling Gradient/°C
In this paper	17.2	6.7
JTG D60-2015 [46]	25.0	12.5
AASHTO [47]	21.0	10.5
BS-5400 [48]	13.5	8.4

As the most unfavorable vertical temperature gradients of paths a and b are the same, the most unfavorable distribution pattern of path a is chosen here for comparison with the temperature gradient distribution patterns of various national specifications. Figure 28 shows the distribution mode of the most unfavorable cooling gradient, and Figure 29 shows the distribution mode of the most unfavorable heating gradient. It can be seen from the figures that the most unfavorable temperature gradient distribution pattern obtained in this article is closest to the provisions in the British BS-5400 specification, followed by the difference with the American AASHTO specification, and the difference with the Chinese JTG D60-2015 specification is the largest. It should be noted that temperature variations within a certain range from the external surface are more pronounced due to the poor thermal storage properties of concrete. For the cooling gradient, the turning points for temperature change occur at 0.45 m from the top surface and 0.4 m from the bottom surface, respectively; for the heating gradient, the turning points for temperature change occur at 0.4 m from the top surface and 0.3 m from the bottom surface, respectively, which is closer to the BS-5400 specification.

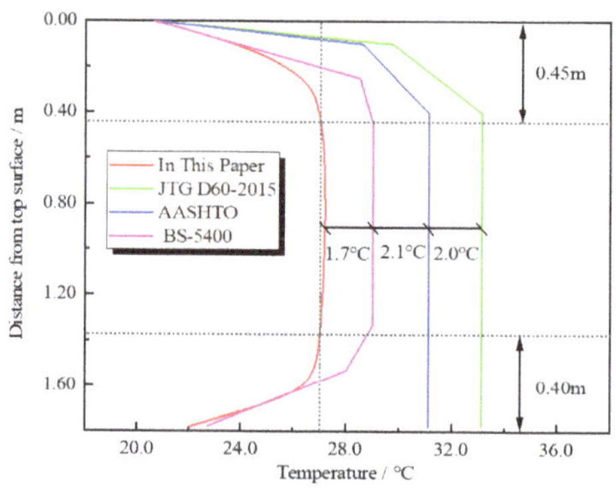

Figure 28. Comparison of negative temperature gradients.

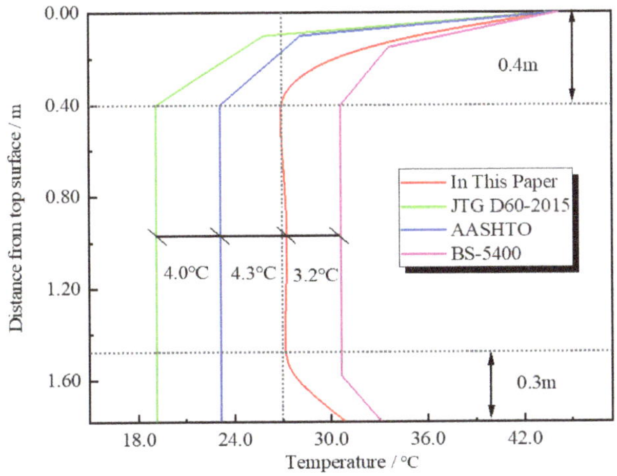

Figure 29. Comparison of positive temperature gradients.

To analyze the C-shaped temperature gradient distribution pattern obtained in this paper, the temperature variation between the upper and lower turning points is minimal, and the temperature distribution pattern resembles a straight line. On the other hand, the temperature distribution of the upper and lower turning sections is closer to an exponential function. Therefore, we used the exponential function presented in FOR (24) to fit the temperature values of the upper and lower turning sections [49,50], and the results of the fitting are shown in Figures 30 and 31.

$$T(x) = T_0 e^{-\alpha x} + C \tag{24}$$

where C is the temperature value at the temperature turning point, °C; T_0 is the temperature difference between the external surface of the box girder and the temperature turning point, °C.

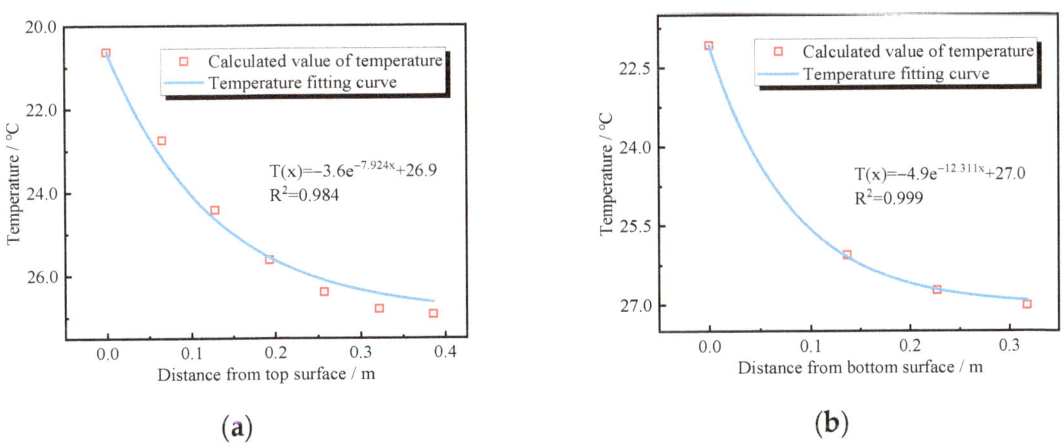

Figure 30. The fitting of temperature values in the transitional section of the cooling gradient. (**a**) The upper transitional section; (**b**) the lower transitional section.

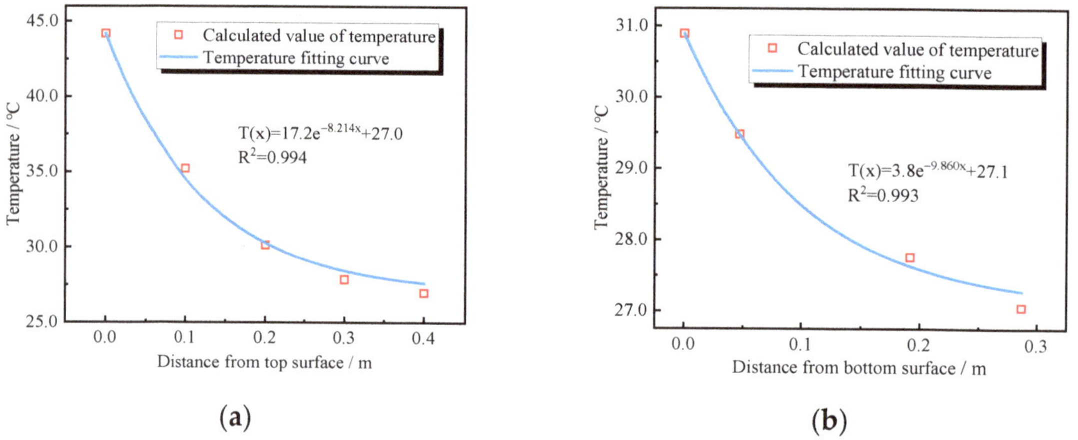

Figure 31. The fitting of temperature values in the transitional section of the heating gradient. (**a**) The upper transitional section; (**b**) the lower transitional section.

The minimum correlation coefficient of the fitted temperature data for the upper and lower transitions of the least favorable vertical negative temperature gradient and the least favorable vertical positive temperature gradient is 0.984, with good fitting accuracy.

Based on the analysis conducted in this study, the recommended distribution patterns for the cooling and heating gradients of concrete box girders in the construction stage are shown in Figure 32a,b, respectively. In these figures, the upper and lower transitional sections follow an exponential distribution, while the middle section follows a linear distribution.

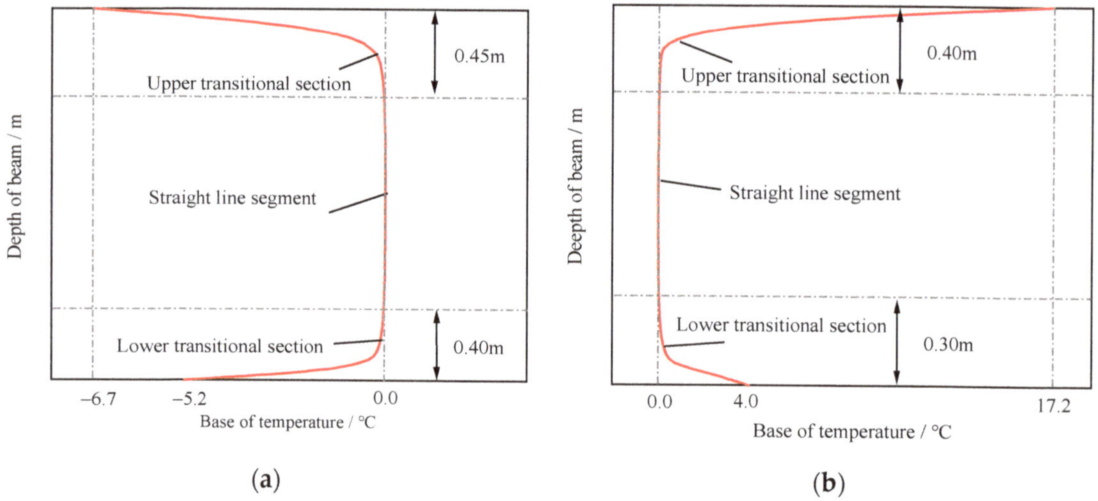

Figure 32. The distribution pattern of the temperature gradient. (**a**) The cooling gradient; (**b**) the heating gradient.

5. Discussion

In the analysis of the temperature of concrete structures, parameters such as material density, thermal conductivity, specific heat capacity, solar radiation absorption coefficient, surface long-wave radiation absorption rate, wind speed, etc. are required. While the concrete density and specific heat capacity can be measured experimentally, the thermal conductivity of the material can also be deduced by inversion from the measured temperature data. However, determining the exact value of the long-wave radiation absorption rate of the concrete and the absorption coefficient of solar radiation on the surface of the concrete structure is challenging, with values ranging between 0.85 and 0.95 and 0.5 and 0.7, respectively. Additionally, measuring the wind speed at the surface of the structure is often difficult. This subsection focuses on the impact of three calculated parameters, namely the solar radiation absorption coefficient, the surface long-wave radiation absorption (or emission) rate, and the wind speed, on the temperature field calculations.

For the study, the control variables method was used to increase or decrease the calculated parameters individually by a margin of 30% while keeping the other parameters constant. The values of each parameter are shown in Table 8.

Table 8. The values of parameters.

Parameter	Reduced by 30%	Unchanged (Original Value)	Increased by 30%
Wind speed (m/s)	1.4	2.0	2.6
Short-wave radiation absorption rate	0.42	0.60	0.78
Long-wave radiation absorption rate	0.63	0.90	1.17

The temperature values of T_1 measuring points on the roof are compared and analyzed, and the comparison results are shown in Figure 33.

From Figure 33, it is apparent that the solar radiation absorption coefficient has the greatest impact on the temperature calculation results, followed by the influence of changes in wind speed. The long-wave radiation absorption coefficient has the smallest impact on the temperature field calculation results while maintaining the same amplitude of change. The difference in the maximum temperature of the structural surface will reach about 13%, and the temperature change can be over 10 °C by increasing or decreasing the solar radiation absorption coefficient by 30%. This is because the solar radiation absorption coefficient determines how much solar radiation reaching the structural surface can be absorbed by the structure, and solar radiation is the primary cause of structural temperature change. Therefore, changes in this parameter have the greatest influence on the calculation results of the temperature field. The influence of wind speed on structural temperature is mainly reflected in the convective heat transfer coefficient between the structural surface and the external environment. The larger the wind speed, the larger the convective heat transfer coefficient, and the faster the heat dissipation of the structure. Decreasing wind speed will weaken the convective heat transfer process and reduce the heat dissipation speed of the structure. The long-wave radiation absorption coefficient mainly affects the radiation heat transfer between the structure and the external environment. However, the temperature change caused by this effect is relatively small. This results in the temperature change caused by changing the long-wave radiation absorption coefficient also being small.

In summary, it is important to choose a reasonable value for the solar radiation absorption coefficient to avoid significant deviations between the calculated temperature field and the actual results for concrete structures.

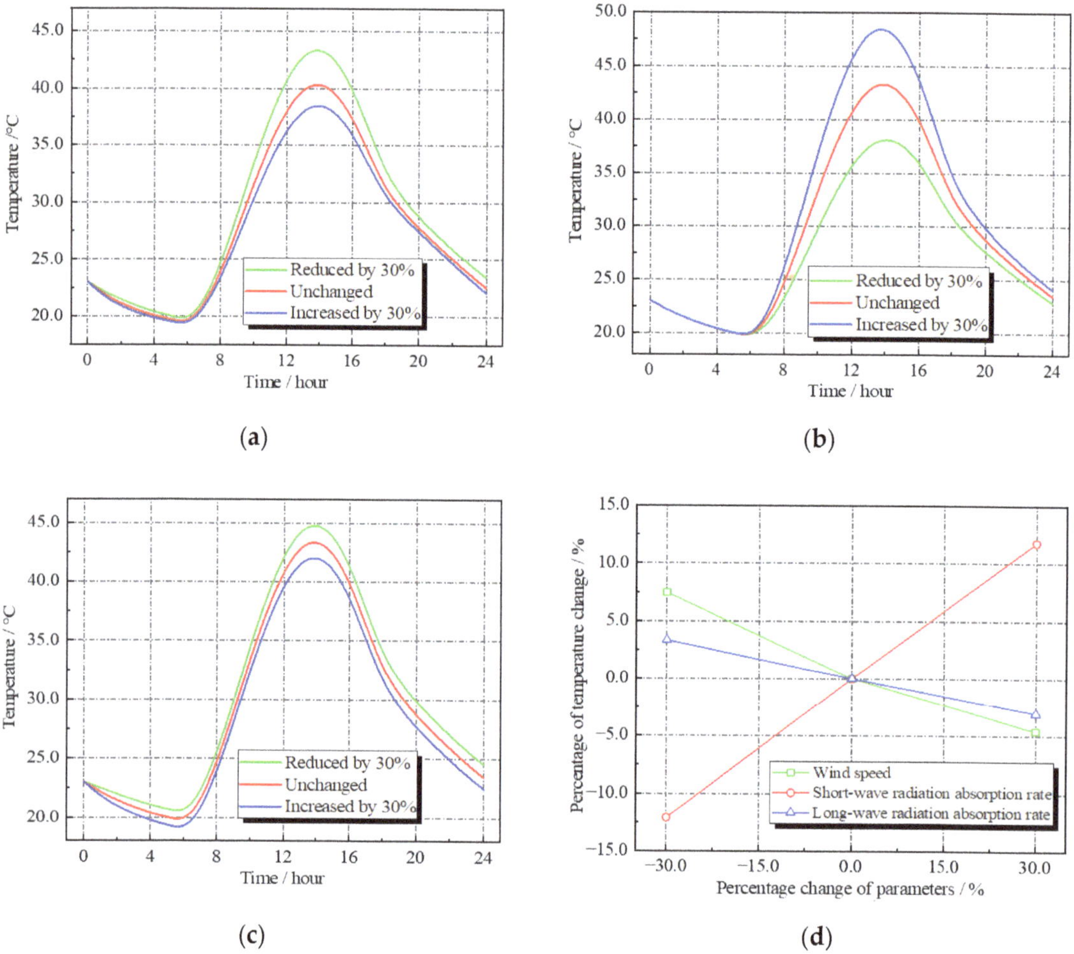

Figure 33. The sensitivity analysis of calculation parameters. (**a**) The influence of wind speed variations; (**b**) the influence of short-wave radiation absorption rate variations; (**c**) the influence of long-wave radiation absorption coefficient variations; (**d**) the percentage of temperature variations.

6. Conclusions

This study involves the development of a new simulation method using the user subroutine function of ABAQUS, which allows for the imposition of complex thermodynamic boundary conditions with ease. The accuracy of the simulation method was verified by comparing the simulated values of temperature with the measured values. Based on this simulation method, the sunshine temperature field of the concrete box girder of a cable-stayed bridge was analyzed. The main conclusions of this study are as follows:

(1) The comparison between the simulated and measured temperature values of the experimental model reveals that the difference between the simulated and measured values is small, with a maximum deviation of 2.9 °C. This result verifies the accuracy of the simple simulation method of the sunshine temperature field used in this paper;

(2) The analysis reveals a significant temperature difference between the internal and external surfaces of the box girder under the influence of sunshine, with the maxi-

mum negative temperature difference occurring around 6:00 a.m. and the maximum positive temperature difference occurring around 2:00 p.m.;

(3) The most unfavorable vertical temperature gradient distribution pattern of the concrete box girder calculated in this paper is not completely consistent with the double-line distribution pattern specified in the current "General Specifications for Design of Highway Bridges and Culverts" in China (JTG D60-2015). The comparison with the vertical temperature gradient values of concrete box girders in different national specifications shows that the Chinese specification is more conservative, and the results of this study are closest to the British specification. Therefore, for the concrete box girders in the construction stage, it is recommended to adopt the C-shaped distribution pattern calculated in this paper for the cooling and heating gradients;

(4) The sensitivity analysis of the parameters used to calculate the temperature field indicates that the solar radiation absorption coefficient has the greatest impact on the temperature calculation results, followed by wind speed changes, while changes in the long-wave radiation absorption coefficient of the structure surface have the smallest impact on the temperature field calculation results. Increasing or decreasing the solar radiation absorption coefficient by 30% can result in a difference of up to 13% in the maximum temperature of the structure's surface and a temperature variation of 10 °C or more.

In this paper, a simple simulation method is proposed to study the temperature field distribution of bridge structures under sunshine, with a focus on the concrete box girder of cable-stayed bridges. The accuracy of the simulation method is validated by comparing the simulated and measured temperature values of an experimental model, and the results show a small difference. The research findings have significant implications for the control of bridge alignment and stress state during construction, ensuring a reasonable initial operating state of the bridge, and enhancing the sustainability of the structure. However, due to space limitations, only the sunshine temperature field of the concrete box girder is studied, and further research is needed to analyze the sunshine temperature field of other materials and bridge members.

Author Contributions: Conceptualization, Q.W. and J.X. (Jianping Xian); methodology, J.X. (Jun Xiao); software, S.Z.; validation, Q.W., J.X. (Jianping Xian) and J.X. (Jun Xiao); formal analysis, S.Z.; investigation, Q.W.; resources, J.X. (Jianping Xian); data curation, J.X. (Jun Xiao); writing—original draft preparation, Q.W., J.X. (Jianping Xian), J.X. (Jun Xiao) and S.Z.; writing—review and editing, J.X. (Jun Xiao); visualization, Q.W.; supervision, J.X. (Jianping Xian); project administration, J.X. (Jun Xiao); funding acquisition, J.X. (Jun Xiao). All authors have read and agreed to the published version of the manuscript.

Funding: This research was funded by the Science and Technology Project of Shaanxi Department of Transportation (grant number: No. 21–59K) and the Science and Technology Project of Shaanxi Department of Transportation (grant number: No. 21–43K).

Institutional Review Board Statement: Not applicable.

Informed Consent Statement: Not applicable.

Data Availability Statement: Data are contained within the article.

Conflicts of Interest: The authors declare no conflict of interest.

References

1. Lee, J.H. Behavior of precast prestressed concrete bridge girders involving thermal effects and initial imperfections during construction. *Eng. Struct.* **2012**, *42*, 1–8. [CrossRef]
2. Imbsen, R.A.; Vandershaf, D.E.; Schamber, R.A.; Nutt, R.V. *Thermal Effects in Concrete Bridge Superstructures*; National Research Council: Washington, DC, USA, 1985.
3. Brancof, A.; Mendes, P.A. Thermal actions for concrete bridge design. *J. Struct. Eng.* **1993**, *119*, 2313–2331. [CrossRef]
4. Tang, H. Study on Key Problems of Cable-Stayed Bridge PC Pylon. Ph.D. Thesis, Southeast University, Nanjing, China, 2006. (In Chinese).

5. Feng, Z.R.; Shen, J.; Wang, X.J. Finite element analysis of thermal stress for cable-stayed bridge tower with cracks. *Appl. Mech. Mater.* **2012**, *178*, 2085–2090. [CrossRef]
6. Zhang, Q.; Ma, Y.; Wang, B. Analysis of temperature field and thermal stress characteristics for a novel composite bridge tower catering for plateau environment. *Bridge Constr.* **2020**, *50*, 30–36.
7. Zuk, W. Thermal behavior of composite bridges-Insulated and uninsulated. *Highway Res. Rec.* **1965**, *76*, 231–253.
8. Mirambell, E.; Aguado, A. Temperature and stress distributions in concrete box girder bridges. *J. Struct. Eng.* **1990**, *116*, 2388–2409. [CrossRef]
9. Rao, D.S.P. Temperature distributions and stresses in concrete bridges. *J. Proc.* **1986**, *83*, 588–596.
10. Churchward, A.S.Y. Prediction of temperatures in concrete bridges. *J. Struct. Div.* **1981**, *107*, 2163–2176. [CrossRef]
11. Hoffman, P.C.; McClure, R.M.; West, H.H. Temperature study of an experimental segmental concrete bridge. *PCI J.* **1983**, *28*, 78–97. [CrossRef]
12. Xia, Y.; Chen, B.; Zhou, X.Q.; Xu, Y.L. Field monitoring and numerical analysis of Tsing Ma Suspension Bridge temperature behavior. *Struct. Control Health Monit.* **2013**, *20*, 560–575. [CrossRef]
13. Chen, L.W.; Liu, G.K.; Zhang, D.M. Study on the influence of temperature effect on the elevation control of main girder of long-span PC cable-stayed bridge. *J. China Foreign Highw.* **2021**, *5*, 100–104.
14. Wu, Q.X.; Long, P.H.; Jiao, C.Y. Temperature Analysis of Steel-Concrete Composite Girder Caused by Solar Radiation. *J. Beijing Univ. Civ. Eng. Archit.* **2016**, *32*, 22–27.
15. Zhang, X.M.; Ren, Z.P.; Mei, F.M. *Heat Transfer Theory*; China Architecture & Building Press: Beijing, China, 2007.
16. Lienhard, J.H. *A Heat Transfer Textbook*; Courier Corporation: Chelmsford, MA, USA, 2013.
17. Ineichen, P. Validation of models that estimate the clear sky global and beam solar irradiance. *Sol. Energy* **2016**, *132*, 332–344. [CrossRef]
18. Engerer, N.A.; Mills, F.P. Validating nine clear sky radiation models in Australia. *Sol. Energy* **2015**, *120*, 9–24. [CrossRef]
19. Duffie, J.A.; Beckman, W.A. Solar Radiation. *Sol. Eng. Therm. Process.* **2013**, *4*, 3–42.
20. Hottel, H.C. A simple model for estimating the transmittance of direct solar radiation through clear atmospheres. *Sol. Energy* **1976**, *18*, 129–134. [CrossRef]
21. Al-Sanea, S.A.; Zedan, M.F.; Al-Ajlan, S.A. Adjustment factors for the ASHRAE clear-sky model based on solar-radiation measurements in Riyadh. *Appl. Energy* **2004**, *79*, 215–237. [CrossRef]
22. Rigollier, C.; Bauer, O.; Wald, L. On the clear sky model of the ESRA—European Solar Radiation Atlas—with respect to the heliosat method. *Sol. Energy* **2000**, *68*, 33–48. [CrossRef]
23. Duan, F. Research on the Sunshine Temperature Field and Temperature Effects on Long-span Steel Bridge. Master's Thesis, Southwest Jiaotong University, Chengdu, China, 2007. (In Chinese).
24. Zhou, L.R.; Xia, Y.; Brownjohn, J.M.W.; Koo, K.Y. Temperature analysis of a long-span suspension bridge based on field monitoring and numerical simulation. *J. Bridge Eng.* **2015**, *21*, 04015027. [CrossRef]
25. Peng, Y.S. Studies on Theory of Solar Radiation Thermal Effects on Concrete Bridges with Application. Ph.D. Thesis, Southwest Jiaotong University, Chengdu, China, 2007. (In Chinese).
26. Li, S.S. *Solar Physics*; Capital Normal University Press: Beijing, China, 1996.
27. Ohvril, H.; Okulov, O.; Teral, H.; Teral, K. The atmospheric integral transparency coefficient and the Forbes effect. *Sol. Energy* **1999**, *66*, 305–317. [CrossRef]
28. Chen, B.; Sun, Y.Z.; Wang, G.J.; Duan, L.Y. Assessment on time-varying thermal loading of engineering structures based on a new solar radiation model. *Math. Probl. Eng.* **2014**, *2014*, 639867. [CrossRef]
29. Severny, A. *Solar Physics*; University Press of the Pacific: Stockton, CA, USA, 2004.
30. Kim, S.H.; Cho, K.I.; Won, J.H.; Kim, J.H. A study on thermal behavior of curved steel box girder bridges considering solar radiation. *Arch. Civ. Mech. Eng.* **2009**, *9*, 59–76. [CrossRef]
31. Duffie, J.A.; Beckman, W.A.; Blair, N. *Solar Engineering of Thermal Processes, Photovoltaics and Wind*; John Wiley & Sons: Hoboken, NJ, USA, 2020.
32. Zhu, J.S.; Meng, Q.L. Effective and fine analysis for temperature effect of bridges in natural environments. *J. Bridge Eng.* **2017**, *22*, 04017017. [CrossRef]
33. Song, A.G. Preliminary study on solar radiation model in sunny days in Beijing. *Acta Energ. Sol. Sin.* **1993**, *3*, 251–255.
34. Elbadry, M.M.; Ghali, A. Nonlinear temperature distribution and its effects on bridges. *Struct. Eng. Int.* **1983**, *3*, 169–191.
35. Kelbek, K. *Effects of Solar Radiation on Bridge Structures*; Liu, X., Translator; China Railway Publishing House: Beijing, China, 1981.
36. Zhao, R.D.; Wang, Y.B. Studies on temperature field boundary conditions for concrete box-girder bridges under solar radiation. *China J. Highway Transp.* **2016**, *29*, 52–61.
37. Zhang, L.L.; Yang, L.; Yang, Z.Y.; Zhao, Y.Q.; Liu, H. Temperature field analysis of long-span concrete box girders. *J. Civ. Archit. Environ. Eng.* **2011**, *33*, 36–42.
38. Li, Y.L.; Huang, X.; Zhu, J.; Zhang, M.J. Thermal effect and anti-crack performance optimization of bridge pylon under extreme weather conditions. *J. Disaster Prev. Mitig. Eng.* **2022**, *339*, 310–319. Available online: http://www.nmglib.com:8901/article/detail.aspx?id=7107321824 (accessed on 8 April 2023).
39. Huang, X.; Zhu, J.; Li, Y.L. Temperature analysis of steel box girder considering actual wind field. *Eng. Struct.* **2021**, *246*, 113020. [CrossRef]

40. Abaqus. Abaqus Analysis User's Guide (V6.11). 2011. Available online: http://130.149.89.49:2080/v6.11/books/usb/default.htm (accessed on 8 April 2023).
41. Pang, Z.Y. The Research of Temperature Field and Effect of PC Curved Box Girder Bridge of City. Master's Thesis, Nanjing Tech University, Nanjing, China, 2015. (In Chinese).
42. Zhang, N.; Liu, Y.J.; Liu, J.; Ji, D.J.; Fang, J.H.; Stiemer, S.F. Temperature effects of H-shaped concrete pylon in arctic-alpine plateau region. *J. Traffic Transp. Eng.* **2017**, *17*, 66–77.
43. Kim, S.H.; Park, S.J.; Wu, J.X. Temperature variation in steel box girders of cable-stayed bridges during construction. *J. Constr. Steel Res.* **2015**, *112*, 80–92. [CrossRef]
44. Wu, L.H.; Woody Ju, J.; Zhang, J.X.; Zhang, M.J.; Li, Y.L. Vibration phase difference analysis of long-span suspension bridge during flutter. *Eng. Struct.* **2023**, *276*, 115351.
45. Zhang, M.J.; Zhang, J.X.; Chen, H.Y.; Xin, X.; Li, Y.L.; Jiang, F.Y. Probabilistic Wind Spectrum Model Based on Correlation of Wind Parameters in Mountainous Areas: Focusing on von Karman Spectrum. *J. Wind Eng. Ind. Aerodyn.* **2023**, *234*, 105337. [CrossRef]
46. *JTG D60-2015*; General Specifications for Design of Highway Bridges and Culverts. China Communication Press: Beijing, China, 2015. (In Chinese)
47. Barker, R.M.; Puckett, J.A. *Design of Highway Bridges Based on AASHTO LRFD Bridge Design Specifications*; John Wiley & Sons: Hoboken, NJ, USA, 1987.
48. *British Standard BS5400*; Steel. Concrete and Composite Bridges. Part 2. British Standards Institute: Milton Keynes, UK, 1978.
49. Zhang, Y.B. Studies on Temperature Effects and Its Influence on Stability for High Pier with Thin-Walled Hollow Sections. Ph.D. Thesis, China Academy of Railway Sciences, Beijing, China, 2011. (In Chinese).
50. Dai, G.L.; Tang, Y.; Liu, Y.; Su, H.T. Numerical simulation and test analysis of three-dimensional solar temperature field of hollow tall-piers with rounded rectangular cross section. *J. Railway Eng. Soc.* **2016**, *33*, 57–62.

sustainability

MDPI

Article

Mechanical Behavior of Refined SCC with High Admixture of Hybrid Micro- and Ordinary Steel Fibers

Qingguo Yang [1,*], Nan Ru [1], Xuefeng He [1] and Yi Peng [2]

[1] School of Civil Engineering, Chongqing Jiaotong University, Chongqing 400074, China; supernan55@163.com (N.R.); hxf202204@163.com (X.H.)

[2] College of Traffic & Transportation, Chongqing Jiaotong University, Chongqing 400074, China; dawsonyp@cqjtu.edu.cn

* Correspondence: yangqg7053@cqjtu.edu.cn

Abstract: The addition of steel fiber to self-consolidating concrete (SCC) may considerably prolong concrete cracking time and improve its deforming performance. Current studies mainly apply high content micro-steel fibers to improve the mechanical performance of SCC whilst assuring its workability, however, there are still very few studies concerning the influence of a mixture of a high content of micro-steel fibers with ordinary steel fibers on the performance of SCC. Thus, this paper conducted experimental studies on micro-steel fiber and ordinary-sized steel fiber hybrid reinforced self-consolidating concrete (MOSCC). Plain self-consolidating concrete (PSCC), micro-steel fiber reinforced self-consolidating concrete (MSCC), and ordinary-sized steel fiber reinforced self-consolidating concrete (OSCC) are proposed for comparison with MOSCC in respects of workability and mechanical performance. Test results show that the hybrid micro-steel fiber and ordinary steel fiber highly enhance the compressive strength, flexural strength, and ductility of SCC as well as maintaining its workability. This paper provides reference to the improvement of the mechanical performance of SCC material and the enhancement of crack resistance of concrete structures.

Keywords: self-consolidating concrete; micro-steel fiber; high admixture; mechanical performance

check for updates

Citation: Yang, Q.; Ru, N.; He, X.; Peng, Y. Mechanical Behavior of Refined SCC with High Admixture of Hybrid Micro- and Ordinary Steel Fibers. *Sustainability* **2022**, *14*, 5637. https://doi.org/10.3390/su14095637

Academic Editors: Kai Wei, Mingjin Zhang, Jian Zhong and Yutao Pang

Received: 11 April 2022
Accepted: 3 May 2022
Published: 7 May 2022

1. Introduction

Self-consolidating concrete (SCC) features high flowability, uniformness, and stability, in which uniform compaction can be achieved due to gravity without vibrating during construction [1]. Owing to the fine workability, SCC is extensively used in structures with complex forms and densely arranged reinforcement. However, SCC has disadvantages such as early shrinkage and cracking, and poor durability. Hence fibers, particularly steel fiber, are considered to mix into the SCC to resolve the above problems whilst maintaining its working performance.

Steel fibers have the potential to improve the mechanical property and deformation performance of concrete. This can be utilized to prohibit the generation and extension of micro-cracks in concrete whilst reducing the self-contraction deformation of concrete by its bridging and cracking resisting function. Researchers first revealed that a single ordinary type of steel fiber can effectively improve the mechanical property and durability of SCC. M. Akbari et al. [2] studied the impact on the compressive strength of SCC by the addition of volume fraction of 0.1%, 0.2%, and 0.3% steel fiber 33mm in length; and the test results show that the concrete compressive strength increased with increment of the volume fraction of fiber. Athiyamaan V [3] added micro steel fiber with a hooked end with the length-diameter ratio of 60 by a proportion of 0.25%, 0.50%, 0.75%, and the test result show that admixture of steel fiber greatly improved the bending resistance of SCC and inhibited the development of cracks, prevented fragile damage to the concrete, and improved its strain performance. Sahith Gali [4] found that the addition of a volume fraction of 0.75% (diameter of 0.75 mm, length of 60 mm) fiber steel with a hooked end can improve SCC, and the experiment results show that admixture of steel fiber improved rupture resistance of SCC, and also improved its shear capacity. Gisele et al. [5] mixed SCC with an addition of 0.25%, 0.5%, and 1% of steel fiber of diameter 0.75 mm respectively, and used V-funnel test and J-ring test to evaluate the slump ratio, flow velocity, and plastic viscosity of the mixture, and concluded that concrete bending was in proportion to the addition amount of fiber.

In their study into improving the properties of SCC with hybrid fibers, Cong Zhang [6,7] improved the mechanical behavior of an SCC beam by mixing steel fibers of 60 mm length, 0.75 mm diameter, and polypropylene fiber of different sizes, and revealed that steel fiber can significantly improve the bending resistance of concrete. Monteiro et al. [8] studied the mechanical properties of SCC with added steel fiber (of length 30 mm, diameter 0.62 mm and length 60 mm, diameter 0.75 mm), polypropylene fiber, and other different admixture of fiber; the test result shows that steel fiber is more outstanding than polypropylene fiber with respect to the crack resistance property of the concrete. Nabeel [9] improved the mechanical performance of polymer concrete by adding steel fibers of 13 mm length (diameter of 0.2 mm) and 6 mm length (diameter of 0.2 mm) into the volume admixture of 2%, while Jiaqing Wang et al. [10] pointed out that the mixture of various steel fibers and polypropylene fibers significantly reinforced rubber concrete. Generally, steel fibers more greatly enhance SCC than flexible fibers such as polypropylene fibers.

Most researchers tend to improve the performance of SCC by adding hybrid steel fibers. Burcu Akcay [11] showed that micro-steel fibers of various lengths and diameters, uniformly scattered in the SCC without any blocking, improved the toughness of the concrete. Deeb et al. [12] added micro-steel fibers (length of 6 mm and diameter of 0.16 mm) at 5% volume admixture together with steel fibers (length of 13 mm, diameter of 0.16 mm) at 1% volume admixture into SCC; the test results showed that a higher content of micro-steel fibers in the mixture can generate SCC with a satisfactory flowability. Dimas et al. [13–15] mixed straight-ended steel fibers (length of 12 mm and diameter of 0.18 mm) and hook-ended steel fibers (length of 5 mm, diameter of 0.55 mm) into SCC, further proving that hybrid steel fibers largely promote the compression strength, tensile strength, and bending deforming performance of SCC. Pajak and Ponikiewski [16] point out that the use of steel fibers improves flexural and residual strength, but may not offer improvements in compressive strength or in elastic modulus. Khaloo et al. [17] also verified that the negative effects of the incorporation of steel fibers on fresh state properties, as well as on compressive strength, are highly dependent on the incorporated volume. Above all, the current studies prove that steel fibers can significantly improve the mechanical properties of ordinary concrete, polymer concrete and SCC. However, the dosage level of steel fibers is limited due to the lower workability of higher admix steel fiber reinforced SCC. Furthermore, improving the mechanical behavior of SCC by increasing the content of steel fibers is less discussed.

Hence, this study aims to demonstrate the feasibility of using a combination of micro-steel fibers and ordinary steel fibers to reinforce SCC with agreeable workability by means of a laboratory testing program and statistically comprehensive methods.

2. Materials and Methods

2.1. Raw Materials

2.1.1. Cement

Portland cement with a strength grade of 42.5 (produced by Red Star construction material company Co., Ltd., Chongqing, China) was adopted. Its physical mechanical properties are presented in Table 1.

Table 1. Cement physical mechanical properties.

Fineness (Remains on the 80 μm Square Sieve) (%)	Consumption of Water with Standard Viscosity (%)	Hardening Time (h:min)		Compressive Strength (MPa)		Flexural Strength (MPa)		Stability (Boiling Method)
		initial hardening	final hardening	3 d	28 d	3 d	28 d	
0.0	27.6	2:25	3:45	28.4	56.2	5.9	9.6	acceptable

2.1.2. Fly Ash

Fly ash is grade I fly ash with a specific surface area of 420 m^3/kg, density of 2.42 g/cm^3, burn loss of 2.48%, and content of 3.82% Fe_2O_3, 2.72% CaO, 0.82% MgO, 29.09% Al_2O_3, and 53.36% SiO_2. For SCC, in order to improve performance, grade I fly ash is prioritized, and grade II fly ash can also be used; the water demand ratio should not exceed 100%.

2.1.3. Aggregate

The fine aggregate is ordinary medium sand, with a fineness modulus of 2.6 and a mud content of 3.3%. Coarse aggregate: crushed limestone, needle sheet particle content of 9.7%. Mix the coarse aggregates of 5~10 mm and 10~16 mm at a mass ratio of 1:1 and use the mixture.

2.1.4. Steel Fiber

The ordinary-sized steel fibers in this research are made by the sheet steel shearing method, as shown in Figure 1a. The micro-steel fibers are ultra-short & ultra-fine steel fibers made by the melting and extraction method shown in Figure 1b. Both types of steel fiber are produced by Daotuo Construction Material Co. Ltd., Chongqing, China. The selected steel fibers are mixed into the concrete volume term. The ordinary-sized steel fiber and micro-steel fiber are denoted as OS fiber and MS fiber respectively, and their mechanical properties are listed in Table 2.

(a)

(b)

Figure 1. The OS and MS fibers. (**a**) Ordinary-sized steel fiber (OS). (**b**) Micro-steel fiber (MS).

Table 2. Mechanical properties of steel fibers.

Types of Steel Fibers	Length/mm	Diameter/mm	Length-Diameter Ratio	Density (t/m^3)	Elastic Modulus (GPa)	Tensile Strength (MPa)
Ordinary steel fiber	35	0.9	39	7.8	210	850
Micro-steel fiber	6	0.2	30	7.8	210	>850

2.1.5. Additive

The water reducer was made with a BASF F10 poly-carboxylic acid powder. After adding the water-reducing agent, a completely unsegregated plaster mixture can be produced, and F10 can reduce mixing, quickly cure, increase the fluidity of the concrete, and improve the slump. After multiple tests to determine the quality of the cementing material, 0.18% was selected as the amount of incorporation. The water used was tap water from the laboratory of Chongqing Jiaotong University.

The target compressive strength of the SCC in this research was 40 MPa. The SCC mixture was designed as shown in Table 3.

Table 3. Mixture design of SCC with steel fibers.

SN	Cement (kg)	Water (kg)	Fly Ash (kg)	Coarse Aggregate (kg)	Fine Aggregate (kg)	Water Reducer (%)	Addition of Steel Fiber (%)
P-0	1	0.41	0.43	1.68	1.68	0.16%	0
OS1	1	0.41	0.43	1.68	1.68	0.16%	1%OS
OS2	1	0.41	0.43	1.55	1.55	0.16%	2%OS
MS6	1	0.40	0.44	0.98	0.98	0.16%	6%MS
OS1 + MS6	1	0.40	0.44	0.98	0.98	0.16%	1%OS + 6%MS
OS2 + MS6	1	0.39	0.44	0.98	0.98	0.16%	2%OS + 6%MS

Note: P-0 is plain self-consolidating concrete, OS is shear type ordinary steel fiber, MS is melt-drawn type ultra-short ultra-fine steel fiber; the volumes of steel fiber are noted by 1%, 2%, and 6% in the manuscript, which represents the mix ratio of steel fibers in SCC. 1%, 2%, and 6% steel fiber contents are represented by 1, 2, and 6 respectively, e.g., OS1 + MS6 stands for 1% ordinary steel fiber plus 6% ultra-short ultra-fine steel fiber mixed concrete.

The fresh concrete and steel fibers are mixed in a concrete mixer. The material container and mixing blade were wetted with a small quantity of water before mixing the weighed cement, sand particles, fly ash, and additives. A fiber concrete disperser can be applied in case of a large amount of steel fibers.

2.2. Sample Preparation

In the fresh state, tests to determine the flow spread and the flow rate using the Abrams cone were determined by ASTM C1611 [18], and passing ability by the J-ring through ASTM C1621 [19]. Fresh state tests were performed within 15 min after the complete homogenization of the mixtures. Considering the time after homogenization, each test was performed with a time-space of 2 min between them. The test results are illustrated in Figure 2a,b. Experimental data are listed in Table 4.

(a) (b)

Figure 2. Concrete fluidity test. (**a**) Slump flow test. (**b**) J-ring extension.

Table 4. Slump extension diameter and J-ring extension.

Group No.	Fiber Length	Content of Fiber	Slump Extension (mm)	T_{500}(s)	J-Ring Extension (mm)
P-0	0	0	610	3.2	570
OS1	35 mm	1% OS	590	3.6	550
OS2	35 mm	2% OS	580	3.9	545
MS6	6 mm	6% MS	575	4.7	525
OS1 + MS6	35 mm + 6 mm	1% OS + 6% MS	565	5.1	505
OS2 + MS6	35 mm + 6 mm	2% OS + 6% MS	550	5.6	495

Through the analysis of several amounts of micro-steel fibers, Khayat et al. [20] gathered results that support the results in this research, indicating that higher amounts of fiber increase the flow rate linearly. Contrary to this study, the flow rate varied with respect to the amount.

2.3. Test Methods

In this research, the mechanical behavior of SCC is presented by cube compressive strength, flexural strength, axial compressive strength, axial compressive stiffness and bending toughness; thus, cube specimens, beam specimens, and cylindrical specimens were prepared for the corresponding tests. Six groups of a specimen of PSCC, OSCC, and MOSCC were initially tested and studied. The influences of the addition of steel fibers on the SCC were compared and evaluated. The amounts of the specimen under each design solution are 18 specimens for respective cube compressive strength, flexural strength, axial compressive strength, and bending toughness tests, and 36 specimens for axial compressive stiffness tests. The total number of specimens was 108. The specimen sizes and numbers are detailed in Table 5.

Table 5. Specimen size and amount.

Test Type	Specimen Size (mm)	Number of Specimens	Total Numbers of Specimens
Cube compressive strength	100 × 100 × 100	3	18
Flexural strength	100 × 100 × 400	3	18
Axial compressive strength	100 × 100 × 300	3	18
Axial compressive stiffness	100 × 100 × 400	3	18
Bending toughness	100 × 100 × 300	6	36

2.3.1. Cube Compressive Strength Test

The most important parameter to elevate concrete material property is compressive strength and flexural strength [21]. The compressive strength was tested by following the Chinese Standard GB/T 50081-2019 [22]. A cube specimen of dimension 100 mm × 100 mm × 100 mm was taken, and three similar specimens made and the form removed after 24 h. The cube specimens and beam specimens were prepared and cured for 28 days. The specimen was placed with its axial core perpendicular to the center of the pressing plate and the pressing plate was adjusted close to and above the upper surface of the specimen without touching. Activating the testing machine for pre-pressing and loading speed in pre-pressing takes 0.5 MPa/s for the physical alignment adjustment and stability adjustment of the compressed. After completion of prepressing, add loading at a speed of 0.1 mm/min on the specimen until it is damaged, and then discharge loading generally. The loading process is illustrated in Figure 3a,b.

(**a**) (**b**)

Figure 3. Cube compressive strength test schemes. (**a**) PSCC. (**b**) MOSCC-OS1 + MS6.

The compressive results of the specimen shall be calculated as per following formula:

$$f_{cc} = \frac{F}{A} \quad (1)$$

where f_{cc} is the compressive strength (MPa) of steel fiber reinforced SCC, F is the ultimate load (N), A is the average value of compression areas of upper and lower surfaces of specimen (mm^2).

2.3.2. Flexural Strength Test

The flexural strength test results were taken according to Chinese Standard GB/T 50081-2019 [22]. The specimen takes dimensions of 100 mm × 100 mm × 400 mm. The test was conducted with the hydraulic universal testing machine, as shown in Figure 4a. Length axle of the specimen is perpendicular to the support circular axle. Loading is added at a constant speed of 50 N/s. Place the specimen on the test machine with its length axle perpendicular to the support circular axle, with spacing between the lower two circular supports at 0.3 m, with constraint being fixed hinge and universal rolling hinged support. Activate the test machine and then pre-press the specimen with a load of 1 KN, check the evenness of the specimen position to ensure test progress is standard and safe. Continue adding load with a loading speed of 0.06 MPa/s before the appearance of initial damage to the specimen. Once a crack appears in the specimen change the loading mode and control the specimen by displacement, and continue loading until damage to the specimen. The loading process was illustrated in Figure 4b.

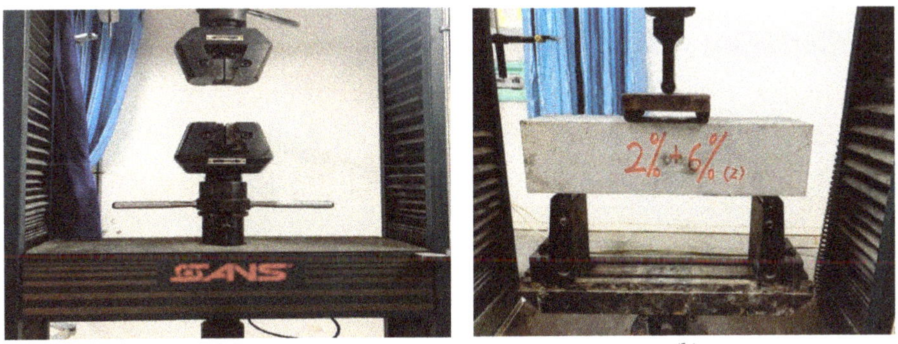

(**a**) (**b**)

Figure 4. Flexural strength test schemes. (**a**) Flexural strength test. (**b**) Flexural strength test specimens.

Specimen flexural strength in MPa is calculated by the following formula:

$$f_f = \frac{PL}{bh^2} \tag{2}$$

where f_f is the flexural strength of steel fiber reinforced SCC (MPa), P is the damaging load imposed on the middle of the specimen at flexural (N), L is the support space (mm), b is the specimen height (mm), h is the specimen width (mm).

2.3.3. Axial Compressive Strength

Non-standard specimen of dimensions 100 mm × 100 mm × 300 mm was used for testing, and the conversion coefficient is 0.95. The loading process is illustrated in Figure 5.

Figure 5. Axial compressive strength test.

The axial core compressive strength of the concrete specimen is calculated as per the following formula:

$$f_{cp} = \frac{F}{A} \tag{3}$$

where f_{cp} is the axial core compressive strength (MPa), calculated result accuracy is up to 0.1 MPa, F is the damaging load of specimen (N), A is the bearing area of the specimen (mm^2).

2.3.4. Axial Compressive Stiffness

The strain sheet is attached to the specimen with a side hole diameter of less than 5 mm and a depth of less than 2 mm. Considering the error caused by unevenly distributed aggregates, tighten the attachment of the strain sheet with a pasting rod at the bilateral centers of the specimen, and the center lines on the sides should be in alignment, as in Figure 6a. After the pasted places are dried, place the specimen on the bearing as per the specification. Prepare to add load on the universal testing machine. The static pressure elastic modulus is tested by the loading manner shown in Figure 7. When loading reaches an initial loading value of 0.5 MPa, stay for 60 s before the next load, until obtaining 1/3 standard loading for the axial core compressive strength of concrete of the corresponding type. Repeat the above steps more than five times and record the bilateral-sided strain values of the specimen from the final time, from initial loading up to the standard loading, as in Figure 6b.

(a) (b)

Figure 6. Axial compressive stiffness test schemes. (**a**) Strain gauge contacts. (**b**) Axial compressive stiffness test experimental setup.

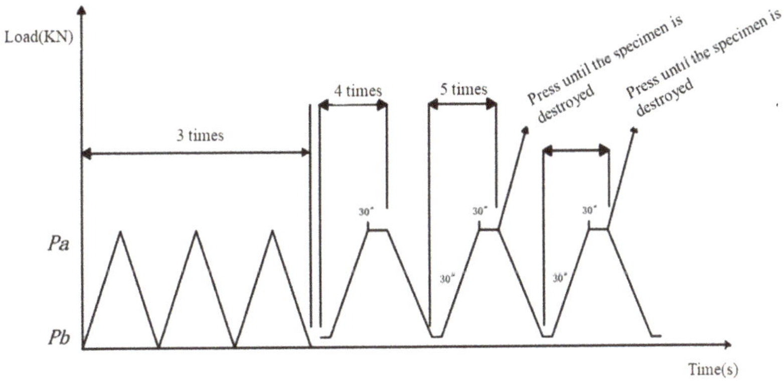

Figure 7. Loading cycle schematic.

Concrete axial compressive stiffness shall be calculated by the following formula:

$$E_c = \frac{F_a - F_0}{A} \times \frac{1}{\Delta n} \tag{4}$$

where E_c is the concrete axial compressive stiffness (MPa), F_a is the standard load (N) at 1/3 axial core compressive strength, F_0 is the initial load (N) at initial load of 0.5 MPa (N), A is the average bearing areas of specimen (mm^2).

$$\Delta n = \varepsilon_a - \varepsilon_0 \tag{5}$$

where Δn is the specimen bilateral sided average value $(\mu\varepsilon)$ in the last time from initial loading (F_0) up to standard loading (F_a), ε_a is the bilateral sided average value $(\mu\varepsilon)$ at loading up to standard load $(F_a)(\mu\varepsilon)$, ε_0 is the bilateral strain average value $(\mu\varepsilon)$, when loading increased to initial load.

2.3.5. Bending Toughness Test

As per the specification test, specimen shall take the dimension of 100 mm × 100 mm × 400 mm. Due to limited test conditions in this paper, a three-point loading method was applied [23]. The bending toughness test was carried out on a WE-100 hydraulic universal test machine, and the specimen lies on a circular support axle, as shown in Figure 8a. By loading the circular support with a constant speed of 50 N/s until rupture of the specimen, strainmeter is used to measure the deflection of the smaller beam, as in Figure 8b.

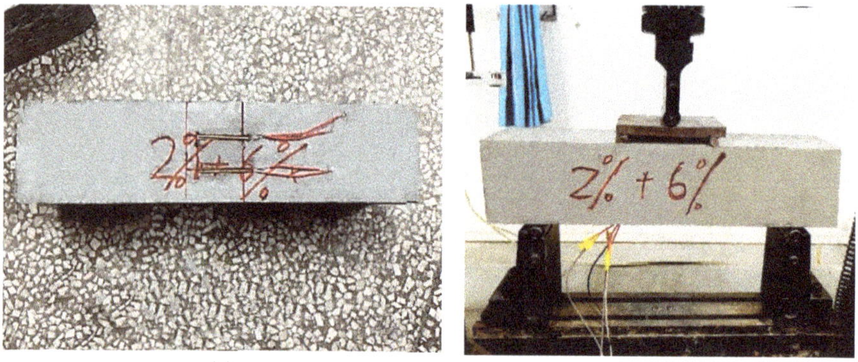

(**a**) (**b**)

Figure 8. Bending toughness test. (**a**) Strain gauge contacts. (**b**) Flexural strength test experimental setup.

The equivalent bending toughness is calculated in the sketch, shown in Figure 9, where the ordinate is the initial crack load, the abscissa is the initial crack deflection, the initial crack point is *A*, and the area *OAD* is the initial crack toughness.

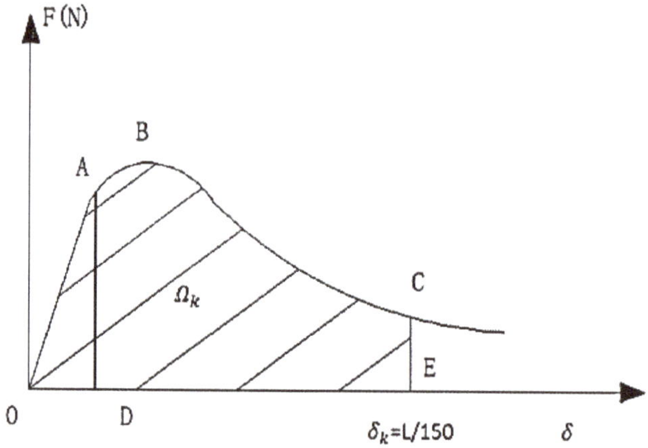

Figure 9. Equivalent bending toughness scheme.

1. Initial cracking strength at bending

Initial cracking point is *A*, the formula for the initial cracking strength calculation is shown in Formula (6).

$$f_{cr} = \frac{F_{cr}L}{bh^2} \quad (6)$$

where f_{cr} is the initial strength (MPa) of rupture resistance of steel fiber reinforced concrete, F_{cr} is the initial cracking load of the steel fiber reinforced concrete, *b* is height of the specimen (mm); *H* is width of the specimen (mm), *L* is the space between supports to the specimen.

2. Equivalent bending strength

The equivalent bending toughness of steel fiber reinforced concrete shall be calculated by Formula (7):

$$f_e = \frac{\Omega_k L}{bh^2 \delta_k} \tag{7}$$

where f_e is equivalent bending toughness, Ω_k areas (mm) in the load-defection curve when mid span deflection is *L/150* (N·mm), δ_k is deflection value (mm) when mid span deflection is *L/150* (mm).

3. Bending toughness ratio *Re*

The bending toughness ratio of the specimen shall be calculated by Formula (8):

$$R_e = \frac{f_e}{f_{cr}} \tag{8}$$

3. Results

3.1. Cube Compressive Strength

Figure 10a–d demonstrate the damages to the cubes of PSCC, OSCC and MOSCC.

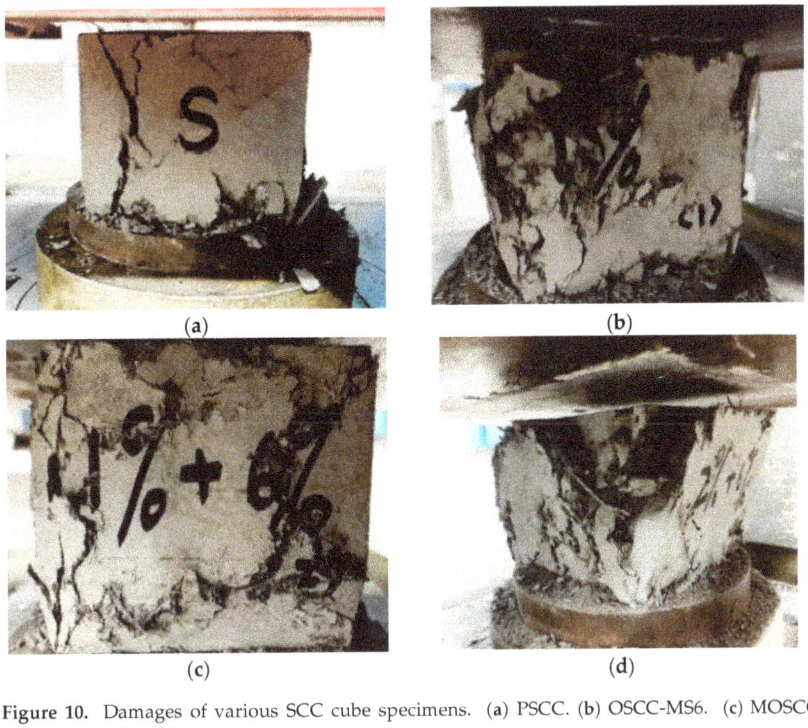

Figure 10. Damages of various SCC cube specimens. (**a**) PSCC. (**b**) OSCC-MS6. (**c**) MOSCC-OS1 + MS6. (**d**) MOSCC-OS2 + MS6.

As indicated in Figure 10, with an increase to the mixture of steel fibers, concrete spalling is slower, its bearing time is longer after spalling, and spalling is uniform. It is known that specimen 2% OS + 6% MS has a longer bearing time than that of specimen 1% OS + 6% MS observation. Namely, with a 1% increase to the long fiber, the hybrid steel fiber concrete has improved ductility. This test result shows that with an increase to the addition amount of fibers, compressive strength shows a rising trend. The compressive strength from S-0 to 1% OS materials increased considerably. However, after steel fiber is incorporated, compressive strength changes slowly and gently. That is to say, increases to steel fiber content and mixture of various fiber mixtures did not bring about an abrupt surge in compressive strength.

Table 6 shows that, compared with PSCC, a single addition of steel fiber improved the compressive strength of steel fiber reinforced SCC by 29.2%, 46.0% and 58.2% respectively, when the volume addition of ordinary steel fiber was 1% and 2%, and the volume addition of ultra-short and ultra-fine steel fiber was 6%. Compared with OSCC, compressive strength is improved by 70.0% and 87.4%, when 1% volume amount of ordinary steel fiber and 6% volume amount of ultra-short and ultra-fine steel fiber, and 2% ordinary steel fiber and 6% ultra-short and ultra-fine steel fiber in 28 d. Hence, a higher content addition of ultra-short and ultra-fine steel fiber mixed with ordinary steel fiber can considerably improve the compressive strength of SCC, as shown in Figure 11. Abukhashaba et al. [24] and Gisele C.S. et al. [5] also found a negative correlation between the amount of fibers and the mechanical strength. With an increase to the volume of steel fibers, the effect of fiber crack resistance begins to dominate. In the process of tension, after the matrix is cracked, steel fibers with greater deformability can bear the tensile force and keep the matrix cracks slowly expanding. This acts until the steel fiber is broken or pulled out from the matrix, thereby improving the compressive strength of the SCC.

Table 6. Test results of compressive strength of SCC.

SN	Fiber Content Added	Compressive Strength (MPa)	Range of Improvement	Variance σ^2
P-0	0	51.3	—	0.62
OS1	1%OS	66.3	29.2%	1.72
OS2	2%OS	74.9	46.0%	2.11
MS6	6%MS	81.2	58.2%	2.87
OS1 + MS6	1%OS + 6%MS	87.2	70.0%	2.40
OS2 + MS6	2%OS + 6%MS	96.6	88.3%	3.80

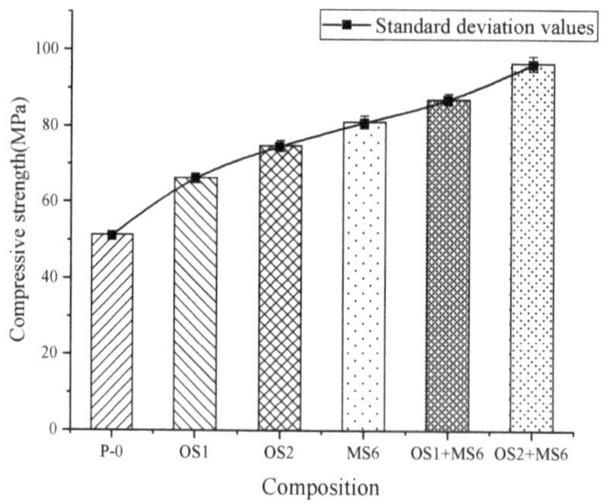

Figure 11. The compressive strength value illustration.

3.2. Flexural Strength

Figure 12a–d shows the flexural damages of specimens made of PSCC, OSCC and MOSCC.

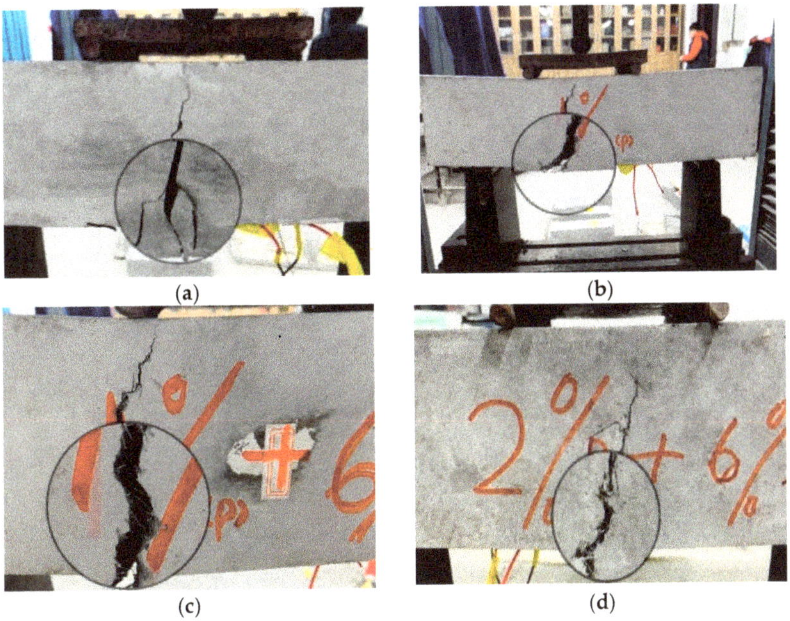

Figure 12. Damages of various SCC specimens. (**a**) PSCC. (**b**) OSCC-MS6. (**c**) MOSCC-OS1 + MS6. (**d**) MOSCC-OS2 + MS6.

These damage pictures show that under ultimate loads, the tensile area at the bottom edges of PSCC developed cracks in a short time under the impact of concentrated stress, with a tip-end effect. The crack extended to the top in a short time and generated a crack with a clear running direction; meanwhile, several blocks spalled in the tensile area. With a clear and loud cracking sound, the concrete block failed quickly and stopped working. The rupture-resistant concrete cube with added steel fiber, under concentrated loads at three separate points, firstly developed the main crack at a tensile area at the bottom of the example. With the increase in loading, the main crack gradually extended from the bottom in tension towards the top of the beam. With the mix ratio, the strength improvement role of these two types of steel fibers is discernible and mixture effects are sound. The flexural strength drawn from the test is shown in Table 7.

Table 7. Flexural strength test results.

SN	Content Addition of Fibers	Flexural Strength (MPa)	Range of Improvement	Elastic Modulus of Bend and Tensile ($\times 10^4$ MPa)
P-0	0	4.1	—	3. 21
OS1	1% OS	6.0	46.3%	3.50
OS2	2% OS	6.5	58.5%	3.58
MS6	6% MS	7.7	87.8%	3.61
OS1 + MS6	1% OS + 6% MS	9.8	139.0%	3.69
OS2 + MS6	2% OS + 6% MS	14.8	261.0%	3.72

Table 7 shows that, when compared with ordinary SCC, for concrete with a single addition of steel fiber, the flexural strength of the steel fiber reinforced SCC is improved by 46.3%, 58.5% and 87.8% when the content of added ordinary steel fiber reaches 1% and 2% respectively, and when the added content of ultra-short and ultra-fine steel fiber volume addition reaches 6%. Compared with PSCC, the flexural strength of SCC is improved by

139.0% with the volume addition of 1% mixture steel fibers and volume addition of 6% ultra-short and ultra-fine steel fiber, and by 261.0% with the volume addition of 2% ordinary steel fiber and volume addition of 6% of ultra-short and ultra-fine steel fiber. Denesh [25] found similar results. Figure 13 demonstrated that the volume addition of steel fiber is in proportion to the flexural strength of concrete. The results are similar to those found by Ponikiewiski [26]. Compared with PSCC, the flexural strength of SCC almost doubled with a content of 1% volume addition of ordinary steel fiber and 6% ultra-short and ultra-fine mixture steel fiber. With a volume addition of 2% ordinary steel fiber and volume addition of 6% ultra-short and ultra-fine steel fiber, the strength improvement soared. This kind of mixing method greatly improved the flexural strength of SCC. The volume addition of steel fiber can improve the tensile elastic modulus of steel fiber reinforced SCC whilst generating minor impact.

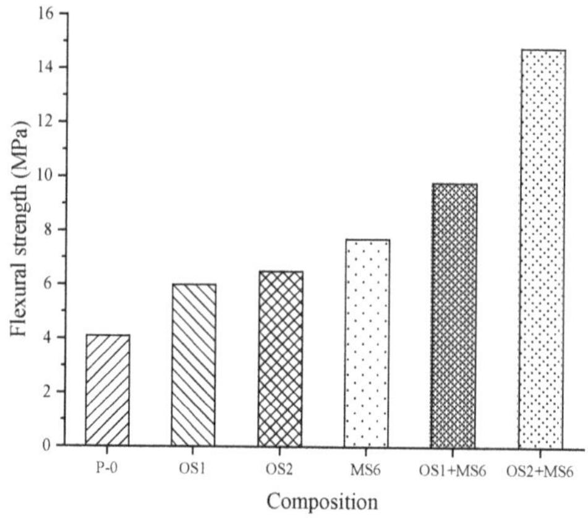

Figure 13. Bend tensile elastic modulus illustration.

3.3. Axial Compressive Strength

Compressive loading on the axial core of rectangular specimens. Figure 14a–d demonstrates the damage to specimens with different additions of steel fibers.

These pictures of damage to the axial core of SCC show that in the axial core loading test, the concrete developed degrees of inclined cracks. In the concrete cube with added steel fiber the cracks started at 2/3 height of the concrete, with 1% OS + 6% MS generating the finest crack, and no discernible fine cracks appeared surrounding the main crack. With the addition of 2% OS + 6% MS steel fiber, the inclined crack ran in a diagonal direction and generated the longest crack. The specimen with added steel fiber demonstrated notable plasticity, and this specimen was not damaged and continued bearing loads, without emitting the clear and loud rupture sound of PSCC at the moment of damage.

Figure 15 shows that the volume modulus of steel fiber is in proportion to the compressive strength of the axial core of the SCC cube, and that any addition of ordinary or a mixture of steel fiber may improve the compressive strength of the axial core of concrete. During the stress process of SCC, the addition of steel fibers can change the stress form of the text block, from concentrated stress to dispersed stress. The restraint of steel fibers can effectively inhibit the generation of micro-cracks and increase the propagation of micro-cracks, prolonging the cracking time of the test block, thereby increasing the energy absorbed by the SCC during failure, increasing its strain energy and relative toughness. Concrete for foundations with hoops and stirrups can be achieved by adding two types of steel fibers in random distributions, and a mixture of steel fibers can further develop these advantages.

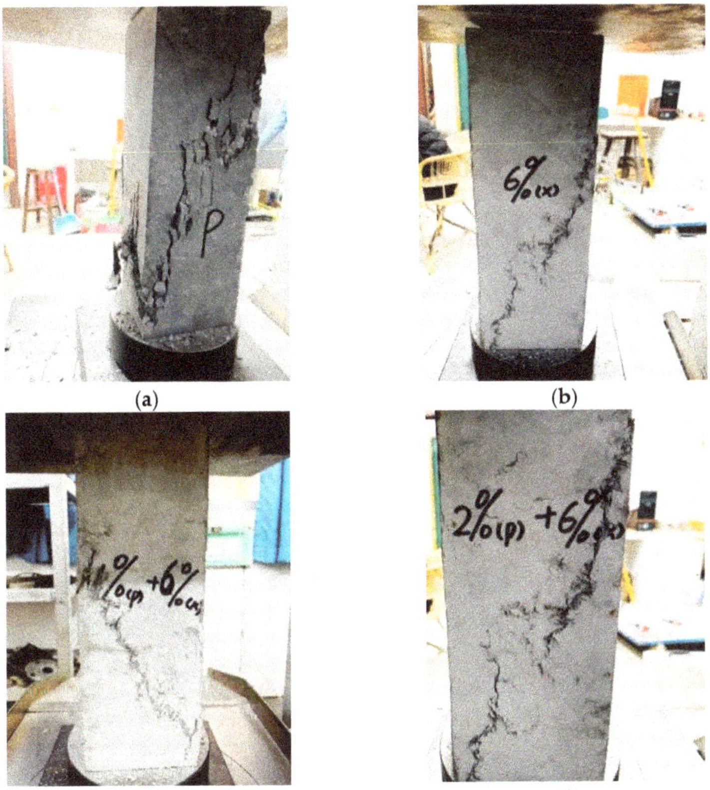

Figure 14. Axial core compressive strength test. (**a**) PSCC. (**b**) OSCC-MS6. (**c**) MOSCC-OS1 + MS6. (**d**) MOSCC-OS2 + MS6.

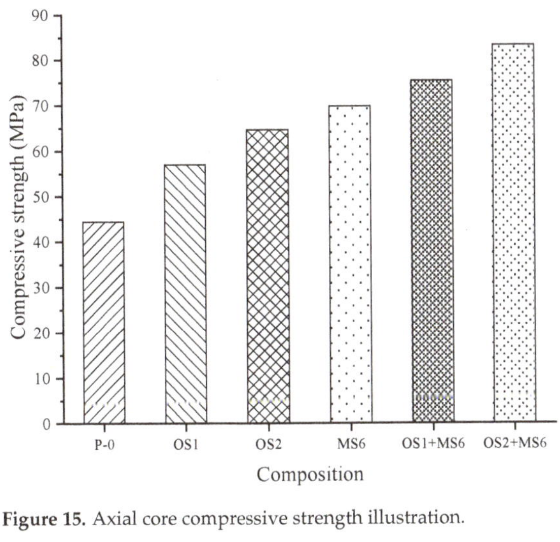

Figure 15. Axial core compressive strength illustration.

3.4. Axial Compressive Stiffness

The test shows that the PSCC cube was damaged shortly after loading the plain, and generated a very long crack which almost penetrated the diagonal of the whole rectangular body. The crack developed top-down along the bearing face; the crack width became very wide when approaching the upper bearing, and was narrow closer to the lower bearing. In the MOSCC cube with a mixture of steel fibers, the cracks are narrow and developed slowly, thus the SCC cube could bear loads over a longer time. The elastic modulus drawn from the tests is listed in Table 8 below.

Table 8. Axial compressive stiffness test results.

Group No.	Compressive Strength of 1/3 Axial Core F_a (KN)	Initial Loading F_0 (KN)	Bearing Area of Specimen A (mm^2)	Last Reading of Strain Meter Δn (×10^{-3} mm)	Average Value of Elastic Modulus E_c (MPa)
P-0	144.7	5.0	10,000	42.9	34,560
OS1	190.0	5.0	10,000	44.7	36,200
OS2	215.6	5.0	10,000	46.4	36,500
MS6	232.6	5.0	10,000	49.8	37,100
OS1 + MS6	222.0	5.0	10,000	57.0	38,070
OS2 + MS6	277.4	5.0	10,000	64.9	41,930

Figure 16 indicates that the axial compressive stiffness of steel fiber reinforced SCC increased with the increment of the addition of steel fibers. When the steel fiber addition amount was 1% OS, 2% OS, 6% MS, and 1% OS + 6% MS respectively the addition of steel fiber had little impact on the axial compressive stiffness, but with 2% OS + 6% MS the axial compressive stiffness has notable improvement compared with cases of other additions of steel fibers. This indicates that the addition of a mixture of two different kinds of steel fibers may result in some degree of improvement to axial compressive stiffness. However, the amount of steel fibers added has less impact on axial compressive stiffness in general.

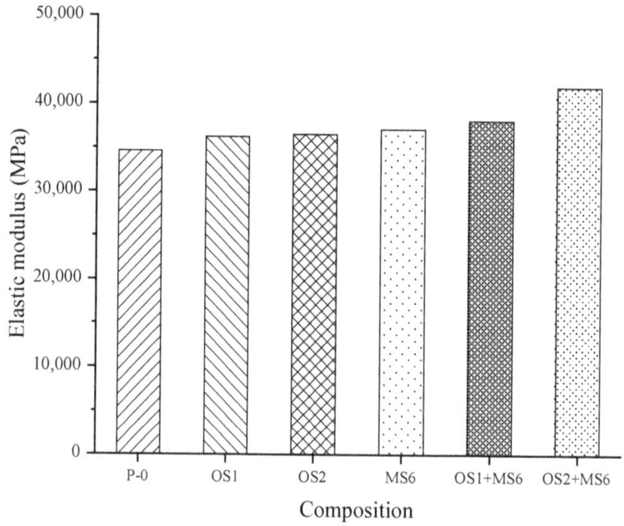

Figure 16. Axial compressive stiffness illustration.

3.5. Bending Toughness

The bending toughness loading defection curve of SCC fluctuated greatly with different volume fractions of steel fiber. The ultimate loads of a specimen increased with the increment of added fiber volume fraction, and the curve becomes fuller and presents greater toughness. The major results of the bending toughness test are shown in Table 9. Compared with PSCC, with the single addition of steel fiber, the compressive strength of SCC improved by 27.5%, 37.3% and 45.1% respectively and its equivalent bending toughness improved by 100%, 161.5% and 230.8%, while its bending toughness ratio improved 60.0%, 96.0% and 132.0% respectively when the volume addition of ordinary steel fiber was 1% and 2% and the volume addition of ultra-short and ultra-fine steel fiber was 6%. Compared with PSCC, the initial cracking strength of SCC improved by 60.8% and 119.6% respectively, and the equivalent bending toughness improved by 323.1% and 453.8% respectively, while the bending toughness ratio improved by 144.0% and 156.0% respectively, in the cases of a mixture of volume addition 1% ordinary steel fiber and 6% volume addition of ultra-short and ultra-fine steel fibers, and mixture of 2% ordinary steel fiber and 6% ultra-short and ultra-fine steel fiber. Hence it is clear that the bending toughness of SCC can be improved by adding steel fiber or a mixture of steel fiber, while SCC with a higher content of ultra-short and ultra-fine steel fiber with a mixture of steel fiber presents the greatest effect.

Table 9. Bending toughness test results.

Test Code	Initial Loading (KN)	Ultimate Load (KN)	Flexural Initial Strength (MPa)	Equivalent Flexural Strength (MPa)	Fexural Toughness Ratio
P-0	17.0	17.0	5.1	1.3	0.25
OS1	21.8	26.3	6.5	2.6	0.40
OS2	23.2	28.9	7.0	3.4	0.49
MS6	24.5	30.1	7.4	4.3	0.58
OS1 + MS6	27.2	32.7	8.2	5.5	0.61
OS2 + MS6	37.3	49.3	11.2	7.2	0.64

Figure 17 shows the data drawn from the three-point bending toughness test in the given study. It shows that a mixture of steel fibers can effectively improve the toughness of center displacement of material and prohibit the generation of micro-cracks in a flexural test. In Dimas [13] experiment in testing bending toughness, the average initial cracking strength was 3.69 MPa, and compound material C1.5% reached the highest tensile strength, and increased by 2.3 times that of UN-reinforced base material, while the effect of the mix is not so great. In a given test, the average initial cracking bend strength reached up to 7.57 MPa. Figure 17 shows that all the curves gradually decline after reaching their top. This is because before cracking, the steel fiber did not slip and move against SCC. Thus, in the loading displacement curve, a linear relation is formed and the skew rate is great. Afterwards, the curve presented many sections of curves, and the displacement at mid-span has a slow increase under the same loads. As the increment of addition increases to the whole base material, and the cohesion strength between the fiber and concrete base body is gradually reduced, which limits improvement to the compressive strength of the material.

The initial cracking strength, equivalent flexural strength, and bending toughness of the SCC specimen are proportional to the volume addition of steel fiber as shown in Figure 18. Pons et al. [27] and Gisele C.S. [5] got results that are compatible with those in this study, assessing that the toughness of the reference mixture was increased due to the addition of steel fibers. This proved that the bending toughness of concrete can be improved by incorporating a mixture of steel fibers. According to Abrishambaf et al. [28] and Thomas [29], this is the expected behavior once fibers are used to improve bending toughness.

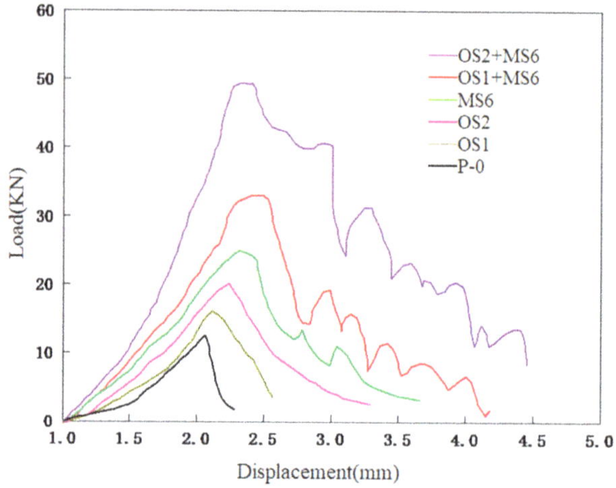

Figure 17. Load–displacement curve.

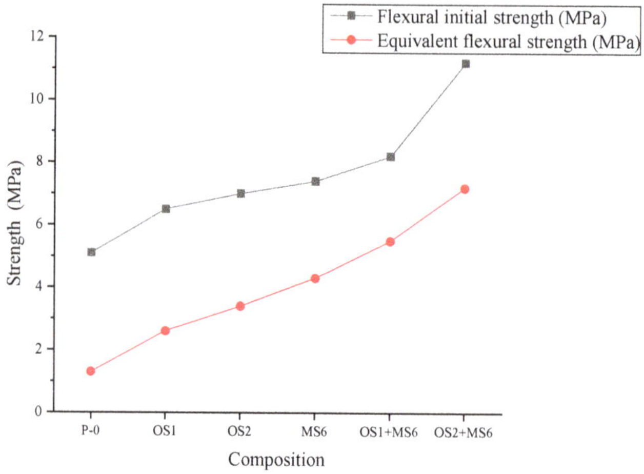

Figure 18. Bending toughness illustration.

4. Discussion

4.1. Addition Amount Recommendation

According to the cube's compressive strength, flexural strength, axial compressive strength, axial compressive stiffness, and bending toughness, it is clear that the addition of steel fiber can effectively improve the mechanical properties of steel fiber reinforced SCC, however, it is hard to determine which mix ratio can most improve the SCC. Therefore, by normalizing the data, the selected criteria should be shortened between 0 and 1 to find the optimal recommended mixing ratio, and by accumulating the specified criteria values, the combined effect of steel fibers on SCC can be determined.

The data normalization formula is shown as follows:

$$D_n = \frac{D - D_{min}}{D_{max} - D_{min}} \tag{9}$$

where D is the original value of test data, D_{max} is the maximum value of the test data, D_{min} is the minimum value of test data, D_n is the normalized value of test data.

The comprehensive influence of steel fiber on SCC is shown in Figure 19. It shows that the incorporation of steel fiber can greatly improve the mechanical property of SCC, and with the increase of the added amount of steel fiber the SCC property is improved constantly. Furthermore, the effect of the fiber mixture is greater. This is because the steel fibers are distributed in multiple directions inside the concrete. MS fibers can effectively prevent the expansion and extension of small cracks in the concrete, and OS fibers can inhibit the propagation of large cracks. The two are mixed and can give full play to their respective roles to prolong the test block cracking time, thereby increasing the energy absorbed by the SCC during failure, increasing its strain energy and relative toughness. Hence higher content of MOSCC can effectively prohibit the development of cracks in SCC and can improve the cracking resistance of SCC.

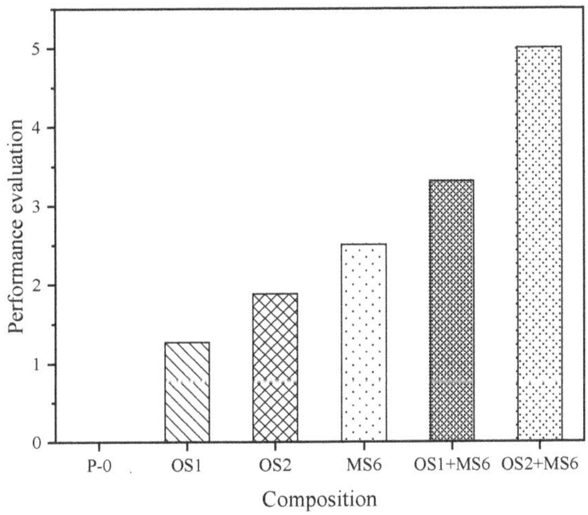

Figure 19. Comprehensive influence of fibers content on SCC.

4.2. Comparison

Burcu [11] applied three different typed steel fibers and incorporated them in two different volume fractions into the mixture, the cement used was ordinary Portland cement with a density of 3.17 g/cm^3, while the density of silica fume was 2.5 g/cm^3, the maximum particle size of aggregate was 16 mm. PSCC is plain self-consolidating concrete, MOSCC (C1.5H) is concrete with 1.0% high strength straight steel fiber (length of 6 mm and diameter of 0.15 mm) and 0.5% high strength hooked-end steel fiber (length of 30 mm and diameter of 0.55 mm). By obtaining the optimum addition and the above-mentioned optimum addition amount, the improvement rate of SCC properties is compared. The specific parameters are shown in Table 10. The table shows that with an increase in the addition of steel fiber, the vertical orientation of the fibers relative to the bending loading direction is more pronounced, thus improving the mechanical property of SCC. However, in Burcu's experiment, the added amount is insufficient to determine whether the optimum state of SCC is achieved, and the improvement mechanism of the steel fiber mixture is unclear. The test results in the experiment provided in this paper show clearly that a higher content of MOSCC improved the mechanical properties of SCC, and that the effect is pronounced. In this paper, considering the economic cost and practical problems, an excessive amount of steel fiber will affect the performance of concrete. The content of ultra-short and ultra-fine steel fibers is increased to 6%, which proves the feasibility of combining micro-steel fibers and ordinary steel fibers to enhance SCC. This test explained the mechanism of improving SCC by adding different amounts of steel fibers, and also verified that SCC containing a higher content of MOSCC not only achieved self-consolidation, which is hard to achieve with a higher steel fiber content, but also verified that compressive performance, rupture performance, axial compressive stiffness, and bending toughness improved constantly with the increasing addition of steel fibers, giving SCC a high cracking resistance capability. At the same time, we can apply the anti-cracking properties of MOSCC to bridge deck pavement.

Table 10. Comparison with Burcu's test.

Test	Burcu's Test			This Resarch		
	PSCC	MOSCC (C1.5H)	Improvement Rate	PSCC	MOSCC (OS2 + MS6)	Improvement Rate
Compressive test	115.3	123.6	7.2%	51.3	96.6	88.3%
Axial compressive stiffness	46.0	45.9	-	34.6	41.9	21.1%
Bending toughness	7.3	14.5	98.6%	5.1	11.2	119.6%

5. Conclusions

(1) MOSCC has a considerably improved strength and deforming property. In the case of volume addition of 2% ordinary steel fiber and volume addition of 6% ultra-short and ultra-fine steel fiber, when compared with PSCC, MOSCC has its 28 d cube compressive strength improved by 88.3%, and flexural strength improved by 261.0%.

(2) The MOSCC cube specimen demonstrated pronounced toughness in the loading process and showed cracking without rupture, while the loading-displacement curve showed a declining stage, and the region surrounded by curves and coordinate axles became fuller.

(3) MOSCC has a considerably improved toughness. In the case of the mixture of volume addition of 2% of ordinary steel fiber and 6% volume addition of ultra-short and ultra-fine steel fiber, compared with PSCC, the initial cracking strength of MOSCC improved by 119.6%, equivalent bending roughness improved by 453.8%, and bend toughness ratio improved by 156.0%.

(4) Damage conditions of MOSCC were changed. With a higher content of ultra-short and ultra-fine steel fiber mixture, the damage conditions of SCC can be improved. When MOSCC is damaged, its cracks will normally not penetrate throughout the whole specimen, as occurred to PSCC. The specimen generated ordinary cracks which caused no damage, and still had initial crack loading, and its damage shows a certain degree of ductility.

To sum up, it can be concluded that MOSCC achieved the workability of the mixture containing a higher content of admixture, and can greatly prevent the development of a concrete macro-crack, and thus it possesses excellent mechanical properties.

Author Contributions: Conceptualization, Q.Y. and N.R.; methodology, Q.Y., N.R. and X.H.; validation, Q.Y., X.H. and Y.P.; formal analysis, Q.Y., N.R., Y.P. and X.H.; investigation, Q.Y., Y.P. and N.R.; resources, Q.Y.; data curation, N.R., X.H., Y.P.; writing—original draft preparation, Q.Y., N.R., Y.P. and X.H.; writing—review and editing, Q.Y., Y.P. and N.R.; visualization, Q.Y., Y.P. and N.R.; supervision, Q.Y.; project administration, N.R.; funding acquisition, Y.P. All authors have read and agreed to the published version of the manuscript.

Funding: This research was funded by China Postdoctoral Science Foundation (grant number: 2021M693918) and Chongqing Postdoctoral Science Foundation (grant number: cstc2021jcyj-bshX0113).

Institutional Review Board Statement: Not applicable.

Informed Consent Statement: Not applicable.

Data Availability Statement: Data are contained within the article.

Conflicts of Interest: The authors declare no conflict of interest.

References

1. Long, J.; Lin, H.; Chen, Z. Effect of fiber on mechanical properties of self-consolidating concrete. *Concrete* **2014**, *295*, 60–63.
2. Akbari, S.M.; Khalilpour, S.; Dehestani, M. Analysis of material size and shape effects for steel fiber reinforcement self-consolidating concrete. *Eng. Fract. Mech.* **2018**, *206*, 46–63. [CrossRef]
3. Athiyamaan, V.; Ganesh, G.M. Analysis of the Alignment of Micro-Steel Fibers in Admixture-Based Self-consolidating Concrete (MSFR-SCC) using NDT and Evaluation of Its Effect on the Modulus of Rupture. *Int. J. Technol.* **2019**, *10*, 5–15. [CrossRef]
4. Gali, S.; Sharma, D.; Subamaniam, K.V.L. Influence of steel fibers on fracture energy and shear behavior of SCC(Article). *J. Mater. Civ. Eng.* **2018**, *30*, 04018295. [CrossRef]
5. da Silva, G.C.S.; Christ, R.; Pacheco, F.; de Souza, C.F.N. Evaluating steel fiber-reinforced self-consolidating concrete performance. *Struct. Concr.* **2019**, *21*, 1–10. [CrossRef]
6. Zhang, C.; Han, S.; Hua, Y. Flexural performance of reinforced self-consolidating concrete beams containing hybrid fibers. *Constr. Build. Mater.* **2018**, *174*, 11–23. [CrossRef]
7. Zhang, C. Effect of hybrid fibers on flexural performance of reinforced SCC symmetric inclination beams. *Comput. Concr.* **2018**, *22*, 209–230.
8. de Monteiro, V.M.; Lima, L.R.; de Silva, F. On the mechanical behavior of polypropylene, steel and hybrid fiber reinforced self-consolidating concrete. *Constr. Build. Mater.* **2018**, *188*, 280–291. [CrossRef]
9. Farhan, N.A.; Sheikh, M.N.; Hadi, M.N.S. Effect of Steel Fiber on Engineering Properties of Geopolymer Concrete. *ACI Mater. J.* **2020**, *117*, 29–40.
10. Wang, J.; Dai, Q.; Guo, S.; Si, R. Study on Rubberized Concrete Reinforced with Different Fibers. *ACI Mater. J.* **2019**, *116*, 21–31. [CrossRef]
11. Burcu, A.; Ali, T.M. Mechanical behaviour and fibre dispersion of hybrid steel fibre reinforced self-consolidating concrete. *Constr. Build. Mater.* **2012**, *28*, 287–293.
12. Deeb, R. Development of self-consolidating high and ultra high performance concretes with and without steel fibres. *Cem. Concr. Compos.* **2012**, *34*, 185–190. [CrossRef]
13. Rambo, D.A.S.; de Silva, F.; Filho, R.D.T. Effect of steel fiber hybridization on the fracture behavior of self-consolidating concretes. *Cem. Concr. Compos.* **2014**, *54*, 100–109. [CrossRef]
14. Rambo, D.A.S.; de Silva, F.; Filho, R.D.T. Flexural behavior of hybrid steel fiber reinforced self-consolidating concretes. *Civ. Eng* **2014**, *67*, 27–32. [CrossRef]
15. Rambo, D.A.S.; de Silva, F.; Filho, R.D.T. Mechanical behavior of hybrid steel-fiber self-consolidating concrete: Materials and structural aspects. *Mater. Des.* **2014**, *54*, 32–42. [CrossRef]
16. Pajak, M.; Ponikiewski, T. Flexural behavior of self-compactingconcrete reinforced with different types of steel fibers. *Constr. Build Mater.* **2013**, *47*, 397–408. [CrossRef]
17. Khaloo, A.; Molaei Raisi, E.; Hosseini, P.; Tahsiri, H. Mechanical per-formance of self-compacting concrete reinforced with steel fibers. *Constr. Build Mater.* **2014**, *51*, 179–186. [CrossRef]
18. *ASTM C1611 11*; Standard Test Method for Slump Flow of Self-Consolidating Concrete. ASTM: West Conshohocken, PA, USA, 2014.
19. *ASTM C1621-17*; Standard Test Method for Passing Ability of Self-Consolidating Concrete by J-Ring. ASTM: West Conshohocken, PA, USA, 2017.
20. Khayat, K.H.; Kassimi, F.; Ghoddousi, P. Mixture design and testing of fiber reinforced self-consolidating concrete. *ACI Mater. J.* **2014**, *111*, 143–152.
21. Pan, L.; Chen, W.; Jiang, C. Concrete strength evaluation based on small size non-standard specimen. *New Build. Mater.* **2017**, *6*, 28–32.
22. *GB/T 50081-2019*; Standard for Test Methods of Concrete Physical and Mechanical Properties. China Architecture & Building Press: Beijing, China, 2019.
23. Rilem, T. Determination of the fracture energy of mortar and concrete by means of three-point bend tests on notched beams. *Mater. Struct.* **1985**, *18*, 285–290.
24. Abukhashaba, M.I.; Mostafa, M.A.; Adam, I.A. Behavior of selfcompacting fiber reinforced concrete containing cement kiln dust. *Alex. Eng. J.* **2014**, *53*, 341–354. [CrossRef]
25. Denesh, K.C. Experimental study on fiber reinforced self compacting concrete. *Int. J. Eng. Res. Technol.* **2014**, *3*, 1209–1212.
26. Ponikiewski, T.; Golaszewski, J. Properties of steel fibre reinforcedself-compating concrete for optimal rheological and mechanicalproperties in precast beams. *Procedia Eng.* **2013**, *65*, 290–295. [CrossRef]
27. Pons, G.; Mouret, M.; Alcantara, M.; Granju, J.L. Mechanical behaviour of self-compacting concrete with hybrid fibre reinforcement. *Mater. Struct.* **2007**, *40*, 201–210. [CrossRef]
28. Abrishambaf, A.; Pimentel, M.; Nunes, S. The effect of fibre orientation on the uniaxial tensile response of UHPFRC: Experimental evaluation and analytical modelling. In *High Tech Concrete: Where Technology and Engineering Meet*; Hordijk, D., Lukovic, M., Eds.; Springer: Cham, Switzerland, 2018.
29. Thomas, R.J.; Sorensen, A.D. Review of strain rate effects for UHPC in tension. *Constr. Build. Mater.* **2017**, *153*, 846–856. [CrossRef]

MDPI
St. Alban-Anlage 66
4052 Basel
Switzerland
www.mdpi.com

Sustainability Editorial Office
E-mail: sustainability@mdpi.com
www.mdpi.com/journal/sustainability

www.ingramcontent.com/pod-product-compliance
Lightning Source LLC
Chambersburg PA
CBHW040818250926
51850CB00007BA/1199
* 9 7 8 3 0 3 6 5 9 7 6 3 8 *